Cambridge Monographs in African Archaeology
78
Series Editors: John Alexander, Laurence Smith and Timothy Insoll

An Archaeological and Geomorphological Survey of the Luangwa Valley, Zambia

Dan Colton

BAR International Series 2022
2009

Published in 2016 by
BAR Publishing, Oxford

BAR International Series 2022

Cambridge Monographs in African Archaeology 78

An Archaeological and Geomorphological Survey of the Luangwa Valley, Zambia

ISBN 978 1 4073 0597 4

BAR Publishing is the trading name of British Archaeological Reports (Oxford) Ltd. British Archaeological Reports was first incorporated in 1974 to publish the BAR Series, International and British. In 1992 Hadrian Books Ltd became part of the BAR group. This volume was originally published by Archaeopress in conjunction with British Archaeological Reports (Oxford) Ltd / Hadrian Books Ltd, the Series principal publisher, in 2009. This present volume is published by BAR Publishing, 2016.

Printed in England

BAR
PUBLISHING

BAR titles are available from:

BAR Publishing
122 Banbury Rd, Oxford, OX2 7BP, UK
EMAIL info@barpublishing.com
PHONE +44 (0)1865 310431
FAX +44 (0)1865 316916
www.barpublishing.com

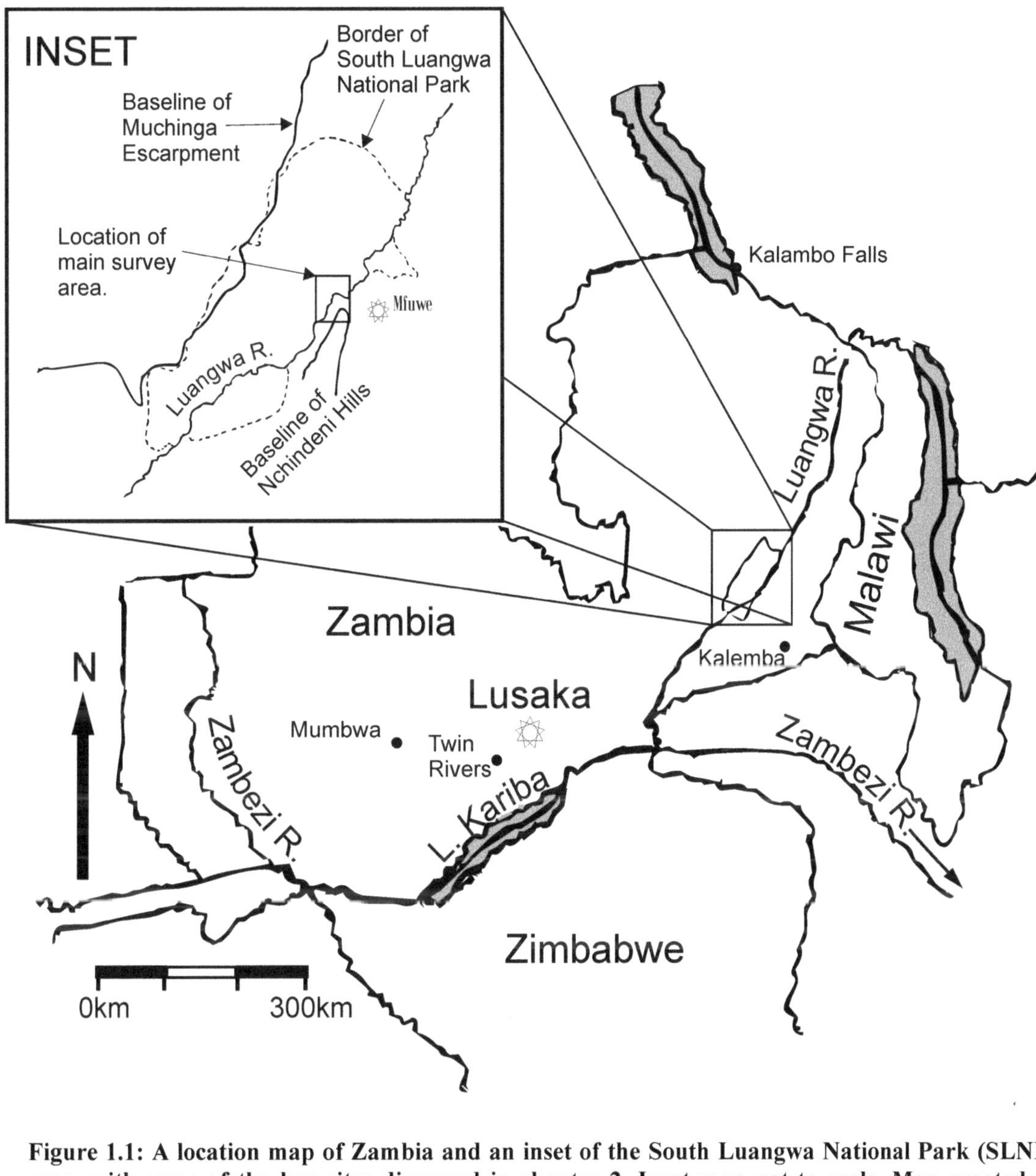

Figure 1.1: A location map of Zambia and an inset of the South Luangwa National Park (SLNP) area, with some of the key sites discussed in chapter 2. Inset map not to scale. Map created in Corel Draw X3 from an original drawing (site locations based on Barham 2000:2; Musonda 1987:148).

Contents

Contents (continued)

Abstract: An Archaeological and Geomorphological Survey of the Luangwa Valley, Zambia.

The aim of this research was to develop an understanding of the geomorphological processes that have affected the archaeological record in the South Luangwa National Park area of the Luangwa Valley, in eastern Zambia, and determine the extent to which genuine behavioural residues survive. Prior to this research little or no investigation had been initiated into Pleistocene and Holocene geomorphological processes, or the archaeological record.

A 12km x 14km area of the park was selected for detailed investigation. Sixteen different geomorphological environments were mapped as separate areas, or zones, and then described and interpreted. The archaeological record was surveyed using an off-site approach, and a stratified sampling technique was employed. The geomorphological zones formed the survey strata, and this enabled the archaeological record to be related to the geomorphological record.

The archaeological material collected during the surface survey derived from the Early, Middle, and Later Stone Ages (Modes 1, 3, and 5 respectively – Modes 2 and 4 are rare or absent in the area), and from the Iron Ages, although diagnostic Early Iron Age material was limited to one artefact. The off-site survey provides artefact density data in the landscape context. Artefact attributes such as raw material, abrasion state, and artefact type, provide information on behavioural patterns as well as the condition of the artefacts themselves.

Geomorphological deposits from the Pleistocene are primarily confined to abandoned alluvial fan deposits and the occasional exposed sedimentary sequence. The alluvial fan deposits are now heavily dissected by Holocene and present day seasonal streams; these deposits contain sequences of archaeological material from the Early to Middle Stone Age, with Later Stone Age material distributed on the surface. The archaeological record for the Early and Middle Stone Age in these areas has been extensively disrupted by both the formational processes of the alluvial fans and subsequent erosion. Raw material selection can be identified for the Early and Middle Stone Age, and there is some limited spatial data for Middle Stone Age activity. Geomorphological controls have restricted the activity of the Luangwa River to a 2km to 4km wide floodplain within the survey area. The deposits here are primarily Holocene, and Early and Middle Stone Age material has either been removed entirely, or is contained in secondary contexts.

Geomorphological processes have least affected the archaeological record for the Later Stone Age and the Iron Age, with the exception of some recent Holocene erosion. These two periods contain the most intact spatial data in the landscape, and distinct patterns can be discerned that are related to environmental phenomena such as hill slope, distance to water sources, and soil type. These patterns are used to produce a predictive model, using a Geographic Information System (GIS), of the land-use intensity of the two potentially contemporaneous groups of people. Producing a predictive model using off-site data has not been attempted before, and the results cannot be statistically tested in the usual manner proscribed by GIS practitioners in archaeology. Qualitative examination of the model reveals that it fits the known artefact distribution. The models provide visual comparisons of the land-use strategies of the two groups of people.

A small number of transects were surveyed outside the main survey area, and on some of these, material that might derive from the Early Stone Age appears to survive in a primary context.

Acknowledgements

This research has benefited enormously from the input and support of many individuals and organisations. Thank you to you all!

The Past Peoples and Environments Project (PPE), of which this project forms a part, was funded by the Arts and Humanities Research Council. Without this funding this thesis could never have been undertaken, and I am very grateful to the AHRC for their support.

Both my supervisors have provided a vast amount of support, encouragement, and assistance, and many thanks go to both. To Larry Barham, for his confidence in me as a student to undertake the project, and in particular for assisting with, and managing, the countless logistical arrangements for the fieldwork in what is a remote and hazardous part of Zambia. Matthew Fitzjohn provided training and assistance in all matters GIS. As both I and Larry were new to GIS at the start of the project, much more input was provided by Matthew than is usually the case for a second supervisor.

In Zambia, much valued assistance was provided by the National Heritage Conservation Commision. Collins Chipote in particular assisted me in the solo trip I undertook for my second season in Zambia, both in regard to obtaining permission to research archaeology in Zambia, and as an excellent guide in Lusaka to organise permits from the Zambian Wildlife Authority (ZAWA) to undertake research inside the South Luangwa National Park (SLNP). Everyone at ZAWA regional headquarters in Mfuwe has my gratitude for providing me with scout cover and keeping me out of harms way. In particular Kephas Kakumbi, my scout for the majority of my second season, was good humoured and good company throughout the surveying. All the staff at Wildlife Camp, my home in Zambia, were excellent company and provided vital logistical support in the form of food, a safe place to camp, and a hire vehicle. The mechanics there were excellent at keeping me on the road, despite the best efforts of my inexperienced bush driving at destroying the 4x4 I hired from them. Steve and Anna Tolan were a valuable source of local knowledge and kindly store all the artefacts and equipment involved in the PPE project. They also provided a rescue in the case of one of my two serious breakdowns (mechanical), and a cup of tea or two and an hours company here and there throughout.

The PPE project is multidisciplinary, and I have benefited from many discussions with individuals who have accompanied the main project, these individuals include (in roughly chronological order): in the preliminary trip, Prof. Ann Wintle and Dr Mayank Jain. In the first season Dr Alf Latham, Dr Pete Ditchfield, Dr Laura Bishop, Dr Andy Plater, and Dr Simon Turner. In the second season Prof Geoff Duller, Dr Stephen Tooth, and Dr Sumiko Tsukamoto.

Dr Liz Maher requires a special mention, she accompanied me to Zambia for part of my second season and, as well as providing a general refresher course in geological field techniques, provided challenging discussions, and introduced me to clast count analysis, without which this project would have been much weakened.

At the University of Liverpool various people have assisted me in my research. Andy Wilson provided instruction in the use of the archaeology departments GPS and GIS enabled PDA (as well as general advice on the functions and errors contained in GIS). Tino DaCruz in the geography department tracked down many Zambian Survey maps covering the South Luangwa region, which are rare and hard to obtain. He also provided assistance in scanning the maps and the aerial photographs used in the project. Jennifer Mirdamadi provided assistance in the Hartley Building of the archaeology department, particularly in helping sort out intransigent technologies such as printers and computers. Assistance was provided from across the pond, by Bill Phillips at the University of Idaho, who provided some of his initial findings regarding the geology and date of the Manzi River section before they were published. Dr S.D. Prince at the University of Maryland, provided an invaluable copy of the original digital files used to create W.L. Astle's (1995) vegetation map of the SLNP.

My PhD colleagues at Liverpool have provided much valued friendship and advice. Those who share my office, Adnan, Alexis, Caroline, Holly, and Lyn have in the last 6 months been witness to my frustration with all things computational, and I apologise to them for having to put up with what at times can only be described as a Victor Meldrew type character talking, muttering, and generally saying unpleasant things to his computer. Others, whose research is also based in Zambia to one extent or another have often shared knowledge about often difficult to obtain resources, as well as friendship, thank you to Ginette, Marcelle, Nick, and Rosie.

Finally, a thank you to all who read and commented on draft chapters of the thesis during the initial write up, and to the reviewers for the Cambridge Monographs in African Archaeology, who's comments helped adapt the thesis to a format more suited to a professional publication.

This thesis is dedicated to all who assisted or were involved in the Past Peoples and Environments project, and to the residents of Mfuwe, who are always welcoming and friendly to us all.

Dan Colton. August 2009

List of Figures

List of Figures (continued)

Figures in appendices

List of Tables

List of Tables (continued)

Tables in appendices

1. Introduction

1.1. Project Location, Background, and Primary Aim

The Luangwa Valley covers about 40,000km^2, and it is 700km long and 90km across at its widest (Astle *et al.* 1969:143). The valley, situated in eastern Zambia, is flanked by Archean granites and metamorphic rocks on its eastern and western margins, and is primarily floored by Karoo age sediments (Dixey 1937a: 52; Theime & Johnson 1981; Utting 1988:1). The Luangwa River that flows through it arises in the north-east of Zambia close to the Malawi border, and flows south-west to join the Zambezi River (Gilvear *et al.* 2000:424) (figure 1.1, front page). No systematic investigations of the archaeological record of the Luangwa Valley have been instigated before the 'Past Peoples and Environments in the Luangwa Valley' project directed by Lawrence Barham. However, the presence of Stone Age tools has been noted by geologists working in the area (Utting 1988:6), and a number of what are presumed to be surface collections of archaeological material have been made (recorded by Clark 1967, see chapter 2). There has also been little investigation into the Pleistocene and Holocene geological deposits of the valley, and no attempt at reconstructing its Pleistocene environments. A broad understanding of the structural formation of the valley is known, and is primarily based on work conducted in the 1930s (Dixey 1937a; 1937b).

The Past Peoples and Environments (PPE)[1] project, of which this survey forms a part, is primarily based in the southern portion of the Luangwa Valley. This survey is specifically located in and around the South Luangwa National Park (SLNP) and its surrounding Game Management Area (GMA). The general location of the park is shown in figure 1.1, it covers approximately 9050km^2 and varies in elevation from 500 – 800m above sea level, and is ecologically diverse compared to many parts of Africa (McIntyre 2004:278). The PPE project aims to record, sample and date sites of all periods to build a chrono-stratigraphic framework for the area (Barham & Jarman 2005:114). One of the key aims within this is to reconstruct the palaeoenvironmental record for the valley, which is critical, as it is possible that it would have served as a refuge and corridor for plants and animals during drier glacial climatic episodes that rendered western Zambia an inhospitable desert (Barham 2001:71-76). The PPE project was already underway when this research commenced, and it was already clear that the Luangwa Valley has dense accumulations of archaeological material in parts of the landscape. Archaeological material is contained within various sedimentological environments, which encompass an actively changing landscape varying from seasonal floodplains to weathering Escarpments. The various archaeological deposits represent the entirety of African pre-history from the first Early Stone Age Mode 1 (Oldowan) artefacts to Late Iron Age settlements, some of which were only abandoned in the early 20th century (Barham & Jarman 2005:116). The primary aim of this research is to develop an understanding of the geomorphological processes in the South Luangwa landscape (the n-transforms) that have affected the archaeological record, and then interpret the genuine human behavioural residues that remain (the c-transforms) (Schiffer 1987:7 & 22). No previous research has been undertaken in the area, and given the broad archaeological objectives of the PPE project, it was important to determine the depositional environment of the artefact bearing sediments, and the nature and condition of the archaeological material in the various geomorphological terrains of the park. This research represents a first step towards constructing a geomorphological framework for interpreting the archaeological record of the SLNP, and hopefully it will enable future researchers to target their efforts on areas and periods of most promise.

[1] The PPE project is an AHRC funded project, which began in 2003, and was preceded by exploratory fieldwork in 2002 at the southern end of the SLNP.

1.2. Methods Employed in the Research

The South Luangwa National Park has one of the highest densities of game in Africa (McIntyre 2004:278), and consequently any research involving fieldwork is a potentially hazardous enterprise. Therefore the methodology had to be designed with a number of constraints in mind, for example, it was necessary to work with an armed scout at all times and to liase with hunting safaris that operate in the GMA. The methodology for this survey was based on some prior knowledge of the South Luangwa environment gained from accompanying the PPE project the year before the PhD was undertaken. Despite this initial experience, the initial survey design still proved unsuitable for the environment, and was radically redesigned for the second season. The result is a unique survey design that allowed a large amount of ground to be covered by a single surveyor. The use of Global Positioning Systems (GPS) and a Personal Digital Assistant (PDA) to display the survey map in the field further reduced the amount of time needed to locate survey units. This survey technique could easily be adapted to a number of different African environments, and could conceivably allow cheap, accurate, and detailed survey work to be undertaken in areas where it would otherwise be too expensive or logistically impractical to put a large team in the field.

To achieve the primary aim of this research an archaeological landscape survey was conducted, and a geomorphological survey built into the project design. An off-site approach (Foley 1981:10) was used for the archaeological survey. Developed in the mid 1970s, off-site spatial analysis sought to create a framework for interpreting archaeological material outside the traditional site-based framework, treating artefacts as the minimal spatial unit which, rather than occurring at discrete locales, occur as continuous distributions of variable density in the landscape (Dunnell & Dancey 1983:270; Foley 1981:10). The off-site approach was deemed particularly suitable given the scattered distribution of artefacts observed previously in the SLNP. It would enable systematic analysis of the density of archaeological material with reference to the various geomorphological terrains of the valley. To achieve an effective spatial analysis it was necessary to create a sampling strategy that would allow the different geomorphological areas to be compared. Stratified sampling, where a sample population is divided into sub-populations (strata) (Banning 2002:115), was selected as there was sufficient justification to believe that archaeological material may vary between two or more different areas. The geomorphological areas in the valley contained varying densities and types of artefacts, these geomorphological areas, or zones, as they are referred to in the survey design, formed the strata on which the archaeological survey was designed. Geographic Information Systems (GIS) have become widely used in archaeological investigation since their introduction to the discipline in the 1970s (Wheatley & Gillings 2002:18), and in this study the GIS 'ArcView' was used throughout the planning, implementation, and analysis stages of the project. ArcView provides a platform for undertaking integrated spatial and statistical analysis of data, especially across landscapes.

A range of geological techniques was employed to develop interpretations of the Pleistocene and Holocene geomorphological record of the valley. These included identifying and recording geological sections in the field, and identifying the boundaries of different geomorphic deposits to construct a geomorphological map. A clast count analysis was conducted to provide data that helped determine the provenance of the deposits. No basic geomorphological data exists for the SLNP and the Luangwa Valley in general, so these methods though simple, provided a robust dataset from which to develop an understanding of the formation of the Pleistocene and Holocene deposits of the region.

2. Background to Archaeological Research in Zambia and the Surrounding Region

2.1. Introduction

Extensive research had not been undertaken in the immediate area of the Luangwa Valley before the initiation of the Past Peoples and Environments project, and the chronology used here is primarily based on published examples from elsewhere in Zambia and beyond the immediate region. An understanding of the chronology is critical to the research as it forms the primary source for indirectly dating the archaeological material, this is because the fieldwork for the archaeological aspect of the project entailed a landscape survey, which by its nature is restricted to using stylistic and typological associations to date finds in the absence of applicable radiometric or other dating methods.

The primary aim of this project is to assess the integrity of the archaeological record in reference to geomorphological effects to determine what remains of the human behavioural record. This primary aim necessitated a methodology that would produce a sample that factors out unconscious or conscious biases of the surveyor introduced by selective collection of material by prospective sampling. This chapter will demonstrate that most landscape orientated surveys conducted in Zambia, and on the continent as a whole, have primarily been aimed at sampling material in a prospective manner to aid the construction of local and regional chronologies. It is apparent that few African surveys have had a theoretical underpinning that allowed a sound assessment of the archaeological record in the context of the landscape, a key exception being the off-site approach developed by Foley (1981) for his research in the Amboseli area of Kenya. It was also necessary to look to projects elsewhere in the world for models on which to base the methodology, particularly as the analysis was to be conducted in a GIS, which is a practice rare in African archaeology.

2.2. The Archaeology of Zambia and Central Africa

2.2.1. Historical context of archaeological research in Zambia

Archaeological research in Zambia has been greatly affected by geopolitics and the varying research interests of individuals and groups within archaeology. This is typified by the experience at Mumbwa caves in central Zambia (figure 1.1). This site was first investigated from the 1920s to the 1940s by at least three different research teams, MacCrae in 1925, an Italian scientific expedition in 1930, and Desmond Clark, shortly after his arrival in Zambia in 1938 (Fagan 1967:8). The archaeology found was described using European Palaeolithic terminologies, reflecting the pioneering stage of African archaeology in the 1920s (Barham 2000:3). Excavations at Mumbwa in the 1930s by Desmond Clark enabled him to start constructing a Stone Age culture-historical framework for the region, later augmented by excavation at other sites such as Kalambo Falls (*ibid*). Further excavations focusing on the Later Stone Age were undertaken in 1973 as part of a doctoral research project (*ibid*: 7). After this, despite the long sequence and relative early fame within Palaeolithic archaeology, Mumbwa Caves were almost forgotten through the 1970s and 80s due to regional political and economic turmoil, lack of academic support, and higher level of interest and excavation in South Africa (see Barham 2000 for a comprehensive summary).

Archaeological research in Zambia as a whole ceased to be a priority in the immediate post independence decades, and active field research dwindled since the 1970s, with a couple of exceptions. These exceptions include Robertson (2000) who conducted a survey in 1977 – 1979 (Robertson 2000:147), Derricourt from 1974 – 1976 (1985:16), and Michael Bisson (1992) who apparently conducted the last season of fieldwork in November 1980 (Bisson 1992:238). Zambia effectively became a country with a great many sites and few archaeologists, a situation exacerbated by severe environmental and infrastructural constraints on survey and reconnaissance (Bisson 1992:234). These problems remain today, and are particularly obvious when Zambia is compared with countries like Kenya and South Africa. Despite these impediments, research has been reinvigorated with further excavations at Mumbwa caves and Twin Rivers from 1993 to 1996 (Barham 2000), along with reanalysis of material previously excavated from the sites (Clark & Brown 2001). Clark's 2001 publication of the final volume of the Kalambo Falls excavations (Clark 2001) has also helped refocus attention on Zambian archaeology.

The recent reactivation of research into the Zambian Middle Stone Age, which has benefited by re-dating with modern techniques (Barham 2000), has allowed it to be tied into recent debates regarding the origins of modern human behaviour (e.g. Barham 2002; 2007; McBrearty & Brooks 2000:518), and how the development of these behaviours may be related to climate change throughout the duration of the Middle Stone Age (Avery 2003; Barham 2001). The Luangwa Valley itself has never been the subject of systematic investigation until the initiation of the PPE project. Research and analysis is ongoing, though publication is already completed for the discovery of a *Theropithecus* femur (Elton *et al.* 2003), cosmogenic nuclide dates for one site (Phillips *et al.* 2005), and for an excavation report of one the Iron Age sites excavated in the GMA (Barham & Jarman 2005).

2.2.2. A chronology for the Zambian archaeological sequence

Although the Luangwa Valley has not had any major investigations into its archaeological record the presence of archaeological material has been noted in a small number of publications. MacCrae and Lancaster (1937:62) report a surface collection in the Viziwa Stream in the northern part of the Luangwa Valley, which appeared to be a mixed assemblage (MacCrae & Lancaster 1937:63). In his Atlas of African Prehistory J.D.Clark plots 19 archaeological localities in the Luangwa Valley (Clark 1967)[2], at least four of which are presumably the surface collections made in 1938 by Lancaster, cited in Clark's 1954 paper on the Sangoan industries of the Luangwa Valley (Clark 1954:202). The atlas was compiled using a combination of data supplied by regional correspondents and Clark's own research (Clark 1967:6), descriptions of these sites appear not to have been published elsewhere, and it is probable they are the result of limited surface collections at various localities rather than systematic fieldwork. It is also possible that the industries identified at each locality would be classified differently today. Prior to the initiation of the PPE project the most recent record of archaeological material in the valley is found in a geological special publication covering the South Luangwa National Park (Utting 1988:6), which provides a brief description of stone tools and their raw materials from a locality near Chipembele Ridge (Rhino Hill) (see figure 5.4 for location of Chipembele Ridge – zone 12).

The paucity of prior research in the Luangwa area requires the archaeological sequence to be constructed with reference to research elsewhere in Africa and Zambia. Taking a global perspective G.Clark (1971) observed that stone tool technology developed following broadly similar stages over much of the world, and he identified and defined five stages, or Modes, of technological development (Clark 1971:31). Though the appearance of new Modes usually occurs in roughly the same order, it is only rarely synchronous in different geographic areas (*ibid*: 30). Hominin dispersals and differing environmental factors inhibiting or disrupting occupation can cause parts of the Mode sequence to be missing in certain areas (*ibid.*). For ease of discussion in this thesis the archaeological sequence is referred to using the Early, Middle, and Later Stone Age terminologies, because Mode 1 technologies continue to be produced into the Later Stone Age. Where possible, in the context of this study, the dates for the development of different technologies are retrieved from the nearest sources of excavated data so that the chronological sequence is more likely to represent the sequence of the Luangwa area. There is an element of unreliability in typologically dating artefacts in this manner as 'typical' or diagnostic artefacts of one technology may continue to be produced in later periods i.e. Early Stone Age (Mode 1) artefacts continue to be produced right through to the Later Stone Age alongside Mode 5 though usually, but not always, with a difference in size. In this study, the problem of technological continuity was overcome by creating separate classifications for material that might derive from two or more periods (see appendix 7.1). These intermediate classes of artefact could then be excluded from the analysis where necessary so that a smaller but more reliable sample could be obtained. Excavations are ongoing as part of the PPE project, and observations made during these excavations (*pers obs* and Barham *pers comm.*) have confirmed that the general sequence matches that expected based on the typologies and reports consulted below.

2.2.2.1. The Early Stone Age (Modes 1 and 2).

Deposits associated with Early Stone Age material have been dated in the Luangwa Valley; Phillips *et al.* (2005) used cosmogenic nuclides to date alluvial deposits containing Oldowan (Mode 1) artefacts found in a layer of cobbles near the base of a dissected terrace on the Manzi River, a tributary of the Luangwa. The date retrieved for the deposition of the artefacts was 0.9 to 1.0 Ma, which is a relatively recent date for the Oldowan (*ibid.*). Comparisons to palaeomagnetic data support this date (Barham *et al.* in prep), although the normal polarity of the sediment allows some wiggle room to match the deposit to earlier sub-chrons that fall within the range of the Oldowan (*ibid.*). However, isothermal luminescence (ITL) and cosmogenic nuclide dates from the upper units of the Manzi section are similar, suggesting that the dates for the lower units are reliable (*ibid.*). This date is approximately 600,000 years too young for Oldowan artefacts compared to the known range elsewhere in Africa (2.6 – 1.7 Ma), and as the unabraded condition of the artefacts suggests that they are *in situ*, it is possible therefore that the Oldowan material at the base of the Manzi sequence represents a localised response to raw material constraints (*ibid.*). The Ethiopian site of Gona has yielded the world's earliest date for any Early Stone Age material, and is dated to 2.5 - 2.6 Ma (Semaw *et al.* 1997:333). In southern Africa the oldest dates for the Oldowan are from Sterkfontein at an estimated 2 - 1.7 Ma (Kuman & Clarke 2000:829). The oldest Oldowan technology of the Luangwa Valley could presumably originate from anytime between these two dates or soon

[2] These include eight localities in the mid-Luangwa Valley of uncertain Lupemban affiliation (Clark 1967: overlay 16), nine certain Sangoan localities (*ibid*: overlay 15), and two uncertain Nachikufan localities (*ibid*: overlay 22).

after. However, there is no reason to suspect that it was produced any earlier than the earliest dates in Ethiopia. A conservative estimate of a maximum of 2 million years before present is suggested as a guideline for the oldest age for Early Stone Age material in the area.

Oldowan (Mode 1) artefacts tend to be simply or opportunistically worked, and the technology primarily consists of cores and flakes where frequently only a few flakes would be struck from each core (Klein 1999:230). There is evidence at Lokalalei 2C, dated to 2.34 Ma, of planning and foresight in procurement and management of raw material, comprehension of basic geometric concepts, and the ability to overcome some technical difficulties in the knapping process. However, the industry is geared toward the production of flakes, and cores clearly fall into the category of a waste product (Delagnes & Roche 2005:466). Some classificatory systems have been complex, for example, in M.D Leakey's (1966) analysis there were six major categories for cores (M.D. Leakey 1966:464). This overemphasises the formality of the industry, as most of these core forms can be created without premeditation, and in fact many can grade into one another during core reduction (Toth 1985:107). So as not to over emphasise the formality of Mode 1 tools in this project, they were simply identified as cores or flakes, and no further subdivisions were used (see appendix 7.1 for definitions of identifying characteristics). The photographs in figure 2.1 illustrate typical Mode 1 cores and flakes retrieved during this research. Mode 2 technologies, or bifaces, typical of the Acheulian tradition are also part of the Early Stone Age repertoire, appearing after the Oldowan. The earliest Acheulian material has been found at Turkana in Kenya and is dated to 1.65 Ma (Klein 1999:333), and a similar early date of 1.5 Ma has been retrieved for Acheulian material at Konso-Gardula in Ethiopia (Asfaw *et al.* 1992:732). Bifaces are rarely found in the area of the Luangwa Valley where this research was undertaken (*pers obs*). However, a small number have been found in streams in and around the National Park (Barham *pers comm*), suggesting that the hominins in the area had the technical ability but, for reasons perhaps linked to raw material availability, did not manufacture handaxes or cleavers in the quantity seen in other areas of Zambia (Barham *et al.* in prep), such as Kalambo Falls (Clark 2001).

Figure 2.1: Examples of Early Stone Age (Mode 1) artefacts. a and b) Cores from this survey – transects GR2 and K2 respectively. (c and d) Ventral and Dorsal side respectively of a flake (abraded) – from transect K2. Scale is cm subdivided into mm.

2.2.2.2. The Middle Stone Age (Mode 3).

The Middle Stone Age of eastern Zambia was initially dated from a small number of sites excavated by D.W. Phillipson. At the Kalemba rock-shelter site near Chipata to the south of the Luangwa area (figure 1.1), radiocarbon dates indicated the Middle Stone Age was 35,000 years or older (Phillipson D.W. 1976:189). The dates retrieved for this material might have been at the limits of the radiocarbon technique used at the time. However, a more certain bracket of 25 – 22,000 bc[3] was given for a more recent level of the site (*ibid*). Cobble layers with Middle Stone Age (Mode 3) artefacts from the Manzi sequence in the Luangwa Valley have been dated using both cosmogenic nuclide and isothermal techniques, and are dated to 45 – 90 ka (Barham *et al.* in prep). Recent research elsewhere in Zambia suggests that the earlier Middle Stone Age is of much greater antiquity, with dates averaging at 260 ka for Twin Rivers in central Zambia (Barham 2002:585). This is concordant with many other dates that have been retrieved in Africa for the appearance of the Middle Stone Age, suggesting it appeared approximately 250-300 ka (McBrearty & Brooks 2000:453). The archaeological record at Mumbwa Caves (central Zambia) provides the only dated sequence available from central Africa and north of the Limpopo River that spans the Middle to Late Pleistocene (Barham 2000:243). The Middle Stone Age occupation at Mumbwa is dated between 130 ka and 40 ka and its Mode 3 industry is characterised by prepared core technology, bifacial and unifacial points, awls, scrapers, and rare backed blades (*ibid*: 41). Two periods of abandonment at Mumbwa Cave occur at 170 – 130 ka and 6 – 2 ka, and are associated with regional scale aridity phases in central Africa (*ibid*: 49). The Middle Stone Age dated elsewhere in Zambia to 25 ka may represent a minimum age (Barham 2002:598), or a genuine expression of the variability of the continuation of Middle Stone Age knapping practices (McBrearty & Brooks 2000:532).

Figure 2.2 illustrates examples of Middle Stone Age material discovered during the survey. The typical Middle Stone Age artefacts recognised in the survey are Mode 3 technologies, which are defined as flakes removed from prepared cores and the prepared cores themselves. The preferential flakes removed from these prepared cores will usually retain a number of scars on the dorsal surface from the core preparation, and are likely to have a dihedral or multi-facetted butt. The cores produced using this method will usually have been worked using a Levallois reduction strategy, the technological classification for which was taken from Chazan (1997:725). Radial cores do not always fulfil the

Figure 2.2: Middle Stone Age artefacts collected during this survey: a) Flake from a prepared core, dorsal surface displayed left and ventral right. b) Prepared core. c) Point from a prepared core with ventral surface displayed left and dorsal to the right.

[3] Dates quoted by Phillipson are uncorrected radiocarbon dates when expressed as bc or ad, and calendar years when expressed as BC or AD.

technological requirements of the Levallois technique, i.e., the knapping surfaces are not always hierarchically related in radial cores, and these radial cores were simply recorded as Middle Stone Age cores at the analysis stage.

2.2.2.3. The Later Stone Age (Modes 4 and 5).

The first elements of the Later Stone Age (Mode 5) to appear in eastern Zambia are dated at Kalemba to 23-22,000bc, but the technology is not fully represented until 15,000bc (Phillipson D.W. 1976:189). The Leopard's Hill Cave sequence to the south-east of Lusaka provided a comparison to the Kalemba sequence (*ibid*: 199). The lowest level of Leopard's Hill is dated slightly later than Kalemba to 21-19,000bc, and the industry found there was described as proto-Later Stone Age (Clark 1970:199), and interpreted as being more evolved towards a Mode 5 technology (Phillipson D.W. 1976:200). After a hiatus of 5,000 years, occupation was resumed at Leopard's Hill (*c.* 14,000bc) at which point a fully-fledged Mode 5 industry was being produced, and this has been subsumed under the 'Nachikufan I' typological grouping (Miller 1969a, quoted in Phillipson D.W. 1976:201). At Mumbwa Caves in central Zambia a later date of 12-6.6 ka was retrieved for the Later Stone Age (Barham 2000:41). However, the sequence from the Late Glacial is missing (15 – 12 ka), so the date reflects reoccupation after a period of abandonment of the site rather than a date for the arrival or development of the Later Stone Age in the area (*ibid*: 243). Throughout the rest of Africa the Middle – Later Stone Age transition appears to have occurred earlier at 50 ka (McBrearty & Brooks 2000:532); sites such as Kalemba are unusual with the persistence of Middle Stone Age Mode 3 technologies as late as 25 ka, but, this could represent a genuine expression in the variability of the appearance of the Later Stone Age (*ibid.*). The sequence at Kalemba demonstrates a steady development from a Mode 3 to a Mode 5 industry (Phillipson D.W. 1976:199), and this is mirrored elsewhere in southern Africa where there appears to be a direct transition from Mode 3 to Mode 5, though the timing and causes of the transition are poorly known at present (Mitchell 2002:112; Phillipson D.W. 2005:104-5). Mode 4 industries, typified by punch struck blades with steep retouch (Clark, G. 1971:31), are not represented south of the Sahara (Phillipson D.W. 1976:4). Although systematic blade production is seen in the Middle Stone Age industries of the Kapthurin formation (Kenya) (Tryon & McBrearty 2006:493), and at Mumbwa Caves and Twin Rivers (Zambia) (Barham 2000:192), the manufacture techniques in the latter sites are difficult to determine (*ibid*: 100), and the blades at Kapthurin were manufactured using a direct hard hammer technique (Bar-Yosef & Kuhn 1999:333). The development of Mode 5 industries in southern Africa appears to stem from the use of backed blade Mode 3 technologies, which gradually reduced in size over time to form the microlithic Mode 5 technology of the area (Phillipson D.W. 2005:104).

Across Zambia numerous separate Later Stone Age industries have been recognised, over which there has been some wrangling regarding their typological groupings (Bisson 1990:116; Phillipson D.W 1976:204). Essentially, there were three groups of tool making traditions identified in Zambia: the 'Zambia Wilton', 'Nachikufan', and the 'Makwe' (Phillipson D.W. 1976:204). The 'Zambia Wilton' is found to the centre and the south of the country (*ibid*: 13). The 'Makwe' is found in eastern Zambia and was first defined by Phillipson based on his excavations at Makwe, Kalemba, and Thandwe, with the typical sequence at Makwe (*ibid*: 190). The 'Nachikufan' was first defined by J.D.Clark (1950); it was named after the type-site of Nachikufu Cave and is found in central and northern regions (Musonda 1987:152). Clark (1950) divided the Nachikufan into three phases. This was later slightly altered to split the second phase into two sub-phases (Sampson & Southard 1973:79). Musonda maintained the simpler three phase system of Clark and provided different dates to that of Miller (up to 2,000 years difference either way (see *ibid*: 86 - 88)). The first phase extends from 18-9,000BP and encompasses all assemblages identified as Nachikufan I, the second phase from 9-5,000BP includes the Nachikufan II material, and the final period corresponds to Nachikufan III at 5,000-100BP (Musonda 1985; quoted in Bisson 1990:116).

Re-evaluation of the Nachikufan typological groupings has revealed that there is no clear differentiation between some of the industries and that the group III industries in particular are not homogenous and can be subsumed within the other groupings (Bisson 1990:117). A consensus has not yet been reached regarding the classifications; however, it is clear that, although regional clustering is present, discrete geographic boundaries cannot as yet be identified (Phillipson D.W. 1976:204). Different cave sequences rarely resemble one another (Sampson & Southard 1973:88), and the variation of the industries may be caused by factors such as chronological change, differing subsistence strategies affected by different environments, or a combination of these along with genuine cultural differences (Bisson 1990:117).

For the purposes of artefact recognition in this project the Later Stone Age material excavated at SL3 'Poachers' Cave' and SL8 provided a guide to the nature of Later Stone Age working in the area; the industry descriptions from D.W. Phillipson's (1973; 1976) work in eastern Zambia also acted as a guide. Figure 2.3 illustrates some of the typical Later Stone Age artefacts recovered on this survey. Many of the Later Stone Age cores in the Luangwa area were water worn cobbles that were first split into two or more pieces using a bipolar anvil and hammerstone technique. This technique involves placing the quartz cobble on an anvil stone and striking the cobble with a hammerstone to split it into two or more pieces (Barham 1987:46). The split cobble would then allow the knapper to exploit the circumference of the cobble as a platform for removing microlithic flakes and blades, thus allowing for a variety of tool forms to be produced (*ibid.*). One of the most common forms of microlithic tools recovered from Makwe and Kalemba are pointed lunates, which are segment-shaped bladelets that have been backed by retouching the curved edge of the microlith (figure 2.3b) (Phillipson D.W. 1976:81 & 135). Though Phillipson recognises a number of types of backed lunate miroliths, they are referred to here generically as segments, as few were recovered during the survey and it was not possible to identify them separately. A total of 27 of the backed lunates recovered from Makwe and Kalemba retained mastic indicating that the microliths were hafted. The mastic was usually found on the backed part of the tool suggesting the sharp edge was the functional edge (*ibid*: 217). Hypothetically, more than one segment could be hafted and a number of different arrangements could be envisaged to create various tool forms beyond those outlined by Phillipson (*ibid.*).

2.2.2.4. The Later Stone Age (Mode 5) to Iron Age transition.

The Later Stone Age persists over much of Zambia to at least the 19th century (Musonda 1987:156). It overlaps chronologically with both the Early and Late Iron Ages, and the nature of the interaction between the two groups has been a focus of interest in Zambian archaeology (Bisson 1990:136-7, 1992:234; Miller 1969b:82; Musonda 1987:156; Phillipson, D.W 1976:206). Three competing theories have been put forward: the first model suggests that the groups remained separate but developed trade links (Miller 1969b:87). This model was revised in response to the discovery of an iron-smelting furnace at the Later Stone Age site of Nachikufu Shelter (*ibid.*). Despite its presence in the sequence there was still no evidence that Later Stone Age people manufactured pottery. In this scenario it was suggested that hunter-gatherers had learnt how to smelt iron and traded to obtain the pottery (*ibid.*). Musonda (1987) presented the second model, suggesting that there was virtually no contact between the two groups in the Early Iron Age, and even minimal exchange networks were not established until the second millennium AD after a long period of coexistence. Most of the pottery found in Later Stone Age sites was therefore scavenged from abandoned Iron Age sites, and possibly collected for their intrinsic beauty (*ibid*: 155). Musonda's interpretation was based on his research in central Zambia where, he suggested, the lack of contact was due to the combination of the environmental diversity of the region in which the two peoples practised different subsistence strategies, and low population densities. Added to this was the likelihood that the Iron Age peoples did not use domestic livestock

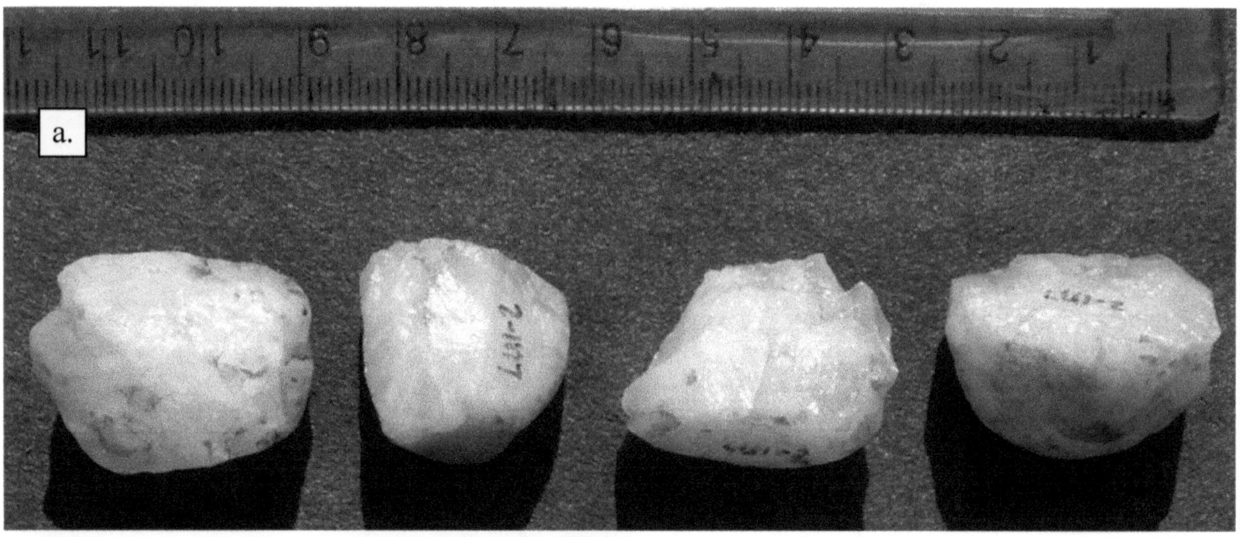

Figure 2.3: a) Examples Later Stone Age mode 5 cores from this survey, produced on small quartz water worn cobbles by first splitting the cobble to produce the necessary platforms for microlithic working. b) Example of a segment recovered in the survey, both faces illustrated. Scale in photographs is in cm, subdivided into mm.

due to the prevalence of tsetse flies, further reducing potential pressures on resources (*ibid.*). It is possible that in the ecologically homogenous forests that occur elsewhere in Zambia relations between the two groups might have been less than friendly, leading to intentional avoidance or actual conflict (Bisson 1992:244). It is possible that the disappearance of hunter-gatherers and the widespread ambivalence towards them in ethnohistories recorded in many Zambian groups is a consequence of long-term tensions engendered by these conflicts (*ibid.*). The third possibility is that the Early Iron Age represented a period of transition from a hunter-gatherer lifestyle to that of permanently settled communities (Robertson 2000:179). Early Iron Age sites from before AD500 were not permanent villages, but were temporary features set up by a portion of a community for use whilst tending crops, while other parts of the group continued to hunt and gather (*ibid.*).

Based on current data it is difficult to determine which, if any, of these suggestions is the more likely in Zambia as a whole and in the eastern part in particular. It is possible that separate scenarios should be considered in parts of Zambia with differing environments and population densities. The development of the second and third theories presented here is based on different excavations in the same region of central Zambia. It would be difficult to support the suggestion that different contemporaneous scenarios could have been possible in one part of Zambia; it is more probable that the two theories result from differing interpretations of similar material (small amounts of pottery found at Later Stone Age sites). Musonda (1987:155) proposed a test to determine if Later Stone Age people were collecting potsherds for curiosity value, he suggests that it would be likely that high proportions of decorated sherds would be found in Later Stone Age contexts if they were collected for their intrinsic beauty (*ibid.*). The spatial analysis conducted in this project addresses this issue with analysis of the pottery types found in association with Later Stone Age scatters. The analysis of the spatial distribution of material in the landscape also provides a comparison of the land-use of the two groups, to determine whether they overlap (section 7.7.5). It is clear that more research is needed to better understand the nature of the transition in each area of Zambia. D.W. Phillipson (1976) argues that the appearance of the Early Iron Age was the result of migration into the area of people from the north-east, along an eastern stream of migration that initially passed through Malawi (Phillipson D.W. 1976:208). This was based on his research in eastern Zambia indicating that the Early Iron Age appeared suddenly and that there are shared typological affiliations with all pottery in eastern and southern Zambia, Mozambique, and Malawi (*ibid.*). This industry is a component of the Chifumbaze complex (Mitchell 2002:264), and pottery is characterised by stamped or incised designs, and rim forms that are most commonly internally or externally thickened (Barham & Jarman 2005:118). As it is likely that pottery was introduced to the area by Iron Age migrations a developmental sequence in the Luangwa Valley from the Later Stone Age to the Early Iron Age would not be expected.

2.2.2.5. The Iron Ages.

In spite of the fragmentary nature of Iron Age research in Zambia regional chronologies have been created for the Iron Age across the country. Based on the four archaeological sequences excavated by D.W. Phillipson in eastern Zambia from 1966 – 1971 (Phillipson D.W. 1976:193), and from other excavated sites in the region, it appears that the Early Iron Age flourished between 310 and 1060AD (Phillipson D.W. 1976:333). The Early Iron Age pottery in Zambia indicates that there were several different cultural groups within Zambia, each with their own sequences of development (Phillipson D.W. 1974:1). In eastern Zambia these included the Kalambo, Chondwe, Kapwirimbe and Kamnama groups (*ibid*: 9). The decorations on Early Iron Age vessels are frequently found on the rim bands, which are often thickened, the most frequent decoration being a horizontal band of diagonal comb-stamping or incisions (*ibid*: 6). In the Luangwa Valley itself dates have been published for an Iron Age site situated on the Chowo River, a seasonal tributary of the Luangwa River (Barham & Jarman 2005:116). The oldest levels of the excavated sequence are radiocarbon dated to Cal AD 400 to 700, and the diagnostic sherds of pottery are characteristic of the Early Iron Age found over the rest of eastern Zambia (*ibid.*). The top 25cm of the sequence contained clay bricks, metal foil, and glass indicating that occupation continued into the early 20th century, and overall the sequence of radiocarbon dates indicated that there was repeated occupation of the site and potentially episodes of lengthy occupation (*ibid.*).

There appears to be no developmental sequence between the Early and Later Iron Age pottery, though chronologically it appears that the two periods may overlap (Phillipson D.W. 1976:194). The Later Iron Age 'Luangwa tradition', which appeared after 1080AD, replaced the four groups of Early Iron Age pottery in eastern Zambia. It is characterised by banded decorations which have usually been comb-stamped in diagonal bands associated with pendant segmented blocks; there is a complete lack of the thickened rims seen in the Early Iron Age (Phillipson D.W. 1974:9). The transition between the Early and Late Iron Age had not been well documented in the excavations conducted in eastern Zambia up to the 1970s (Phillipson D.W. 1976:195), and this certainly remains the case in the Luangwa Valley area. Despite having no predecessor in the region, the Later Iron Age is relatively widespread, occurring in all of eastern Zambia and much of the centre of the country; within this transition there appears to be a typological discontinuity between the periods (Phillipson D.W. 1974:16). It has been hypothesised that this represents a large population shift, in which an external group of Iron Age farmers introduced the Luangwa tradition to the area at the beginning of the second millennium AD (*ibid*:24), probably by dispersal and movement from the by the

Eastern Highland Group of Bantu speaking peoples (Phillipson D.W. 1976:213).

The term 'Luangwa tradition' is slightly misleading as it suggests that the Luangwa Valley has been researched with regard to the Iron Ages, and that the pottery found there formed the type industry for the region. However, the valley just happens to lie in the centre of the region in which it is found, and no pottery has been described to confirm that the Luangwa tradition occurs there. This of course does not suggest that it is not present within the area, but is a consequence of the lack of archaeological investigation in the valley. Figure 2.4 illustrates a sample of the diagnostic pottery rim sherds from this survey.

Figure 2.4: Examples of Late Iron Age pottery discovered in this survey (transect GA1.4.2). Similar to pottery excavated by Phillipson in eastern Zambia at Kamnama (see Phillipson D.W. 1973:17).

2.3. Brief History of Landscape Survey in Archaeology

Before proceeding to discuss archaeological survey in Zambia and Africa, it is necessary to place them in the broader context of the development of landscape archaeology as a source of behavioural data. Survey has been an integral feature of archaeological research since the origins of the discipline (Banning 2002:2; Orton 2000:68). European curiosity had led to the documentation of sites in past landscapes since the 16th century (Banning 2002:2; Orton 2000:68). As early as 1755 pioneer field archaeologists surveyed and mapped antiquities (Phillips 1980:3), and from 1801 the Ordnance Survey of Great Britain included antiquities on its detailed maps (*ibid*: 5). These very early works were not explicitly conceived as archaeological surveys and were often an add-on to geographical research or military expeditions (Banning 2002:2). Surveying as a specific enterprise in archaeology began to develop after the 1920s (Banning 2002:3; Orton 2000:68), however, the aims of early surveys were limited to prospecting for artefacts and sites using techniques such as fieldwalking (Clarke 1922) and aerial reconnaissance (Crawford 1928; 1929) – though the earliest aerial photograph taken of an archaeological site appears to date to 1899 (Barber 2006:19). As early as 1932, researchers such as C. Fox (1932) produced site distribution maps of the UK to analyse temporal patterns of behaviour in the landscape. These early studies were conceptually limited and methodologically uncritical, basing conclusions on visual appreciation of distribution maps, with no assessment of how survey intensity may have correlated to site density, and with little discussion to support the reasoning behind why some hypotheses were given preference over others (Hodder & Orton 1976:3).

Until the 1960s most surveys were conducted with the aim of locating sites suitable for excavation (Ammerman 1981:63) and were often seen as a mundane precursor to the real business of excavation (Orton 2000:68). Up to the late 1970s a 'defence' justifying the reasons for going to the effort of conducting a survey was still regarded as necessary in many projects, especially in American archaeology (Ammerman 1981:63; Schiffer *et al.* 1978:1). However, by the 1940s some surveys had begun to move beyond merely prospecting for sites; Gordon Willey's (1953) survey of the Viru Valley in Peru is widely regarded as the turning point away from the earlier site-based approach (Ammerman 1981:65; Banning 2002:4-5; Bower 1986:22; Chennall 1975:3; Orton 2000:28). In 1946 at the beginning of the Viru Valley project it was far from clear how such a survey should be conducted (Ammerman 1981:65), and at the time Willey considered that he had been dealt a bad hand in getting the survey element of the project (Willey & Sabloff 1974:149). Willey's own justification of the survey, that it would provide data from the landscape that reflected human behaviour (Willey 1953:1), arguably did not do justice to the work that he undertook. He not only concerned himself with the distribution of sites in space, but also with site functions, population sizes, and socio-political organisations (Banning 2002:5). An explicitly designed sampling procedure was not followed, and should not be expected, given that statistical theory in general was still developing (Orton 2000:68). Willey was aware that a random selection of survey sites was desirable, and that there were factors that might skew his sample of sites (Willey 1953:6). Willey's study (1953), combined with Binford's (1964) first experiments with probabilistic sampling in archaeology, paved the way for the improved spatial samples and the environmentally orientated studies of the 1970s problem based research strategies typical of the New Archaeology (Banning 2002:5; Orton 2000:69; Wobst 1983:43).

Archaeological survey during the 1960s benefited from a number of developments in the theory and practice of archaeology (Bower 1986:22). On the theoretical side, there was an increasing dependency on formulating questions in regional terms which required data to be gathered using survey techniques (Bower 1986:22; Dunnell & Dancey 1983:267). On the practical side the ability to satisfy the needs of cultural resource management in America, combined with increasing costs of excavation, placed an emphasis on survey work in general (Bower 1986:22). The main theoretical development was that of the New Archaeology; this adopted a positivist approach to the subject, with the hope that it would produce law-like statements about human behaviour (Conolly & Lake 2006:7). It was no longer regarded as satisfactory to simply suggest or assert that features appeared to be grouped together, or looked aligned along a particular landscape feature (Wheatley & Gillings 2002:6). Scientific approaches were used in the New Archaeology requiring hypotheses that could be objectively and statistically tested against the patterns seen in data which, in the case of landscape studies, were often in the form of distribution maps (*ibid*:7). These theoretical developments were concurrent with practical pressures on research; in the USA these were primarily caused by the growth of cultural resource management that followed the legal changes requiring environmental impact assessments to be made on building work that might affect the cultural landscape (Orton 2000:69). Though cheaper than excavation, the increasing costs of conducting survey further encouraged the development of methods such as probabilistic and non-probabilistic sample selection in an attempt to reduce the amount of time spent in the field (Schiffer *et al.* 1978:2).

The probabilistic sampling, or random sampling, of the New Archaeology was first explicitly introduced to the discipline by Lewis Binford (Ammerman 1981:78; Dunnell & Dancey 1983:267; Orton 2000:68). Landscape sampling as defined by Binford, is not merely the substitution of total coverage with partial coverage of an area, but is the science of controlling and measuring the reliability of information gathered through the theory of probability (Binford 1964:427). Binford's aim was to redirect the motives of archaeological research away from collecting 'samples' of cultural items in regionally defined areas, where sites discovered were treated as 'mines' for such items, towards using samples to analyse the population of cultural items within regionally defined universes, whilst being able to evaluate the results as to their reliability as an accurate representation of the population investigated (Binford 1964:433). Probability sampling was sold as an uncomplicated panacea for improving data collection and inferences made (Ammerman 1981:78). However, the technical aspects of sampling are far from a simple matter (*ibid.*; Wobst 1983). It can be argued that the development of spatial studies was slow and that it was hindered by the difficulty of getting to grips with the interface between archaeology and statistical theory (Hodder & Orton 1976:1; Orton 2000:69). The 1970s and 1980s are perhaps best seen as a period of experimentation and consolidation in probabilistic landscape surveys (Orton 2000:70) when factors specific to archaeological survey, such as the obtrusiveness and visibility of sites (Schiffer *et al.* 1978:6-7), were defined and developed for inclusion in sample design and analysis.

The debates regarding the application of sampling methods in archaeological survey that appeared in the 1970s and 1980s have effectively continued to this day. One key question of interest here is the suggestion by some researchers that only total survey can answer the most interesting questions regarding the archaeological record of a landscape (Orton 2000:70). Bintliff (2000:201), for example, entirely rejects the notion that sampling blocks of the landscape can yield useful information. However, surveys that cover the entirety of a landscape are rarely possible for logistical and financial reasons, as was certainly the case in this study, and is further examined in chapter five below with regard to the sampling scheme for this project (section 5.4).

With the development of landscape sampling in archaeology new techniques appeared, both technical and theoretical, for gathering archaeological data via landscape survey (Orton 2000:71-73). One of the key developments of interest here is the development of the "off-site" approach ("non-site" in the USA), which was initially developed by D.H. Thomas for his survey in the Reese River Valley of Nevada, in the south-western United States. He argued that the artefact should be the "basic operational unit" of investigation, rather than the site (Thomas 1975:62). This approach sought to understand the distribution of all archaeological material in the landscape, rather than the distribution of sites, due to the fact that in most regional contexts the archaeological record is continuous, with highly variable density characteristics, where sites represent only a part of this continuous record (Dunnell & Dancey 1983:272; Foley 1981:31; Thomas 1975). This is particularly true in highly mobile hunter-gatherer and herder societies (Foley 1981:29). An increased awareness that the activities of hunter-gatherer groups were not centralised around sites, and that they often left widely scattered evidence of their landscape use, also played a role in the development of the approach (Banning 2002:7; Foley 1981:12-27; Plog *et al.* 1978:387). The essential theoretical component to the approach is an understanding that the accumulation of material over time in the landscape will result in time averaging of spatial distributions (Foley 1981:23). If the landscape is stable a behavioural pattern will emerge relating to general areas or habitats seen as suitable. However, if sedimentation or erosion has been variable over a landscape then the distribution would be meaningless (*ibid.*). It is therefore vital to understand the geomorphological history and processes of an area when analysing off-site data. Though the off-site approach was developed as a method for studying hunter-gatherer societies, it is also equally applicable to sedentary societies who also leave archaeological signatures outside the site context (Keay & Millet 1991:129). Since its development the off-site approach has been applied in numerous studies in Europe and the Americas (Bintliff 2000; Gillings & Sbonias 1999; Keay & Millet 1991; Shennan 1985). An off-site approach was used in the design of the archaeological survey for this project. Foley's project provided a model for its development, so the field methodology is analysed in greater detail in section 2.5, and the specific methodology used here is developed in section 5.4.

2.4. Zambian Archaeological Surveys

2.4.1. Case studies

During the late 1960s and the 1970s six studies in Zambia were undertaken that had a spatial element to them (Phillipson, D.W. 1973; Phillipson, L 1975; Vogel 1978, 1984, and 1987; Robertson 2000). The first, undertaken between 1967 and 1969, was designed to investigate the variation of a cultural entity within a restricted geographical area, for which the Middle Stone Age (Mode 3) of the Upper Zambezi Valley was chosen (Phillipson, L 1975:1). In this study the key features that were used to identify a Middle Stone Age site were outlined and were designed to be sufficiently broad to allow all Middle Stone Age sites to be found (Phillipson, L 1975:2). It is not possible to reconstruct the survey methodology, as beyond a general outline sketch map, no indication is given as to how far the survey extended to either side of the Zambezi, or what search pattern was used for the survey. It is apparent from the published papers that sites in very close proximity to the Zambezi were given the most emphasis (Phillipson, L 1975, 1976, 1977). Although the initial stated aim of the project was investigation of the variation of the Middle Stone Age in a geographically defined area, it is clear that this investigation used survey in the traditional manner as a site location tool, rather than using the archaeological record as a source of behavioural data. Sites were located for excavation, the cultural sequence was then described, and the nature of the material was compared between the sites. From this perspective, while the project met its aims, it did not attempt to develop an understanding of the distributions of the sites or the possible natural and cultural factors affecting that distribution.

A second survey was completed at roughly the same time by D.W. Phillipson (1973). This survey was designed to better understand the prehistoric succession in eastern Zambia. The area of interest was 25,000km^2 of plateau between the escarpment of the Luangwa to the north-west, and the Mozambique-Malawi border to the south-east (Phillipson D.W. 1973:3). The region was subjected to a four-week reconnaissance to locate rock shelters that may have had good archaeological sequences (*ibid*: 6). However, no mention is made of how this reconnaissance was undertaken, and with such a large area to cover it seems unlikely that all parts of the region were sampled in the time allowed. Within both these studies no systematic sampling methods were used, the aim of the research being to find and identify archaeological sequences in the areas covered and, as a result, no attempt to understand the landscape surrounding the sites was initiated in keeping with the aim of developing a culture-historical sequence in a preciously unstudied region. As such, it was a successful exercise but rooted in the pre-New Archaeology practice.

The third study examined here was conducted by Robertson from 1977-79 in central Zambia (Robertson 2000:148). Though this study was again directed at establishing a local sequence that could be linked to wider discussions, in this case the Later Stone Age to Iron Age transition, it is apparent that the survey undertaken was relatively extensive, using teams of school students to cover accessible fields and areas of clear land away from dense bush (*ibid*: 149). While there is some limited interpretation of the likely reasons for the patterns of settlement seen in the area (*ibid*: 150), Robertson concedes that some of the main patterns in the distribution of sites are purely a reflection of where land had been cleared for modern farming, resulting in only one area yielding good spatial data (*ibid.*). The search pattern is not described, so it is not possible to know factors such as spacing of walkers, or number of fields and total area sampled in the survey area defined.

In contrast to the studies above, Vogel conducted three landscape-orientated studies that were designed to address questions of spatial organisation of archaeological sites in the landscape (Vogel 1978, 1984, 1987). Two of these spatial studies were conducted based on the locations of known and published sites, so no field survey was necessary (Vogel 1984:71; 1987:160). The 1978 research did involve some field survey, but this appears to have been limited to the immediate vicinity of the known sites (Vogel 1978:148). The 1984 survey of published sites was designed to assess the spatial relationship of Iron Age sites to develop an understanding of the settlement systems in operation within the area between two tributaries of the Zambezi River near Victoria Falls. Vogel (1984:74) accepted that the lack of actual fieldwork was a potential problem, and suggested that the geography of the settlements was worth more detailed examination. However, it appears that this was never undertaken, presumably due to the economic difficulty in conducting research in Zambia at the time. Future investigation using a landscape survey could test whether Vogel's observations of spatial patterning genuinely reflects behavioural choices of the Iron Age communities, or that of research interests of different archaeologists working in the area at the time.

The two most recent field studies, conducted in the late 1970s and in 1980, have been more systematic and list the methodologies employed to conduct the surveys. The first of these was conducted in three seasons from 1974 – 1976 by Derricourt (1985) where the aim was to survey and record archaeological material threatened with loss following the construction of a dam that would flood the Itezhitezhi lake area on the Kafue River (*ibid*: 10). The survey work commenced with plotting the relevant area on 1:50,000 scale maps to define access routes and zones for concentrated searches. Two teams of four people carried out the work, which was described as a "fast but conscientious search for every likely locality for all classes of finds, sites, or other features" (*ibid.*). From a base point reached by vehicle, each team would spread out and explore the area using transects (*ibid*: 17). However, the widths of each transect, the spacing between fieldwalkers, and the exact amount of coverage they provided is not given. As a result, it is difficult to quantify the amount of ground covered or the effectiveness of the survey in the context of its aim of determining how much material was present in the area. In the initial part of the survey the teams were inexperienced so it is possible that artefacts were missed (*ibid.*), and overall in the survey a high proportion of Later Stone Age and Iron Age material was collected with very little Early and Middle Stone Age (*ibid*: 118). It is possible that earlier material was not recognised by the surveyors; however, the proximity of much of the survey to the Kafue River is also a probable cause, as overbank deposits from the river will have buried older archaeological horizons, or lateral movement of the channel might have removed them.

The second of the more recent spatial studies, conducted in 1980 by Michael Bisson (1992:238) was, apparently, the last field survey undertaken in Zambia before this project. The primary goal of the project was to survey all archaeological sites in the Luano drainage basin. However it seems that virtually all sites discovered were Later Stone Age and Iron Age (Bisson 1992:234). As a result, the analysis was focused towards determining the relationship between Later Stone Age and Iron Age sites with regard to the development of socio-political organisations in the Iron Age (*ibid.*). The study focused on an 11km long deforested stretch of the Luano, a tributary of the Kafue, where widespread cultivation had left the archaeological record more visible (*ibid.*). The survey covered a strip of land one kilometre in width either side of the stream; a trained crew walked a zig-zag pattern within parallel transects 30m wide. Artefacts were then recorded on a 1:10,000 scale map provided by a mining company that had surveyed the region (*ibid*:238). This approach allowed computations to be made of the relationships between site size and their placement in the landscape relative to topographic and vegetative zones, allowing modelling of Later Stone Age and Early Iron Age site location strategies (*ibid*:240). It demonstrated that Later Stone Age and Iron Age groups had to share resources at times, but with seemingly little contact evident in the material culture in the archaeological sites (*ibid*: 244). It is likely that co-existence would have been difficult as clearing the land for agriculture and competition for game would have put strain on hunter-gatherers in the area (*ibid.*). The conclusions drawn from this allowed a more certain assessment regarding the relationships between Later Stone Age hunter-gatherers and Early Iron Age farmers than had previously been possible based on the few sites that had been excavated with no landscape survey attached to them (*ibid.*). As a result, it is the only survey of its type to have been undertaken in Zambia.

2.4.2. Overview

Most of the surveys conducted in Zambia were primarily designed to collect data to aid the construction of regional chronologies for the Middle and Later Stone Age, and the Iron Ages, and none of the surveys published has been directed at the Early Stone Age. By the 1970s, when landscape surveys were becoming more sophisticated (Orton 2000:69), Zambian archaeological research was beginning to be affected by political and economic problems. It is therefore unfair to judge these projects by modern standards. Up to this point a relatively small number of surveys had taken place in Zambia; these were prospective studies aimed at locating sites with a stratigraphy deep enough to aid the construction of local and regional chronologies. None of these surveys produced data that could be used to develop an understanding of the distribution and nature of the archaeological record in the landscape. The data gathered would merely reflect the judgment of where the researchers expected archaeological material to be, rather than being a genuine record of the distribution of material in the landscape. The two most recent studies, conducted in 1974-5 (Derricourt 1985) and 1980 (Bisson 1992), did aim to understand the archaeological landscape using a recorded sampling design. For the purposes of this project Bisson's Luano area project (Bisson 1992) is the only one that provided a potential model on which to base the methodology for this project. To expand the knowledge of approaches used in areas with similar archaeological material and in environments that might present similar challenges to those faced in the Luangwa Valley, it was necessary to examine projects undertaken elsewhere in Africa.

2.5. Comparative Examples from Elsewhere in Africa

2.5.1. Introduction

In general, research in the rest of Africa has also suffered from a lack of landscape studies that have had methodologies designed to understand the archaeological record from a spatial perspective. At various times many countries in south-central Africa have been politically and economically unstable resulting in little opportunity for continuous archaeological research across much of the continent. Where surveys have been undertaken the exact methods of survey were often not recorded, particularly in the case of early surveys, where to discover the methods used the original excavation diaries have to be consulted, or the methodologies inferred (Bower 1986:27). However, there have been many surveys, usually located in the more politically stable eastern part of Africa and South Africa that have used a landscape sampling approach to retrieve archaeological data, with varying degrees of success (Bower 1986: 29-37; and e.g. Asombang 2004; Barnett 2005; Mazel *et al.* 1984; Mitchell 1994; 1996; Tappen 1995; Willoughby 1993). As it would be impossible to review all of them an admittedly eclectic set of examples that demonstrate the various approaches taken and problems faced in African field survey are analysed here. These are Garth Sampson's (1985; 1988; 2001) work in the Seacow Valley in South Africa, Susan and Roderick McIntosh's (1993) survey in Senegal, Peter Robertshaw's (1994) in Uganda, and Frederic Pearl's (2004) and Bruce Dickson *et al.*'s (2004) work in the Mukogodo Hills in Kenya. Gilbert Pwiti's research in Zimbabwe (Pwiti 1996) will be looked at first as it is a relatively local example to Zambia and, rarely for African projects, includes some analysis using GIS (the use of GIS in this survey is discussed in chapter 4). As noted above, Foley's (1980) off-site survey is one of the most useful projects from the point of view of this research, and its main aims and methods are described here.

2.5.2. Case studies

Pwiti's research focused on the Iron Age of an archaeologically unexplored area of northern Zimbabwe. Ambitiously for one project, he set out to develop a culture-historical framework for the area, understand human settlement preferences in relation to economic and climatic change, test theories regarding socio-cultural change in the area, and finally to use GIS to attempt to understand some of the environmental factors in site location preferences, with the ultimate aim of providing models that may be applicable in other areas of the country (Pwiti 1996:16). The survey was undertaken in three different areas of 375km^2 each, taking a sample of 6% of each land area, resulting in 22km^2 being surveyed in total (*ibid*: 47). The survey units were chosen using a combination of simple random and multistage sampling methods (*ibid*: 45 & 48), the survey units (transects) were then walked by parties of eight people spread 10 metres apart. This survey data was then combined with subsurface sampling of some of the sites found (*ibid*: 57). This was a well-designed survey with a clearly outlined methodology, using sampling methods to eliminate bias introduced by the researcher; as a result Pwiti was able to develop arguments around all of his key aims. GIS was used in the project to map the sites found in relation to the soils and the vegetation of the area (*ibid*: 142-146). The simple analysis provided interesting results that indicated a clear relationship between sandy to sandy loam soils, mosaic vegetation of *Colophospermum mopane* and *Brachystegia allenni*, and preferred site locations (*ibid*: 143). The use of GIS here may be regarded by some as simplistic, even by the standards of the mid 1990s, however, it remains remarkable as GIS was still relatively new to western archaeology, and has been very rarely used in Africa since. Pwiti's work was therefore at the vanguard. The example demonstrates the effectiveness of combining a well-designed methodology with GIS where prior information is limited. As a result, a clearer determination can be made between environmental attributes and the spatial nature of human activity.

Sampson undertook what is probably one of the largest single surface surveys of prehistoric material in Africa (Sampson 1985; 1988; 2001). The initial survey was undertaken from 1979 – 1981 utilising a team of eight archaeologists who spent 15 months surveying 5000 km^2 of the semi-desert Seacow Valley on the South African plateau (Sampson 2001:28). During the course of this survey 16,000 Stone Age surface sites were located. A second survey was undertaken to sample the pottery of the 987 sherd-yielding sites of the upper part of the valley and create a map of the stylistic boundaries of the pottery types, and thus the hunter-gatherer groups of the area (Sampson 1988). The exact method by which the surveys were undertaken is not given, and systematic survey design was openly rejected, as it was perceived to be unnecessary for the survey goals (Sampson 1985:25). The search paths appear to have been random walks in the country, and there was a concerted effort to target areas of high site density (Sampson 1985:24-27). Whilst the survey was not intended to find all sites in the area, a 100% recovery on the search paths was aimed for in an attempt to develop an understanding of the relative changes in densities in various terrains.

Although a large amount of statistical evaluations were undertaken (Sampson 1985:25; 1988:173), the survey tactics could not guarantee a reliable sample; their success therefore would seem coincidental. In some of the analyses mentioned it is difficult to perceive how the statistical tests could be successful. For example, the Acheulian sites have a clear clustering to the higher ground to the south of the survey area (Sampson 2001:31), and despite the claims that it is not an effect of the survey design (Sampson 1985:25), it is difficult to rule this result out as being an artefact of the rather unusual survey ethos, which clearly resulted in a much higher density of 'search paths' in the south of the region visible in Sampson's maps (Sampson 1985:24-27: figures 13-15).

The data from the survey was incorporated into a GIS for analytical purposes in 2001 (Sampson 2001:34). The technology was not widely available when the project was initially designed, so the survey was not specifically planned with a GIS analysis in mind. This can lead to problems in the analysis stage due to researchers not fully appreciating the limitations of the technology (Wheatley & Gillings 2002:1). However, Sampson's analysis succeeds in demonstrating a probable relationship between Acheulian sites and outcrops of hornfels, and a negative relationship with water sources (Sampson 2001:34). Whilst the concentrations of sites are argued as not equating to factors such as thinner soils or the intensity of the survey (*ibid*: 35), it has not been made clear how this has been determined, and as a result Sampson's work cannot provide much assistance in meeting the requirements of the Luangwa project.

Susan and Roderick McIntosh undertook a survey in 1988-9 in the tumulus zone of northwestern Senegal where there are 6,982 earthen mounds, clustered in 1,432 sites over a 32,000 km^2 area. They are certainly Iron Age phenomena, however the identity of the builders and the exact chronology of the development of the tumuli are unknown (McIntosh & McIntosh 1993:75-76). The aims of the research were clearly laid out: the collection of data on geographic variability of location, size, and density of the monuments with respect to landforms, soils, and vegetation. It was also hoped to determine the relationship of the mounds to sites of habitation, which were unknown at the time. The survey methodology was designed to address these aims, and a stratified sampling technique was chosen in which the strata were delineated on the basis of landform types found in the region along with vegetation type and density of tumuli (*ibid*: 77). The plan was to walk transects 50 m apart for five kilometres, recording sites as they were encountered. This methodology was revised when it was discovered that there were shallow surface scatters over the entire area surveyed, probably due to the short-term mobile nature of settlement activity (*ibid.*). This made it too time consuming to record all of the sites individually. Instead, intensive surveys were conducted around known tumulus clusters to map their exact size and distribution. An area of one kilometre surrounding each cluster was surveyed for signs of habitation using radians extending outward from the centre of the cluster.

The mounds themselves had representative samples of materials gathered to give additional material, however, for a number of sites the sample size remained too small for meaningful statistical analyses to be conducted (McIntosh & McIntosh 1993:77). Although this could be deemed as poor sample design, i.e., not allowing for a large enough sample, there were constraints on the project time that precluded the gathering of larger samples as the low density of material would have required large scale excavation (*ibid*: 99). Even if time had allowed collecting more data from some sites to make up the difference between low-density sites and those with an adequate sample, the results would have been unfairly skewed in favour of lower density sites (*ibid.)*. The survey technique did allow for valid conclusions regarding the distribution of tumuli in the landscape, and correlations could be made with prominent natural features such as dune crests. The analysis of the larger pottery collections allowed re-evaluation of prior assumptions regarding the relationships of the tumuli to other historical events in the region (*ibid*: 105). This example demonstrates the flexibility of sampling techniques, which, when combined with clear project aims, can allow a methodology to be adapted when initial plans are found to be unworkable. This case study was useful to this project methodologically as it provided a model for developing a survey that could determine the relationship between archaeological material and differing geomorphological areas in a landscape with continuous distributions of artefacts.

Robertshaw (1994) attempted a site survey of Bunyoro – Kitara sites (Ugandan Iron Age), and his aim was to locate sites using transects spaced one kilometre apart, however, the vegetation was so dense in most of the survey area that it was difficult to see any ground material. The plan was abandoned in favour of searching in farmed fields, where ground visibility was better (Robertshaw 1994:112). The fields were surveyed in an *ad-hoc* (*ibid.*) manner and, although a selection of field localities was chosen (slopes, valley floors, and ridge tops) in a form of stratified sampling, no formal sampling methodology was applied. As testing of hypotheses regarding settlement patterns was not an aim of this investigation, the lack of a survey design was not regarded as a problem. It is clear from this example that before surveying can be undertaken the surface conditions of the land area must be taken into account and if possible the methodology tested. It should also be noted that the spacing of transects one kilometre apart is exceptionally high for any archaeological investigation of the landscape, particularly considering the visibility of the ground surface. Even with good visibility it is difficult to conceive that an Iron Age village would be large enough to be intersected by these transect widths.

A survey of particular interest when considering the design of this project is that of Pearl (2004) and Dickson *et al.* (2004) whose project attempts to understand the distribution of sites in relation to the local geomorphology of the Kipsing and Tol River valleys in central Kenya. The survey was conducted in two field seasons in 1996 and 1999 (local unrest due to cattle raiding erupting into open warfare delayed the second season by two years), and its aim was to locate and record archaeological sites, link them to known sites and relate them to Quaternary geological deposits in the region. A north and south survey tract was chosen for the 24km^2 area of their survey area, in which parallel transects approximately 30m apart were surveyed. Dry season fires that had cleared vegetation provided good visibility allowing 30% coverage of the area to be attained. Despite this clearance, it is likely that sites of low obtrusiveness were missed (Dickson *et al.* 2004:170). Site distribution was assessed in relation to the geomorphology by mapping the alluvial deposits of the area, and then assessing site distribution in relation to the mapped deposits. The geomorphological analysis allowed differentiation between natural and cultural site formation processes, providing a basis to understand genuine land-use patterns of Middle and Later Stone Age people (Pearl 2004:42). The Middle Stone Age sites were uniformly distributed, and the Later Stone Age sites were concentrated in the southern end of the survey area (Dickson *et al.* 2004:181; Pearl 2004:42). No Early Stone Age sites were discovered, and this is put down to lack of exposures of suitable age (Pearl 2004:29), however, it is possible that some of the sites listed may contain Early Stone Age material in a secondary context. This is suggested as two examples of cores recovered (figure 11 in Pearl 2004:32), though classed as Middle Stone Age, are technologically Mode 1 and could derive from the Early Stone Age. Additionally, the lower core recorded in figure 11 of Pearl (2004:32) appears abraded. When this observation is combined with the fact that the Kipsing River is approximately 100m away it seems possible that some material is in a secondary context. Though the study is successful in linking the sites to the geomorphological landscape, it was weakened by not taking into account the abrasion of the tools when considering the formation processes of the sites. Despite this reservation over the site classifications, the study demonstrates the effectiveness of using a multi-disciplined approach that combines archaeological and geological survey with a good sampling design, allowing the development of theories relating to human behavioural decisions in the landscape.

Foley's research in the Amboseli, Kenya, is another project of particular interest in designing this survey. It was aimed at developing a behaviourally based model of regional archaeological evidence for hunter-gatherer activities that could take into account the continuous nature of the archaeological record on the landscape surface, and to then develop a field methodology to test the model (Foley 1981:2). The survey developed and used the off-site approach in a 600km^2 area for which a stratified sampling strategy was selected (*ibid*: 34). The area was split into blocks measuring two by four kilometres, and a two stage sampling procedure was used to produce a basic map of the distribution of artefacts, followed by a series of experiments to look at interesting or anomalous variations in density (*ibid*: 35). For the first and most crucial stage of the investigation, six sampling points were located in each survey block using random number tables of which only four points were used, with two in reserve. After experimentation, an innovative transect design was decided on in which four 50m x 5m (250m^2 each and 1000m^2 total area) transects were walked in a radial pattern around the selected point in the landscape (*ibid*: 37). A system of codes was used for the speedy recording of environmental, taphonomic, and artefact data in the field (*ibid*: 39).

The data was analysed manually by plotting artefact densities at the sample localities and then extrapolating values using a contouring system to demonstrate the spatial patterning of the density of various artefact groups. The contouring system developed for the data analysis of the project enabled an understanding the continuous spatial variation of artefacts, which was one of the key aims (Foley 1981:134). This approach was combined with a detailed taphonomic analysis to factor out any natural disturbance (*ibid*: 161). With this method, Foley was able to successfully disentangle the taphonomic information from the behavioural information that could then, for example, be used to model the likely population levels of the valley in relation to the regression of the Amboseli Lake (*ibid*: 180). The analysis of the archaeological data in relation to the taphonomic and behavioural interpretations would conceivably have benefited from the use of a GIS. This would have allowed the complex information to have been displayed more succinctly allowing, for example, artefact condition, density, and soil type to be shown which, due to the manual nature of data interpolation, had to be displayed separately. If the data was held and presented using a GIS, a composite map incorporating two or more features at once could be created that would make patterns easier to interpret. At the time of publication, GIS, as a new technology, was virtually unknown in archaeology, thus precluding it as an option. In conducting a similar survey today GIS, as a very powerful analytical tool, should at least be considered for use.

2.6. Discussion and Summary

Most published surveys in Zambia are prospective surveys designed to aid the construction of chronological sequences. Bisson's (1992) research is the only example of a landscape survey in Zambia that employed a defined methodology for a systematic survey. Although Derricourt's (1985) survey was arguably systematic, it is not possible to assess its effectiveness due to the lack of detailed methodology recorded for the survey design.

Archaeologists working in Africa have not developed the same impetus behind spatial research that has occurred in Europe and America. Some surveys have not been well designed in comparison with those outside the continent or have not taken into account the frequently difficult working environments of African survey, and have as a result, failed to deliver the results the researchers might have originally hoped for. On the other hand, there have been many surveys that demonstrate the effectiveness of employing a well-defined sampling approach with clearly identified aims, allowing them to move beyond the construction of historical frameworks into the realm of assessing site distributions in the landscape (e.g. Dickson *et al.* 2004; Foley 1980; McIntosh & McIntosh 1993). More typically, the field methodologies are not recorded or, as in the case of Sampson (1985; 1988), are nonexistent. However, African surveys should not be overly criticised as it has to be acknowledged that before the 1980s, and even today, the pioneering state of research and the socio-economic and logistical problems in many parts of Africa mean that detailed investigations are simply not feasible. The result is a disparity in the amount and intensity of research undertaken across the continent and in comparison to Europe and North America. Zambia demonstrates why direct comparisons are not necessarily valid. Large areas of Zambia have had relatively minor archaeological investigations, for example the Luapula and Copperbelt provinces briefly investigated by Derricourt (1980) and Bisson (1992:237) respectively, as well as the Luangwa Valley itself which has never been examined archaeologically. As a result, including the ongoing investigations of the PPE project in South Luangwa National Park, there have been approximately five individuals, who have conducted approximately 12 months' fieldwork on the archaeology of these areas, which cover an area of 181,930 km^2. In comparison, the modern Greek county of Boeotia, which encompasses an area approximately 1.5% (2,580 km^2) of the area of the Zambian regions listed above, has been subjected to a ten-year programme of field survey totalling more than 20 months of research (Bintliff 1985:196; 2000:202). A survey in logistically awkward areas where there is little prior research and data, such as this one, will therefore need to have a much simpler methodology than those undertaken in Europe and America or of the types recommended by archaeological statisticians (e.g. Orton 2000). A strictly statistically bound survey was not achievable for this project, and an effective survey design could only be created if it was driven by the practicalities of the landscape and environment (see chapter 5).

The disparity of research area and survey time allowed in African surveys should not preclude the possibility of developing a systematic survey or accurately recording methodologies used in the fieldwork. Neither is it likely to reduce the effectiveness of a well-designed sampling methodology, provided that it is sensitively designed to the often awkward working environments in Africa as well as constraints on time, money, and manpower. Therefore, though surveys in Africa should not necessarily be assessed directly in comparison with European and American surveys, the European and American methodologies can be used as models on which to base research work, provided that the constraints in any one area are well understood.

This project needed to move beyond the discovery and recording of types of archaeological material to develop an understanding of the impact of natural geomorphological processes on site preservation. Only then can human agency be discussed in terms of site location preferences. To achieve this, a systematic methodology was required. Although the survey was designed with a reasonable knowledge of the working conditions in the Luangwa Valley, the plan was designed to take into consideration any unforeseen developments in the field. It was clear from all the examples researched and presented above that flexibility and a good working knowledge of survey practice is key to a successful survey. Even with this knowledge the survey as initially designed required significant modification during and after the first season's research, and a complete redesign for the second season.

3. Geology, Geomorphology, and Past Climate of Zambia and the Luangwa Region

3.1. Introduction

As stated previously, there has been little previous geological or geomorphological research in the Luangwa Valley. The most extensive studies were undertaken in the 1930s by Dixey (1937a & b), and his research, though dated, is used to provide an outline of the Luangwa Valley's formation and sedimentary history, and allows the area to be set it in its regional context. More recent research has primarily been restricted to economic geological concerns such as coal, oil, and diamond prospecting, along with some research into palaeontological deposits of the Karoo sediments. However, there has been some research into the geomorphology of the area, primarily focusing on the relationship of the vegetation to the geomorphology (Astle *et al.* 1969), and the recent action of the Luangwa River (Gilvear *et al.* 2000). Understanding the long and short term geological and geomorphological history of the area was vital to this research as, only with this working knowledge, no matter how basic, was it possible to devise a strategy for further developing an understanding of the past and present processes that may have affected the archaeological record.

It is also necessary to set the Luangwa Valley in its climatic context, and to outline the nature of past climate of the area, as these changes would have had a direct bearing on geomorphological processes. There is currently no published data on the past climate and vegetation of the valley itself, therefore evidence regarding past climatic regimes is taken from terrestrial records around the southern African region and sediment cores from lake basins such as Lake Malawi and Lake Tanganyika, and the ocean floors surrounding the continent. An absolute and accurate climatic model for past climate variability and vegetation in the valley is not possible, but a tentative model of these past climate changes is presented.

3.2. Background to African Geology and Rift Valley Formation

Africa is a massive fragment of continental crust that has been moving northward relative to degrees latitude since the break up of the super-continent Pangea; it has shifted through 60° of latitude and rotated 15° counter clockwise in the past 300 million years (Pavitt 2001). However, Africa has been stable in comparison to other continents; this stability, with little large-scale continental deformation or uplift, has allowed long periods of erosion. Much of Africa is, as a result, characterised by ancient erosion surfaces that give the continent relative flatness (Shaw 1997:468). Large parts of Africa consist of cratonic shields; areas of crust that have remained virtually unchanged since Pre-Cambrian times (500Ma +) (Prothero & Schwab 1996:476). Interspersed between these are orogenic (mountain forming) belts of various ages, for example the Palaeo to Mesoproterozoic Irumide belt (DeWeale & Mapani 2002), which forms part of the Muchinga Escarpment that defines the northwest margin of the Luangwa Valley. The stability of the continent has allowed time for considerable amounts of subsidence of the crust. This periodic downwarping in the continent's history has resulted in the development of basinal structures which in turn allowed deep sequences of sediment to accumulate (Clark 1982:13). Between these basins swells developed in the crust, creating strain that caused rifting in certain areas (*ibid*: 13).

In South Africa one of these downwarped basins, the Karoo basin, was affected by a large amount of sedimentation, forming the Karoo Supergroup. The Karoo basin reaches a depth of 7000 metres and forms the classic geological sequence of the Karoo. It was deposited from the Late Carboniferous to the Early Jurassic, and at different times affected the entire southern portion of Africa, with the result that strata belonging to the Supergroup can be found as far north as the equator (Clark 1982:6). This roughly coincides with the time that Africa formed part of the continent of Pangaea, which had begun to break up by the Triassic (Cairncross 2001:530). The Karoo thus documents the climatic and positional change of Africa through the glaciations of the southern latitudes, to warm swamp environments, and eventually to desert landscapes (Clark 1982:7).

Although Africa is still gradually moving northward, it had reached its general modern day position by the Miocene (Gamble 1997:48), at which point (approximately 20 Ma) the uplift and rifting that forms the East African Rift Valley began in southern Ethiopia and northern Kenya (Sepulchre *et al.* 2006:1420). The western arm of the Rift Valley began to develop in the central Tanganyika Basin approximately 12 Ma, with further rifting down to the Malawi rift from 5 – 2 Ma, creating a 6000km long north south orientated elevated area (*ibid.*). This uplift is likely to have affected the Luangwa Valley by uplifting its north-eastern end, which abuts the Malawi rift, and indeed older rifting from the Oligocene (~30 Ma) may have had the effect of cutting the Luangwa off from the Rufiji River, causing a flow reversal of the Luangwa (Stankewitz & de Wit 2006:81). The Karoo plateau in South Africa has also been raised in the past five million years to an elevation of 1400m (*ibid.*). The Luangwa Valley appears to form part of this Miocene rift structure. However, though it is aligned with the Y-shaped rift structures of the rest of the region, it is likely that it is a far more ancient structure (Dixey 1937b:89). The area also lacks the volcanism of the more active East African Rift, which causes problems for chronology building, as no radiometric (K/Ar) dates can be retrieved in the absence of volcanic deposits. The PPE project has attempted to resolve this issue by applying the relatively new technique of cosmogenic nuclide dating to fluvial deposits in the valley (Barham *et al.* In prep).

3.3. The Luangwa Valley

3.3.1. Sedimentary and tectonic history

The Luangwa Valley is an approximately parallel-sided flat-bottomed trough, which is up to 90km wide and approximately 700km long (Astle *et al.* 1969:143; Drysdall & Weller 1966:42;). In the southern region, where this survey is based, the valley is asymmetrical, as its north-western margin is sharply delineated by major faults which form the boundary with the uplifted Muchinga Escarpment, whereas on the south-eastern side the margin is irregular where the valley floor abuts the Nchindeni Hills (Drysdall & Weller 1966:42).

In spite of the lack of research into the geomorphology and geology of the Luangwa Valley it is possible to briefly summarise what is known about its formation and present structure. During the 1930s an extensive survey was carried out of the Karoo age rocks to the northeast of the area being researched in this study. This research was conducted by Frank Dixey (Dixey 1937a; 1937b) on the plateau between Lake Nyasa (Lake Malawi) and the northernmost tip of the Luangwa Valley. Dixey (1937a: 72-73) gives a brief synopsis of the structural formation of the Luangwa Valley. It has a relatively complex history involving movement that has occurred twice in a similar direction. Initially the valley was formed in the sub-Karoo surface; this was followed by Karoo sedimentation (Later Carboniferous to Early Jurassic (Clark 1982)), and then synclinal folding and faulting. At this point there was regional peneplanation during which the valley was eroded to its present depth, and following this there was some additional folding and faulting and sedimentation in the Cretaceous. Finally, there was Miocene peneplanation and uplift, with some slight additional faulting associated with the uplift and creation of the East African Rift Valley. Most of the Cretaceous sedimentation was removed, and the Luangwa Valley was re-eroded to its present depth, with only a few outliers remaining of Cretaceous sediments. The old scarps and dip slopes of the original valley were effectively re-exposed revealing the ancient pre-Karoo landscape (Dixey 1937a: 73). The valley is therefore aligned along original basins of depression, or synclinal folds, which were then faulted normally into a horst and graben structure during the early part of the Karoo period (figure 3.1). This movement had been completed by the beginning of the Cretaceous (Dixey 1937b: 89). The peneplanation also removed overlying material on the north and south uplifted blocks of the valley, with the result that the Archean basement rocks have been exposed (Cole 1963:293).

Though Dixey's interpretation of the complex development of the rift has never been subsequently confirmed by further research, the Luangwa and Zambezi valleys are still accepted as developing as part of a pre-Karoo rifting system (Castaing 1991:60; McConnel 1972:2553; Shoko 1996:617), or from rifting contemporary with Triassic Karoo sedimentation after 1200m of sediment had already been deposited (Osterlen & Blenkinsop 1994:178). It is likely, however, considering that the uplift causing the Malawi rift (5 and 2.5 Ma) created a five kilometre deep basin bounded by uplifted flanks (Ebinger *et al.* 1993:17821), that the abutting Luangwa Valley rift would have been reactivated, possibly to a significant degree. The present form of the East African Rift Valley had been gained by 1.5 – 2.0Ma (Gamble 1997:48), and it is probable, based on the research presented here, that the Luangwa Valley has been stable throughout the Pleistocene and Holocene. Recent geological research in the valley has almost entirely been limited to the economic geology (e.g. Cairncross 2001; Van de Velde & DeWaele 1994), though some attention has been given to the mapping and description of the Archean complexes, such as the Irumide belt (e.g. DeWaele & Mapani 2002). As a result, little is known of the distribution and structure of Quaternary deposits.

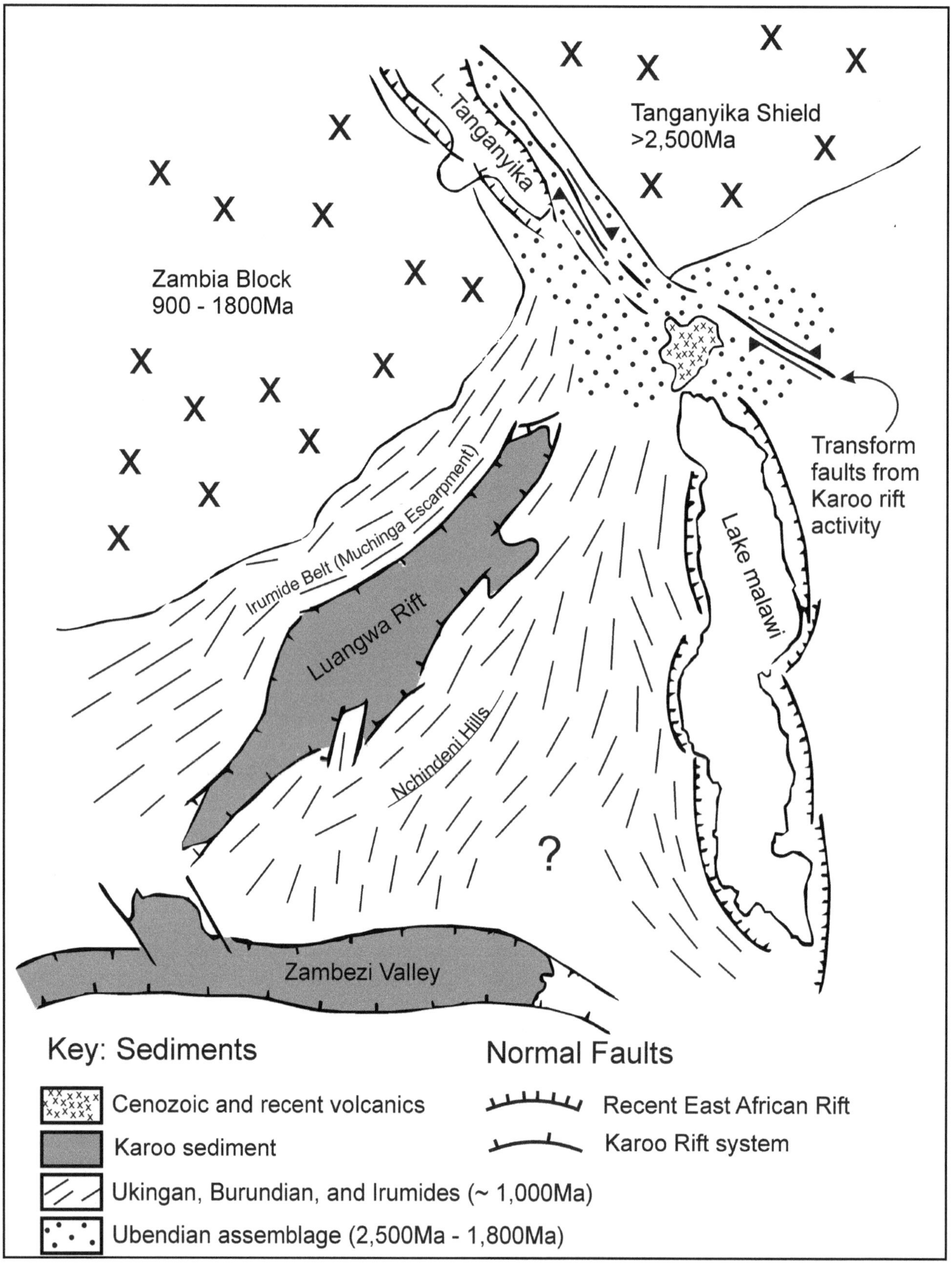

Figure 3.1: A schematic summary of the key geological formations and fault systems surrounding the Luangwa Valley. Original drawing created in Corel Draw X3 (not scaled), based on McConnell (1972:2558 - 2559 figures 2 and 4), Castaing (1991:61 figure 6, and p67 figure 11), and Dixey (1937:53 figure 1).

3.3.2. Geomorphological research

Very few attempts have been made to study the Pleistocene geomorphology of the Luangwa Valley, and it is apparent that some deposits that are clearly Pleistocene in date have been misidentified in the past as belonging to the Karoo succession (see section 6.4.2). These sediments are cobble deposits that are found at numerous localities across the south Luangwa region, and Dixey (1937a:70) surmises, based on his observations in the north Luangwa region, that they must be Karoo in origin. Utting (1988:6) identifies the cobble material as originating from recent erosion and redeposition. However, as his work is meant as a basic guide to the geology of the area, no research into their origin was initiated. He does state that judging by their nature and geographic distribution their depositional history must have been complex (Utting 1988:6). In the 1940s a survey of land resources of the valley was begun but, due to the inaccessibility of the area at the time, it was not completed (Astle *et al.* 1969:143). Astle *et al.* (1969) used a combination of aerial photography and fieldwork to classify the landscape of the park into nine different land-systems to aid management decisions. These land-systems incorporated both ancient and recent sediments (Astle *et al.* 1969), but, no attempt was initiated to develop an understanding of the formational processes of the features observed.

One recent study conducted on the planform change of the Luangwa River itself (Gilvear *et al.* 2000), is unique both to the valley and within its field, as little research has been conducted on the geometry of rivers in tropical zones (*ibid*: 421). A record of flood peak flow and aerial photographs of the region taken over the past 50 years has allowed a relatively detailed analysis of the river's activity. Formed by a combination of anastomising channels and meandering single stream lengths, the Luangwa is a sandy river characterised by a low gradient 2km wide floodplain (<0.0001) whose channel ranges in width between 100-200m. Cut-offs and oxbows are relatively common, as is a form of anastomising, with anabranches up to 14.7km long. The rate of erosion on meanders can be as high as <33ma^{-1} (Gilvear *et al.* 2000). Their research revealed that over the past 50 years there has been great change in the planform of the Luangwa; in one reach of the river a previously abandoned channel has been reactivated (*ibid.* 427), while other sections have had meanders cut off to form oxbow lakes. In general there is evidence for straightening of the channel planform suggesting that the Luangwa was at one time more sinuous than at present (*ibid.* 428). Flood peaks have, on average, been increasing for the past 50 years, particularly since 1989 (Gilvear *et al.* 2000:424)[4]. This increase of flood peak activity and the changes in the planform of the river will undoubtedly have an effect on the archaeology. Much may have been eroded since the Stone Ages, and even from the more recent Iron Ages, which date to at least the 19th century. The Luangwa was clearly a key geomorphological agent in the area; determining the effects of its activity was one of the goals of this research.

The present increase in flood peaks in the Luangwa, and the suspected straightening of the stream planform, is likely to be due to increased rainfall, possibly related to global warming (Gilvear *et al.* 2000). This increased activity today might have analogies with past climate changes that may have increased erosion rates. Work undertaken in the Chipata district to the south-east, approximately 80km away, has revealed that there have been high-energy land forming events in the past, such as landslides and debris flows, that are not produced under the present climatic regime (Thomas 1999; Thomas & Murray 1999). These are probably related to climatic fluctuations at the end of glacials marked by increased rainfall, and at first may have been episodic, triggering instability on previously semi-arid land surfaces that had sparse vegetation cover (Thomas 1999: 292; Thomas & Murray 1999:130). These events have been difficult to date precisely, but roughly date to 90 ka+, (possibly relating to the last glacial cycle 110-90 ka MIS 5d - c), and 21-24 ka, corresponding to the Last Glacial Maximum (MIS 2) (Thomas & Murray 1999:126). There are also deposits that are assumed to relate to the Pleistocene – Holocene transition (12,500 – 9,000 years ago) (Thomas 1999: 293). These events appear to correspond to regional climatic events in Africa (Thomas & Murray 1999:124-125); the increase in rainfall that triggered these events is likely to have increased the discharge of the Luangwa, and thus its erosion rate and planform. Presuming that vegetation had been reduced in a similar scenario, the landscape in the valley would have been unstable, producing similar land forming events.

The probable nature of the past vegetation and climate is outlined in the following sections as a necessary aid to the interpretation of the past geomorphic processes that have occurred in the valley (chapter 6).

[4] Severe flooding in January to February 2007 damaged or removed remaining archaeological deposits excavated as part of the PPE project in 2003 – 2006, including the loss of Mode 1 deposits at two sites, designated SL6 and SL12 (on the Manzi River and West Mwamba River respectively) (Barham *pers comm.*).

3.4. The Vegetation and Climate of Zambia – Past and Present

3.4.1. Rainfall and vegetation of Zambia today

Rainfall is the most important element of the Zambian climate today (Archer 1971b:22; Iwasaki & Shinoda 1989:43), and it imposes a strong rhythm on most of the rural population (Archer 1971a:20). Rainfall, in combination with temperature, defines Zambia's three seasons: the cool dry season (April to August), the hot dry season (August to November), and the warm wet season in which the annual rainfall is entirely confined (November to April) (Archer 1971b:22; White 1983:87). This climatic regime is modulated by the strength of the African-Asian monsoonal circulation, and the position and seasonal migration of the Intertropical Convergence Zone (ITCZ) (Trauth *et al.* 2003:297). The rainfall is caused by the southward movement of the ITCZ in the summer (Archer 1971a:20), with the zone of maximum rainfall following the latitudinal position of the overhead sun, with a time lag of four to six weeks (Trauth *et al.* 2003:297).

In Zambia the most notable feature of the distribution of mean annual rainfall is the general decrease from north to south (Archer 1971a:20; Iwasaki & Shinoda 1989:46). The northernmost parts of Zambia receive 1500mm of rain, whereas southern Zambia averages 635mm to 1015mm, the lowest figures deriving from the Zambezi Valley (Archer 1971a:20). The Luangwa Valley is a notable exception to this general pattern as, due to its relative low altitude, it experiences similar rainfall conditions to southern Zambia. The north-east trending valley receives less than 815 - 830mm per year (Archer 1971a:20; Astle *et al.* 1969:146), whereas the escarpments on either side of the valley receive 1220mm per year (Archer 1971a:20). This difference in altitude also has a considerable effect on the temperature of the valley; for example, the average temperatures in October are higher (37.5°C) than on the bordering escarpments (25-30°C) (*ibid.*). The Luangwa Valley is therefore hotter and more arid in today's climate than the surrounding plateau of the same latitude.

Rainfall is the critical factor that affects the type of vegetation present in differing climatic regimes across central Africa (deMenocal 1995:57; Schefuß *et al.* 2003:418; Hopley *et al.* 2007:427). In the context of humans in the landscape the relationship between rainfall, soil nutrients, vegetation, and herbivore biomass is critical in governing the amount of energy available to human foragers (Barham 2001:69). In general, rainfall combined with the underlying geology controls the leaching of soil nutrients – where rainfall is high there is greater leaching of nutrients from the soils. A result of this process can be seen, for example, in the Guineo-Congolian rainforest where 1600-2400mm of rain falls a year; the poor geological substrate in combination with leaching results in the forage having low nutritional value (usually large woody trees with high canopies) (*ibid*: 70). Therefore, though the plant biomass of the region is high, it offers low quality and difficult to obtain forage for mammalian herbivores. Herbivore biomass is thus low, and as a result so is the amount of protein available for human consumption, which controls population density (*ibid.*). A similar scenario would have been likely on the Zambian plateau region to the south of the rainforest down to a line roughly corresponding to the 1000mm isohyet. This area supports open grassy woodland today, but once supported a distinctive closed-canopy forest type, the Zambezian dry evergreen forest (Huckabay 1989:89). This forest consisted of the emergent *Entandrophragma delevoyi* and the dominant *Marquesia macroura* and *Cryptosepalum pseudotaxus*, as well as a number of other species endemic to the region (*ibid*: 90). It represents part of a transition from the Guineo-Congolian rainforest to Zambezian woodland (White 1983:89). The isolation of this woodland to a few areas today is not due to them being a relic of a past climatic era, as in today's climate the forest can be seen to be self-sustaining and fire-proof providing it is not cut and burnt (Huckabay 1989:93). Their absence over most of Zambia relates to the destruction of the forest by Iron Age farmers in the past 800 years (*ibid.*). It is likely that before its destruction this woodland, with its closed canopy, would not have supported much grazing, with the result that herbivore biomass would be low similarly to the rainforest.

The open woodland of the Zambian plateau today primarily consists of miombo woodland. It is characterised by *Brachystegia*, *Julbernardia*, and *Isoberlinia* species, and covers half of Zambia (Mäckel 1971:24). Conversely, hotter drier areas such as the Zambezi and Luangwa valleys have more mixed vegetation. The Luangwa Valley is dominated by mopane woodland consisting primarily of single species *Colophospernum mopane* (Astle 1995; Mäckel 1971:24), accounting for 55% of the valley floor vegetation (Astle 1995). Though the rainfall in the valley (815 – 830mm) is lower than the tolerance for drier miombo woodland (< 850mm (Barham 2001:71)), there are areas of miombo that make up approximately 15 to 20% of the vegetation in the valley (estimate based on Astle 1995). These areas are primarily on higher ground near the Muchinga Escarpment, where presumably more rainfall is received. Miombo woodlands also border the larger wet season rivers, where water will be retained in the localised sandy soils throughout the year. There are also large areas up to *c.* 300km^2 that are covered in open or sparsely wooded grasslands (estimate based on Astle 1995). This mixture of woodlands, in combination with open areas of nutritious grazing, supports dense concentrations of otherwise dispersed species of large mammals (Barham 2001:71). It is not currently possible to determine whether the vegetation is an artefact of farming practices of the past 800 years, as is the case with the plateau. Despite this possibility, it is likely that reduced rainfall in the past on the valley floor would have supported grasslands with open woodland currently seen in drier, undisturbed areas today, such as on Kalahari sands in the east of Zambia (Barham 2001:71), and in northern Zimbabwe (Thomas & Shaw 2002:785 & 786, fig 2c). Therefore, although at first it may seem counterintuitive, the relatively dry Luangwa Valley would have been easier for hunter-gatherers to exploit in comparison with the higher plateau regions due to the increased density of game. However, this is only true for present conditions in the area, and an understanding of past climate change is critical for developing an understanding of how conditions may have varied over time.

3.4.2. Introduction to the past climate records of central Africa

The effect of past climate on rainfall during the past two million years of human occupation of the Luangwa Valley is clearly key to developing models of the changing interaction of vegetation, geomorphological processes, and the Luangwa River and its seasonal tributaries in the valley. Constructing a chronology for changing rainfall patterns over time is problematic as most continental records of past climate only extend to a maximum of the last 100 to 200 ka spanning the late Pleistocene and Holocene (e.g. Partridge *et al.* 1997; Scholtz *et al.* 2007; Stokes *et al.* 1998). However, there is now published data from Buffalo Cave in South Africa for rainfall and vegetation change for the Pliocene/early Pleistocene (1.99 to 1.53 Ma) (Hopley *et al.* 2007) and, potentially, one of the Lake Malawi cores will provide an unbroken climate record for the past 1.5 Ma (http://malawidrilling.syr.edu/ accessed 21/10/07). Deep-sea sediment cores can yield further data regarding past rainfall and vegetation (Dupont *et al.* 2001; Gingele *et al.* 1998; Schefuß *et al.* 2003), with some cores extending the chronology to *c.*5 Ma (deMenocal 1995:53).

3.4.3. Climate variability in Africa from 2 million years ago to the present

Plio-Pleistocene climate change has been driven by Milankovitch cycles, which are long-term changes in the Earth's rotation and orbit (Klein 1999:55-56). Before 2.8 Ma the African climate varied with precessional periodicities (23 – 19 ka) (deMenocal 1995:55). From 2.8 – 1 Ma high latitude climate oscillated between moderate interglacial and glacial conditions on the 41 ka obliquity cycle, where ice sheets grew during sustained glacial cycles; the aeolian dust records of deep-sea cores suggest African climate variation was also sensitive to these remote high latitude changes, and became periodically cooler and drier (*ibid*: 57). One million years ago the climatic extremes in the northern hemisphere increased intensity, and the dominant variation shifted to the 100 ka eccentricity cycle (deMenocal 1995:53).

Despite the alterations to the high latitude climate cycles caused by changes in the intensity of Milankovich cycles, it is apparent from most records that African low latitude climates remained linked to precession throughout the two million years in question here. A high-resolution climate curve spanning 1.99 to 1.52 Ma, retrieved from coring flowstone in Buffalo Cave in South Africa, reveals that rainfall was dominated by the precessional cycle. However, vegetation responded to the longer obliquity cycle at this time (Hopley *et al.* 2007:430). This demonstrates that independent orbital periodicities can dominate co-occurring rainfall and vegetation proxies. However, the two parameters are still coupled at the precessional frequency due to the effect of rainfall on tropical vegetation, particularly when precessional monsoon forcing (see 3.4.4 below) is at its maximum (*ibid*:427).

An oceanic sediment core to the west of the Congo region provided a record of C_3 and C_4 plant ratios for the central part of the African continent from 1.2 Ma to 450 ka, illustrating that vegetation during the mid-Pleistocene was still responding to the precessional cycle (Schefuß *et al.* 2003:419). More recent rainfall proxies from the Pretoria saltpan (Partridge *et al.* 1997:1131), and the freshwater output of the Congo River recorded in clay minerals of the Congo Fan (Gingele *et al.* 1998:23), record the influence of precession on rainfall for the past 200 ka. The precessional cycle is also believed to have caused the hyper-arid events recorded in recently extracted sediment cores from Lake Malawi (Scholz *et al.* 2007:4). These indicate that there were severe periods of aridity 74-78 ka (MIS5a), 85-110 ka (MIS5b-c), and 127-135 ka (MIS5d), which in the oldest two events saw the lake reduced 95% (550-600m) below its present level (*ibid*: 3). The longer-term high latitude changes throughout the past two million years do remain a critical factor, if superimposed, affecting African climate (deMenocal 1995:57; Gingele *et al.* 1998:23; Hopley *et al.* 2007:419; Schefuß *et al.* 2003:420). The aridity phases recorded in the Malawi core would have affected the adjacent Luangwa Valley, and would have severely reduced the precipitation in the catchment for the Luangwa River.

3.4.4. What is the link between precession, rainfall, and the longer climatic changes?

The link between rainfall and precession, and consequently vegetation, is governed by low latitude insolation forcing of African monsoon strength (Schefuß *et al.* 2003:419). The Earth's orbital precession affects the seasonal distribution of insolation at low latitudes (deMenocal 1995:57; Partridge *et al.* 1997:1132), greater insolation raises sea surface temperatures in the south Atlantic, which increases tropical evaporation and atmospheric water content, increasing precipitation (Schefuß *et al.* 2003:420). However, this simplified pattern does not fully explain climate changes in the region. After the 100 ka rhythm of the late-Pleistocene was initiated from 1 Ma, and fully in place by 650 ka, all investigated climate records record the longer term climatic cycles. This imprinting of the longer-term Milankovich cycles is also likely to be due to the relationship between these cycles and sea surface temperatures of the Atlantic and Indian Oceans, and thus the humidity and aridity of southern Africa (deMenocal 1995:57; Dupont *et al.* 2000:119; Dupont *et al.* 2001:195; Gingele *et al.* 1998:24; Schefuß *et al.* 2003:420). The long-term glacial – interglacial cycles affect sea surface temperatures, and during the mid to late-Pleistocene tropical Atlantic sea surface temperatures are reduced by up to 5°C in glacial periods. This would have resulted in large scale aridification of the south central African region, thus affecting the distribution of vegetation significantly (Schefuß *et al.* 2003:420). Pollen records for the past 150 ka from the Congo fan reveal that during glacial periods the rainforest was significantly reduced, and more open grass rich vegetation prevailed (Dupont *et al.* 2000:114-115). Precessional rainfall variation was still a critical factor in climate change of the area (deMenocal 1995:57), altering the amount of rainfall at the shorter timescale, though at a reduced influence during the cooler long-term glacial episodes that preconditioned terrestrial conditions to aridity (Gingele *et al.* 1998:23; Partridge *et al.* 1997:1131). However, the exact nature of these relative influences of low latitude insolation and high latitude affects on sea surface temperature, and how these vary through time, is unknown (Thomas & Shaw 2002:795).

3.4.5. Shorter-term variability

The link between the orbital forcing parameters and climate change is relatively easy to recognise due to their long-term cyclic nature. A great deal of shorter-term variability may exist, but is either not present or not recognised in older records. For example, climate records for the past 100 ka in the Indian Ocean indicate that low-latitude climate responds to the Dansgaard-Oeschger (D-O) millennial scale cooling and warming events recorded elsewhere in the world (Leuschner & Sirocko 2000:252). The periodical enlargement of the Kalahari Desert into a sand sea covering 2.5million km^2 (20-26 ka (MIS2), 38-42 ka (MIS3), and 95-115 ka (MIS5c)), the Mega Kalahari, does not relate to the precessional cycle, but does correlate to D-O cycles (Stokes *et al.* 1998:318). Furthermore, the oldest of these cycles correlates to one of the hyper-arid conditions seen in the Malawi data (85-110 ka (Scholz *et al.* 2007:3)), and in the Pretoria record (Partridge *et al.* 1997:55). This confirms the presence of these arid events across the south-central African region. However, limitations of most records at the millennial scale mean that these links are difficult to determine with any accuracy (Thomas & Shaw 2002:795).

The D-O cycles that lowered sea surface temperatures off the west coast of Africa resulting in reduced evaporation and rainfall, are relatively short lived spanning 5 – 20 ka (Stokes *et al.* 1998:318). This suggests that the geomorphological responses to short lived climatic changes are rapid. Vegetation is also likely to respond quickly to these short-term changes, and low pollen production and low charred-particle deposition during the megadrought phases in the Malawi core reveal that the surrounding watershed of the lake was converted to a semi-desert, with cool conditions and less than 400mm/year rainfall (Cohen *et al.* 2007:16425). There are few other direct records relating to the African continent at present, however data from the C^{13} content of stalagmites in Villars Cave, France, indicates that in European vegetation there is also a relatively short time-lag between climate amelioration and vegetation response (Genty *et al.* 2003:835).

3.4.6. The Last Glacial Maximum and the Holocene

Some relatively detailed climate records have been produced for south central Africa for the most recent glacial to interglacial cycle, representing the Last Glacial Maximum (LGM) (MIS 2, ~21 ka) and the Holocene (*c.* 10 ka to present (MIS 1)). During MIS 2, the LGM, the rainforest was reduced to small refuges, and its current extent was primarily covered in dry forest and savannah grasses (Dupont *et al.* 2000:117). Data from Lake Malawi indicates that water levels were low at the LGM (Gasse *et al.* 2002:410), though were not as severely reduced as during megadroughts, resulting in lake levels being reduced by only 20 – 200m (Cohen *et al.* 2007:16425). This appears to be the case for most lakes across Africa, indicating that arid conditions prevailed across the continent (Partridge *et al.* 1997:1131; Gasse *et al.* 2002:411). Throughout the Holocene wetter conditions increased the water level in Lake Malawi, but the level fluctuated throughout the past 10,000 years, and remained lower than today's level until 4,900BP (Gasse *et al.* 2002:410). Evidence in the Kalahari region also suggests a cool dry LGM, followed by wetter conditions from 18 – 14 ka (Thomas & Shaw 2002:793). The Holocene warm period in Africa was interspersed with millennial scale cooling and warming periods, similar to those outlined above. Cooling periods were focused around 10.2, 8.0, 6.0, 4.6, 3.0, and 1.9 ka, with two more recent cooling periods centred around 1500 and 1800AD where temperatures dropped by 3° - 4°C, with a warm period (similar to present temperatures) between 400 and 1000AD (deMenocal *et al.* 2000:2199 - 2201).

Presuming that past climate systems have changed on both a long-term and a short-term scale it is likely that rainfall, vegetation, and geomorphological processes have been significantly affected in Zambia and the Luangwa Valley. In the next sections the nature of the response of vegetation will be considered, and a tentative chronology for climate and vegetation change for the Luangwa Valley will be outlined.

3.4.7. Vegetation change in central and southern Africa

Most records of vegetation change for central and southern Africa derive from three main sources: pollen from marine and terrestrial sources (Dupont *et al.* 2000:95), organic components of aeolian dust in sediment cores (Schefuß *et al.* 2003), and ^{13}C isotopes in stalagmite cores (Hopley *et al.* 2007). The last two sources have provided estimates of the relative amounts of cold C_3 and warm C_4 vegetation types on the continent. C_4 plants are favoured over C_3 plants in warm humid environments, whereas C_3 plants are favoured over C_4 plants in cool dry conditions (Collatz *et al.* 1998:441). Nearly all trees, and cold season grasses and sedges use the C_3 carbon fixation pathway in photosynthesis, whilst warm-season grasses and sedges use the C_4 method (Schefuß *et al.* 2003:418).

Buffalo Cave provides one of the few direct records for the early Pleistocene vegetation change in southern Africa, documenting the period from 1.99 to 1.52 Ma (Hopley *et al.* 2007). The key finding of the research revealed that the vegetation altered with obliquity, indicated by a 41 ka cycle of arid to humid conditions. Furthermore, overprinting this there was an increase in aridity from 1.8 to 1.7 Ma in southern Africa with a resulting increase in C_4 savannah grassland. This corresponds to records in eastern Africa, suggesting a pan-African shift toward aridity (*ibid*: 428-9). In the mid Pleistocene period covered in Schefuß *et al.*'s research (2003), 1.2 Ma – 450 ka, the C_4 plant record corresponded to the precessional forcing of sea surface temperatures and rainfall, where maximum C_4 plant occurrence corresponded to the maximum insolation and sea surface temperature (Schefuß *et al.* 2003:419). Additionally, overlapping data in the form of pollen records from the Congo Fan, reveals that between 1.05 and 0.6 Ma *Podocarpus* forest was common in the Congo basin during glacial stages, which indicates that conditions were cooler but still relatively humid (Dupont *et al.* 2001:195). In the late Pleistocene interglacials were warmer and humid (*ibid.*). A detailed examination of the pollen records from the same area for the past 150 ka indicates that the rainforest expanded in the interglacials and contracted in the cooler glacial periods, there is also marked variation within the interglacial stage five, where the rainforest significantly contracted in the cooler substages of 5b (95 – 85 ka) and 5d (115 – 105 ka) (Dupont *et al.* 2000:114-115), probably reflecting the influence of precessional forcing noted in other records discussed above.

The effect on vegetation by the millennial scale changes outlined above is not known at present, the effect on sea surface temperatures in some cases would have been minimal, as there may only be a relatively minor change of a degree or two centigrade. However, in the largest of the Holocene events, at 8.0 ka, there was a reduction of 8°C, similar to the reduction at the LGM (deMenocal *et al.* 2000:2199). Older relatively short-term events of the mid to late-Pleistocene have clearly had a significant effect on vegetation, as documented by the megadroughts in Lake Malawi, and expansion of the Kalahari. In these periods the southern portion of the African continent would have been sparsely vegetated.

3.4.8. The Luangwa Valley itself

There are no direct records from the Luangwa Valley. However, its climate, and thus vegetation and geomorphological landscape would have responded to the changes recorded elsewhere on the continent. The general effects of the longer-term glacial and inter-glacial cycles are that during periods of aridity the miombo woodland that is near its tolerance levels where present in the valley today would not have survived. Water in the valley would have been scarce, and even during the rains it would have been restricted to the larger seasonal tributaries and the Luangwa itself. In today's climate the Luangwa becomes extremely narrow and shallow in the height of the dry season, therefore during arid glacial cycles the river could potentially dry out completely on a seasonal basis. The mopane vegetation dominant in the valley today would presumably have survived a certain amount of rainfall reduction, particularly on moisture retaining soils, however, the woodlands might have become concentrated nearer the Luangwa and by any tributaries still seasonally wet. It is possible in this case that the Luangwa Valley might have become a refuge for wildlife and vegetation whilst the plateau and western parts of Zambia dried out

(Barham 2000:227). Mopane and miombo woodlands are not frost resistant (Huntley 1982:106; White 1983:94), and in cooler glacial episodes this might have been the critical factor in affecting the distribution of these woodlands. In central Zambia today there is an average of one to ten frost days a year (Archer 1971b:22), and presumably frosts would occur in the lower elevations in eastern Zambia during glacial episodes. It is likely that the driest parts of the landscape would have become denuded of vegetation with the result that the sediment would have become unconsolidated and unprotected. It is therefore likely that similar colluvial scenarios to those seen in the Chipata district would have occurred (Thomas & Murray 1999:128), and when interglacial conditions restarted, the denuded landscape would have been prone to landslides, debris flows, and other semi-arid geomorphic processes. Meanwhile, the open woodland that is present today due to deforestation may have been present on the Zambian plateau during glacials, and certainly further afield would have replaced the Congo rainforest, as evidenced by pollen records in the Congo Fan (Dupont *et al.* 2000).

The response of the valley during the hyper-arid phases recorded in the Kalahari Desert and Lake Malawi would have been more extreme. The sands of the Mega-Kalahari and the Lake Malawi valley geographically sandwich central Zambia and the Luangwa Valley. In these hyper-arid phases the rainfall of the valley would have been similarly affected. If the rainfall were halved, as it was in the Lake Malawi basin, which at its lowest elevation receives the same rainfall as the Luangwa Valley today (though a great deal more in the highlands surrounding the lake, up to 2,400mm/yr (Cohen *et al.* 2007:16422)), the response would presumably have been similar. Much of the Luangwa Valley would therefore have been reduced to a semi-desert state, and presumably there would have been a decline in much of the large game in the area. The Luangwa could possibly remain in a diminished form, as it is fed by rainfall on the surrounding escarpments that seasonally fills dambos, and its source area on the Nyika Plateau in Malawi, which may have continued to receive more rainfall than the valley.

3.5. Summary

In the Luangwa Valley there has been little research into the geology of the area. However, it is clear that the Luangwa rift is an ancient feature that has presumably undergone some disruption with the Miocene uplift that created the East African Rift Valley and uplifted the Karoo basin in South Africa. This rifting had ceased by two million years ago, resulting in tectonically stable conditions throughout archaeological time. Climate, on the other hand, would have been a key factor of change in the valley, responding to both long term arid glacial and humid interglacial cycles, as well as the rapid changes seen in recent records. These climatic alterations would have affected the human populations in the valley. The arid periods would have seen a reduction in the amount of vegetation and, possibly more crucially, a variation in the vegetation types, which would have reduced the number and variety of animals, in turn would have reduced the population sizes that the valley could support. Elsewhere in Africa climate changes have been linked to the arrival of savannah adapted hominids (Hopley *et al.* 2007:429), or the development of suites of behaviours not previously seen as in the Mode 3 technologies first appearing in the interglacial MIS 7 or MIS 8 (253-186 ka) (Barham 2001:78-79). At Mumbwa Caves in central Zambia the hiatus in human occupation between 130 – 170ka corresponds to glacial MIS6, indicating a response to deteriorating climatic conditions over much of central Africa that caused the expansion of the Mega Kalahari (Barham 2000:242). Clearly the past climate would have had consequences for past human populations, and while it was not possible in this project to tie specific archaeological events to specific climatic changes in the Luangwa Valley it is, however, possible to discern geomorphological processes (outlined in chapter 6) that must have occurred in the past under different climatic regimes when the valley was drier, and less vegetated. Contained within these deposits are archaeological deposits indicating that the valley was occupied in at least some of these arid cycles, for example the upper portion of the Manzi sequence (Barham *et al.* in prep.), and demonstrated by un-abraded artefacts in the cobble deposits in the valley (chapter 7).

Chapters 2 and 3 have provided an introduction to the archaeological, geological, and climatological record of the Luangwa Valley. Much of these records have been transposed from other areas due to the paucity of research in the Luangwa Valley itself. This will therefore be the first study to attempt to interpret the nature of Pleistocene landscape development in the area, and how this relates to the archaeological record.

4. The use of Geographic Information Systems in this Survey

4.1. Introduction

Geographic Information Systems (GIS) is a general term for describing methods of computing that combine geographic spatial information with data using software programs that have been designed specifically for the purpose. GIS is used in many disciplines including the geographical sciences, urban planning, economics, and in planning corporate ventures such as store location (Wheatley & Gillings 2002:9). GIS are versatile programs that store, analyse, and present information about geographic space; they can be used anywhere that there are spatial coordinates or maps, and are essentially amalgamations of various software packages, i.e. mapping, statistics and graphical display. Statistical software is usually built in, allowing complex relations to be determined that may not be visible on standard maps or graphs (Conolly & Lake 2006:15; Wheatley & Gillings 2002:9). GIS allows researchers in archaeology to combine layers of information to gain a better understanding of a place, creating a virtual landscape using base maps to locate where things are. These layers could be topography, sources of water, vegetation, settlement areas of various periods, or other zones of human activity (Aldenderfer 1996:4).

Geographic Information Systems are a relatively new technology in archaeology, and have been described as '*the most powerful tool to be applied to archaeology since C^{14} dating* (Westcott & Brandon 2000: back cover)'. However, there have been criticisms that it is used without intellectual vigour and is overly dependant on simple presuppositions about spatial patterns (Conolly & Lake 2006:10). It is necessary to remain cognisant of these theoretical implications to ensure the success of a research endeavour (*ibid*).

4.2. Theoretical implications of using GIS in archaeology

4.2.1. Processual and post-processual perspectives

Geographic Information Systems has appeared relatively recently in archaeology and arguably its application has mirrored the theoretical development of archaeology as a discipline, though over a greatly compressed timescale (Conolly & Lake 2006:6-8). GIS are spatial analytical tools and, as a result, one of the key concerns with its use in archaeology is how archaeologists interpret human use and understanding of space and landscapes (Wheatley & Gillings 2002:3-20; Conolly & Lake 2006:5). The quantitative revolution in archaeology of the 1960s allowed archaeologists to move beyond mere visual appraisal of spatial patterns in the archaeological record, and on to explore their shape, form, and nature (Wheatley & Gillings 2002:6). A parallel development to the New Archaeology of the 1960s occurred with the use of GIS in archaeology in the 1990s, when some archaeologists began to move out of the pioneer phase of its use (Conolly & Lake 2006:7) toward developing methods of combining statistical tests in GIS with geographic phenomena, such as hill slope and elevation (e.g., Kvamme 1990:378). This particular form of argument has often been termed as environmental determinism, because it regards cultures as functions of, or entirely shaped by, environmental pressures such as water supply or ecological change (Gaffney *et al.* 1995:211).

By the early 1980s it was realised in mainstream archaeology that viewing past social groupings as systems in the manner advocated by the New Archaeology removed the human aspect of past societies (Wheatley & Gillings 2002:8). With post-processual archaeology, notions of space are intimately tied to cultural activity (Ingold 1993:167), with space considered to be a meaningful medium for human action instead of being an unchanging backdrop. Space is also an entity that can be ordered and altered through human agency, including cosmologically based beliefs (Wheatley & Gillings 2002:8). Some GIS practitioners in archaeology were concerned by the mid 1990s that they had accidentally or purposefully reintroduced processualism (Conolly & Lake 2006:8). However, since then many GIS projects have engaged with post-processual frameworks (Conolly & Lake 2006:8 and e.g. Bell *et al.* 2002; Fitzjohn 2007; Llobera 2003:29, 2007; Whitely 2004).

Advanced theoretical studies that engage with post-processual thought (e.g. Whitley 2004), provide innovative ideas for the use of GIS to examine and resolve archaeological problems, however they have limited archaeological applications, as historical data, and ethnographic and oral records, are needed to provide a large amount of detail that is not often available to archaeologists, and certainly not to Palaeolithic archaeologists.

4.2.2. The theoretical setting of this project with regard to GIS

Despite the distrust of purely functional interpretations engendered by proponents of post-processual archaeology, the use of environmental determinist models is still a valid approach to address certain questions, though it can represent a methodological trap for the unwary (Gaffney 1995:371). This is particularly true where GIS is concerned, as it is a spatial program designed for geographic purposes to interpret the environment, and was not specifically designed for archaeological use. Purely determinist models in GIS are of use in eliminating environmental patterning in a data set, to develop a clearer picture of what cultural factors may be present in the record (van Leusen 1995:370). The success of predictive modelling (section 4.3.3 below) suggests that such determinist models are valid in their own right, and an underlying key to their success is due to the fact that archaeological sites do tend to recur in environmental settings favourable to human settlement, in that ancient peoples were influenced by characteristics of the natural environment which are frequently reflected in modern patterns of environmental variation (Warren & Asch 2000:6).

This project encompasses the past two million years of occupation in the South Luangwa Valley. Over this time it is clear that there have been numerous changes to the climate and the geomorphology of the area that will have affected the archaeological record by displacing artefacts and damaging or destroying sites. The locations of unaltered sites or undisturbed artefacts may therefore be a reflection of the unsystematic nature of geomorphological processes, rather than the result of patterned human behaviour. A large proportion of the fieldwork and interpretive work for this project was therefore geared toward assessing the impact of these environmental factors. Once these factors are understood and screened out, the remaining pattern of archaeological material should reflect human decision making in the landscape. In the context of this study, which is undertaken in an archaeologically unexplored landscape, and is the first of its kind in Zambia, the simplest method of starting to build an understanding of human use of space is to understand the archaeological record in relation to geographic features, such as large rivers, hills, and raw material sources that can be demonstrated as having been present in the past. GIS allows rapid mapping and analysis of these patterns that would not be possible by manually plotting relationships on maps. An essential assumption of this project is that the structure of the environment, i.e., the locations of water and useful materials such as stone for tool production, would have been significant factors in decision making in the landscape, and will be evident in the archaeological record in the area. It was hoped that after this was achieved it would be possible to begin to engage with possible cultural interpretations of the data and, as will be demonstrated in chapter 7, this has been possible. The following section explores some of the applications for which GIS has been used and outlines the data sources that were used to aid the research, as well as outlining the key procedures used in this project.

4.3 Applications of GIS in Archaeology

Archaeologists have been able to adapt GIS to fit many niches; at the smallest scale it has been used to map the changes in dental topography of mantled howling monkeys (Dennis *et al.* 2004), and positions of cut marks on bones (Abe *et al.* 2002), through larger scale intra-site level analyses (D'Andrea *et al.* 2002; Huggett 2000), and eventually up to continental scales of analysis (Crumley 1993; Field & Lahr 2005). By the early 1990s five broad themes or applications of GIS in archaeology had emerged (Aldenderfer 1996:9). Aldenderfer's breakdown is still useful for giving an outline of the broad areas in archaeology where GIS is used; the themes outlined do not have firm boundaries, and many projects undertaken use their data to develop ideas within two or more of the different themes. The first two themes: regional data management and simulation are not relevant to this project and are not considered here. A comprehensive summary is provided in Aldenderfer (1996) chapter one, and Connolly and Lake (2006) pages 33-36. The three most relevant to this project – the management of remote sensing data, regional environmental analysis, and location modelling – are discussed below with reference to the aims and needs of this project.

4.3.1. Management and investigation of remote sensing data

This category involves the processing and display of data gained by means other than direct excavation or the recording of artefacts. Aldenderfer places an emphasis on satellite imaging (Aldenderfer 1996:11). However, aerial photography and data retrieved from non-invasive archaeological techniques such as geophysics, are also commonly integrated with GIS (Chapman 2006:30 & 35; Conolly & Lake 2006:68; Madry 1990:175; Wheatley & Gillings 2002: 74 & 80, and e.g. Barton *et al.* 2002:172; Crumley & Marquardt 1990:78; Forte 1995:232). Of key interest here are remotely sensed images that had the potential to aid the project at all stages from planning to analysis. A number of options were explored, however images from Landsat, IKONOS, QuickBird were not suitable because of cost or the limitations of image resolution.

One of the most well known digital satellite imaging systems is Landsat. The resolution of the images produced varies between 15 and 50m depending on the product purchased and location (ResMap 2005 http://www.resmap.com/products/index.html accessed 29/06/09). This resolution renders Landsat useful for identifying large-scale geomorphological features, however it is of limited use in identifying smaller landscape features and archaeological entities (Ur 2003:104).

In addition to the limitation of resolution, cost is another factor that can make Landsat imagery prohibitive to archaeologists. The most recent Landsat 7 ETM+ images can cost up to $1,500 U.S. (Challis *et al.* 2002:140). However, earlier archived images from the Landsat 5 and 7 system can cost as little as $12.00 to $1,495.00 U.S., depending on the product required (ResMap 2005 http://www.resmap.com./products/index.html (see product sub-menus) accessed 16/11/05). These products were considered for use in this project for broader analyses of vegetation patterns in the valley, but access was gained to a relatively detailed vegetation map, as well as to aerial photographs that together precluded the need for Landsat data. High-resolution satellite imagery was another option considered, and is currently available commercially from two systems: the IKONOS system offers 1.0m resolution, whilst QuickBird offers 0.61m resolution imagery. Quotes from the operating companies revealed that to task either of the satellites to image a part of the Luangwa Valley would cost $5,000 U.S, for a minimum area of 100km^2, or $208,000 for a block covering the entire park.

Information from the NASA Space Shuttle can provide data giving values for elevation anywhere on the planet. Though elevation data can be retrieved from contour lines on paper maps, which can be incorporated into GIS by digitising the map (Wheatley & Gillings 2002:66-69), this practice is only feasible for relatively small areas, as digitising large amounts of contour data is exceptionally time consuming. Fortunately, there is a simple and free solution in the form of the Shuttle Radar Topography Mission (SRTM). The SRTM utilized dual radar systems (SIR-C and X-SAR), acquiring two images at the same time that can be used to produce elevation data (USGS 2008 http://seamless.usgs.gov/website/seamless/products/srtm3arc.asp accessed 29/06/09). SRTM data has been used to generate a digital topographic map of the Earth's land surface with a resolution of 90 metres. The SRTM data is accurate horizontally to 20 metres, while the vertical accuracy is actually significantly better, being closer to +/- 10 metres (*ibid.*). Data for any required area can be downloaded free from http://seamless.usgs.gov/, so it was an exceptionally cost effective method for gaining elevation data that, for this project, was suitable for making broad-scale geomorphological assessments of larger features in the valley (see chapter 6.2.3).

4.3.1.1. Corona imagery

A cheaper form of satellite imagery, available at very low cost to individuals, is in the form of declassified American and Russian surveillance satellite imagery. Corona imagery has been used with success by archaeologists prospecting for new archaeological sites; for example Ur (2002, 2003) was able to create vector files in a GIS, based on features and potential sites visible on Corona images, which could be used as a base map in the field to design a survey (Ur 2002:58). Challis *et al.* (2002) and Philip *et al.* (2002) used similar techniques for identifying potential sites. Challis *et al.* (2002) mapped geomorphological features, whilst Philip *et al.* (2002) identified archaeological and geomorphological features that had not yet been noticed by previous surveys.

Images taken by the Corona system are, however, limited by a number of factors; there can be cloud cover obscuring large proportions of the image (http://eros.usgs.gov/products/satellite/declass1.php accessed 29/06/09), and there is an emphasis on areas of the globe that were of strategic importance, particularly the former Soviet Union and the Middle East (Challis *et al.* 2002:140). There are relatively few images covering the area of interest to this project, and of the higher two-metre resolution imagery there are only six images that include the Luangwa Valley. Though there is no cloud cover, the Luangwa Valley with its relatively low elevation, flatness, and high proportion of alluvium, sands, and gravel, appears to be highly reflective to sunlight and as a result is overexposed in the images. Interestingly, this is also an issue with earlier digitally sensed Landsat 1 - 4 images, copies of which were examined in the University of Liverpool Geography Department's map room. As a result, both Corona and early Landsat images were ruled out as being useful to this study. Though satellite imagery has been of enormous use to other research projects working with GIS in landscape studies it was unfortunately, with the exception of the SRTM data, not a practical or economic option for this project.

4.3.1.2. Aerial photography

Aerial photography is perhaps one of the best-known techniques for remotely identifying archaeological sites, the methods for which have been developed since the early 20th century, and studies now frequently integrate aerial photographs with extensive analyses (Barker 1998:52-55). For example, Barton *et al.* (2002) successfully used GIS to link geomorphological processes with the archaeological record in the Polop Alto Valley in Spain. Areas of erosion visible from aerial photographs were analysed with reference to Early to Late Palaeolithic artefact density. This was achieved by creating polygons (shape files representing geographic space) of the collection units and comparing them to polygons representing erosion. The areas with little or no erosion were then used to identify areas with artefact densities that genuinely reflected cultural spatial patterning in the landscape before Holocene erosion had affected the area (Barton *et al.* 2002:172-6).

If aerial photographs do not already exist for an area, commissioning a flight to take them can be prohibitively expensive, especially in such remote areas as the Luangwa Valley. Fortunately, aerial photographs are routinely taken of the valley at a scale of 1:50,000 for ecological and management purposes. Copies of these purchased for the South Luangwa Past Peoples and Environments Project were used to map large and medium scale geomorphological features; they were also of use in the planning stages of the final field season. Aerial photographs proved useful for more general geomorphological mapping as discussed in section 5.6.2, and the method of incorporating the aerial photographs into GIS is described in section 5.6.3.

4.3.2. Regional environmental analysis

Regional environmental analysis involves the examination of large-scale phenomena and how they relate to the environment, and this has obvious applications to this project. Archaeologists using GIS are able to analyse large areas of spatial phenomena more simply than is possible with non-GIS means, and large regions can be examined systematically (Aldenderfer 1996:12). Though it is relatively easy to link historical events to environmental phenomena, they must be correlated with care so as not to reduce an argument to simple environmental determinism and allow scope for human agency (Crumley 1993:377).

One of the most comprehensive landscape studies is that of the Kythera Island Project (KIP). Bevan and Conolly (2004) analysed survey data retrieved from approximately one third of the 280km^2 area of the island. They studied the archaeological landscape by systematic survey to explore land-use, site formation, and artefact taphonomy, as well as undertaking some ecological modelling (Bevan & Conolly 2004:123). The field survey was undertaken over four seasons, and designed to build up 1km wide transects aligned north south, surveyed by 4 – 7 surveyors at 15m intervals (Bevan & Harlan 2003 http://www.ucl.ac.uk/kip/survey.php accessed 29/06/09). With this survey data, and geographic and geological data from the maps, it was possible to understand the relationship of differing agricultural practices and slope angle, relationships of site discovery, survey intensity, and surface visibility, site definition, and also site location preferences (Bevan & Conolly 2004:124-135). The key experiment of interest here involved modelling the terrain and factoring in hydrology to determine how they may have affected settlement location preferences. The results revealed correlations with Bronze Age Neopalatial sites and areas of low to medium relief changes, which are usually near watershed boundaries (*ibid*: 135). However, though these correlations exist, they are not enough in themselves to explain why these locations may have been chosen (which would be the environmentally determinist explanation), as human decisions are made for various reasons, from regional (whether the general landscape was suitable to live in), to the local (whether a particular patch of land was suitable for a settlement) (*ibid.*).

Another project of interest is the survey undertaken by Barton *et al.* (1999, 2002) in the Polop Alto valley in Spain. In a similar scenario to this project, the researchers were faced by a landscape that was dominated by artefact assemblages that could have derived from various periods. Their survey was conducted over three seasons by four to eight walkers spaced at 15m intervals, and field boundaries were used to demark the randomly selected survey units (Barton *et al.* 2002:162). By systematising their method of determining the ages of assemblages in each survey area they were able to rank artefact assemblages according to the probability that they derived from a particular chronological interval. This ranked value was called a 'Temporal Index' value (*ibid*: 176). As this did not permit modelling of land-use over time a 'settlement intensity index' was created to give an estimate, at the regional scale, of the proportion of the assemblages from each survey unit that accumulated in each time period (*ibid*: 178). Eventually, information regarding the intensity of land-use (rather than settlement patterns), and how this varied across the 30km^2 survey area throughout the past 200ka was gained (Barton *et al.* 1999:629). This survey was completed on a relatively small scale compared with the size of area analysed within the Luangwa Valley. Additionally, the survey units in the Polop Alto valley were near each other, allowing a density map to be created of the whole valley with few gaps; a practice that was not possible in the time allowed in this project due to the size of the area under investigation in South Luangwa, which as a result required survey units to be spaced apart (see chapter 5). Nevertheless, a similar methodology was required, one that could be employed to compare the density of artefact accumulations throughout the survey area. Whereas in the Polop Alto it was possible to outline up to nine factors in assigning temporal index values to assemblages (Barton *et al.* 1999:618), it was not a suitable approach in this project as there are not as many typologically distinct artefacts for the different periods in the area (see table 2 in *ibid.*).

4.3.3. Location/predictive modelling

Location modelling involves predicting the value or probability of occurrence of variables in un-sampled locations using one or more independent variables (Conolly & Lake 2006:179). This is achieved by projecting known patterns and relationships onto unknown times and places (Warren & Asch 2000:6). In archaeology predictive modelling has been focussed toward determining the probability of archaeological sites and settlements occurring in un-sampled locations (Conolly & Lake 2006:179). Location modelling has strong links with regional data management as it often uses this methodology to predict the location of sites for the purposes of conservation and management (Brandt *et al.* 1992; Dalla Bona 2000; Fry *et al.* 2004), or as an aid to planning potential excavation or survey (Potts *et al.* 1996). There are two key assumptions made when using predictive models: first, that settlement choices made by ancient peoples were strongly influenced by the natural environment, and second, that these environmental factors survive, in some form, in modern day maps (Warren & Asch 2000:6). Predictive models are created by using two key methodologies. The first involves statistical derivation of the critical attributes that affect the archaeological distribution followed by logistic regression methods to construct the model (Connolly & Lake 2006:183; Wheatley & Gillings 2002:171). The second and more simple method involves applying weightings to physical features on maps based on statistical analyses of known archaeological distributions and/or theoretical concepts of distributions (Wheatley & Gillings 2002:170).

The weighted feature approach is used to create a model of Later Stone Age and Iron Age activity in the landscape in the final part of this project, and is undertaken to aid an understanding of the contrasting patterns of land-use of the two groups of people who co-existed in the landscape until the 19th century. In the weighted map approach features are assigned values based on whether there is a high or low probability for sites being on or near them, these features can include soil quality and type, slope angle, vegetation, or distance to water (Brandt *et al.* 1992:271; Dalla Bona 2000:75; Fry *et al.* 2004:99; Wheatley & Gillings 2002:167). The key methodological issue with this approach is the selection procedure for applying the weighting. In some studies this has been done purely on a theoretical basis where the weights are created using a deductive process (Brandt *et al.* 1992:271; Wheatley & Gillings 2002:166); however, these weights might be no more than subjective notions about where archaeologists believe human activity is likely to occur (Brandt *et al.* 1992:271). It is better, where possible, to base the models on available empirical data to at least introduce a level of objectivity into the analysis (*ibid*), an approach sometimes referred to as the inductive approach (Wheatley & Gillings 2002:166). For example, Brandt *et al.* (1992) in the Regge Valley in the Netherlands used six variables in creating their predictive model including soil texture, geomorphology, and distance to water; the result was a composite map where high values represented land with a high probability of containing sites. On testing the model with archaeological data a good correlation was revealed between the model and reality (Brandt *et al.* 1992:276). Well-designed weighted feature models that are tested either statistically or in the field can yield good predictive results regarding the location of archaeological material (Brandt *et al.* 1992:278; Dalla Bona 2000:82; Fry *et al.* 2004:105; Warren & Asch 2000).

Predictive models make the assumptions that settlement choices made by prehistoric peoples were strongly influenced by characteristics of the natural environment (Dalla Bona 2000:75). As noted above (section 4.2.1), the charge of environmental determinism has been laid against GIS practitioners in archaeology, and this is particularly true when considering predictive modelling, which has been described as one of the most controversial applications of GIS in archaeology (Wheatley & Gillings 2002:179). For the purposes of this research it is accepted that there are unresolved methodological and theoretical issues regarding the use of predictive modelling, particularly with regard to including cultural factors affecting land-use of past peoples. A more critical problem for this project is that the majority of predictive modelling enterprises, if not all, have been directed at predicting the localities of sites (e.g. Bevan & Conolly 2004:132; Brandt *et al.* 1992; Dalla Bona 2000; Duncan & Beckman 2000; Fry *et al.* 2004; Warren & Asch 2000; Wescott & Kuiper 2000). Contrastingly, this project is directed towards off-site material. The site centred nature of predictive modelling is regarded as a serious methodological issue by some, as it ignores a vast body of anthropological and archaeological evidence and thought indicating that sites represent only one aspect of human activity in the landscape (Ebert 2000:131). Only recently has it been specifically acknowledged that off-site information is important to predictive modelling (Lock & Harris 2006:51), but even then only as a method of better defining the factors that affect the location of sites (*ibid.*). Considering that the modelling output is a continuous surface where each grid-cell of the map represents a value in the landscape, rather than specific locales that might indicate the location of sites (Wheatley & Gillings 2002:167), it is at first surprising that off-site models have rarely been attempted. The reason for this is that the site based concept is regarded as a necessary prerequisite of predictive modelling (Lock & Harris 2006:51), as it is only on the basis of the site concept that statistical analyses can be conducted to test the model predictions. As a simplified example, if 16 of the 25 observed sites fall in the area where sites are predicted, then the model is 64% accurate (Conolly & Lake 2006:184). As a result of the requirement to use sites, off-site data has not been incorporated into predictive models, which made it difficult to know how to proceed in creating models of land-use in this project. It is argued here that though it is not possible to test a model produced from off-site data that does not mean that they are not valid or should not be attempted.

Using an off-site survey strategy restricts the statistical operations that can be undertaken in the research design and results analysis stages of research (Foley 1978:59; Dunnell & Dancey 1983:272, and see section 5.4), and this restriction will extend to the testing of a predictive model produced using off-site material. Usually predictive models are used to predict the probability of sites occurring in an un-surveyed area, so the question raised is, What will an off-site predictive model be predicting? The answer is that proximally artefact density in the landscape is the focus of prediction, but ultimately this acts as a proxy for land-use intensity. Land-use intensity has been approximated using artefact density using GIS based analysis by Barton *et al.* (1999; 2002 see 4.3.2 above), though they were not using a predictive methodology. Foley (1981), on the other hand, used artefact density as a proxy for land-use intensity, and used a contouring technique in his analysis (section 2.5.2) that could be described as an early attempt at predictive modelling, but without the use of GIS. Foley's method of drawing contours in the landscape to extrapolate probable density between survey units using various attributes of artefact data (e.g., raw material (Foley 1981:142, fig 7.7)), and ultimately summarising intensity of human activity based on analyses of artefact density (Foley 1981:192, fig 9.3), is effectively a predictive land-use model produced by hand. The aim of the predictive models developed in chapter eight is the same. Land-use intensity of Later Stone Age and Iron Age peoples is graphically represented based on observations of the distribution of archaeological material in relation to environmental factors that can be demonstrated to have influenced the intensity of activity in various parts of the landscape. The predictive models created are tentative models to illustrate and compare the different land-use strategies of two groups in the area. As a result of the inability to statistically test them their effectiveness can only be assessed subjectively.

4.4. Summary

The theoretical debate over the interpretation of space and cultural activity within archaeology will continue; it is possible to carry out both environmental studies and humanistic studies using GIS, though they will usually have distinct sources of data, ecological or social, resulting in the studies answering different questions (Wheatley & Gillings 2002:234-6). Environmentally based studies are usually unable to generate some of the more complex cognitive explanations to which some researchers aspire (e.g. Whitley 2004). However, once the environmental factors that may have affected the archaeological record have been removed, an environmentally based study can be used to develop theories regarding the behavioural choices of human groups in their landscapes (e.g. Barton *et al.* 1999). This is an approach of particular use when considering the aims of this project and those of Palaeolithic or Stone Age archaeology in general, as there will inherently have been more time for environmental factors to alter archaeological remains. Each research project is different, and needs a separate theoretical approach to meet its particular needs. In this case, it is vital to engage with an environmentally determinist argument to understand the geomorphological development of the valley, as only then will it be possible to determine what genuine human behavioural residues remain. The potential disruption to the archaeological record, particularly in the earlier periods of occupation in the Luangwa Valley, may preclude higher level cognitive interpretations regarding behaviour because, as noted above, cognitive interpretations need a large amount of high-resolution data.

This project was unable to use a 100% survey covering the area of interest, so a sampling methodology was employed that would provide data from a sample of the landscape. It was also not feasible to undertake a survey similar to some of the intensive small scale, or long-term large-scale surveys outlined here and in chapter 2. A compromise approach was needed which could meet the project aims. Both European and African archaeological surveys that might, or did lend themselves to an analysis with GIS, were used to aid the survey design, with consideration being given to the constraints on time and the nature of the working conditions in South Luangwa. The next chapter outlines the field methodology used for this project, which is based on the off site approach initially developed by Thomas (1975), and first used in Africa by Foley (1981). Conceptually, Foley's project provides a template for the research undertaken here.

5. Methodological Approach for the Archaeological and Geomorphological Survey

5.1. Introduction

For the first season's fieldwork in South Luangwa it was hoped that archaeological transects could be surveyed at various distant points across the South Luangwa National Park to sample a wide range of environments in the valley. It was also hoped that most of the geomorphological observations could be completed at the same time as the archaeological survey. Recording the archaeological and the geomorphological information at the same time seemed the most economic approach, considering the logistical constraints and limitations on time in the field. This approach was revised at the end of the first season, as it was realised that the geomorphology of the area varied more greatly over relatively short distances than was initially expected. A better understanding of the positions and extent of these differing geomorphological areas was therefore needed, and for this a geomorphological map was created during the first season and improved in the off-season before the second season using a combination of field notes and aerial photography. Furthermore, a more systematic and targeted approach was needed to understand Pleistocene and Holocene geomorphological processes in the region, and the methodology for this and the geomorphological map are presented in section 5.6. The archaeological survey strategy was redesigned to encompass a smaller more manageable area of the landscape so that the differing geomorphological environments could be sampled. In the final sections of the chapter the finalised archaeological survey strategy is outlined.

5.2. Aims and Objectives

The key aim of this project was to develop an understanding of the geomorphological processes that have affected the preservation and distribution of the archaeological record, and to determine the degree to which behavioural information relating to early and modern human use of the landscape survives in the South Luangwa archaeological record. Essentially, the main aim was to factor out the non-cultural factors that have altered the archaeological record. In designing the fieldwork to meet these aims three key objectives were identified:

1) To determine the relative quantity and distribution of surface archaeological material in the region, and how the distribution varied throughout time.

2) To outline what geomorphological processes affected different parts of the landscape, and in what manner. Once these natural processes have been identified, the effect on the distribution of archaeological material in the landscape can be determined.

3) To understand how the archaeological material may relate to resources and geographical features in the landscape such as fresh water, lithic raw material sources, vegetation types, and topographic features. This information will provide the basis of an understanding of behavioural choices (c-transforms) of humans in the landscape.

5.3. The Research Environment of the Luangwa Region

The survey area is located within the naturally defined boundaries of the Luangwa Valley, which is bounded by the Muchinga Escarpment rising from the valley floor to the north-west, and the Nchindeni Hills to the south-east (fig 5.1 below). The ability to survey any area is controlled by the accessibility (Orton 2000:80), and within the Luangwa Valley there are four key conditional variables. First, the area under investigation is very large at 9050km^2, making extensive surveying or accessing remote parts of the park difficult. The low number and density of all-weather roads magnified this issue, and where these roads are present they are time consuming to negotiate even in a 4x4 vehicle. The roads are mainly un-surfaced dirt tracks that are seasonally damaged and made impassable during the rainy season, and as they are not always re-graded in the dry season, the long travel times can be exacerbated. Second, for the first season there was only one vehicle available to the PPE team, therefore the research aims of this project had to be compatible with the PPE project aims, which limited time in the field. Third, it is not possible to work independently, due to the risks of being accidentally shot by a hunting safari, or attacked by an animal, as the SLNP and its surrounding GMA has a large population of wild game including elephant, hippo, Cape buffalo, lion, and leopard, not to mention a number of species of poisonous snake. An armed scout was therefore needed at all times when working in the field, and required by law within the park if the vehicle was exited. Only one or two individuals can survey at any one time as usually people are required to walk behind the scout single file, and cannot fan out at intervals as is usually the case in archaeological survey. During the first season, only one scout was available for the whole team. Combined with transport this restriction resulted in the entire team being used for a short intensive survey time. Manpower was therefore constrained by both safety considerations and available survey time. For the second season a separate vehicle and scout were available for an extended three-month season that, with me as a single surveyor, eliminated these logistical and safety related problems.

Fourthly, the vegetation of the region, which comprises a mix of woodland and open plain, complicated the accessibility in different areas. The largest component of the vegetation on the valley floor is a near single species woodland of the deciduous *Colophospermum mopane* (Astle 1995). The remainder of the vegetation comprises mainly semi-deciduous miombo woodland (mixed woodland primarily made up of *Brachystegia* of various species), along with plant communities of *Acacia*, *Combretum*, *Diospyros*, and *Terminalia* (Astle *et al.* 1997a:109). The density of these plant communities varies from open plains with occasional trees, to dense bush and wooded areas that are particularly common in the riparian and riverine woodlands close to the Luangwa River (Astle 1995). The varying density and type of vegetation was one of the key factors affecting accessibility to different areas of the landscape; this was not so much due to high density vegetation blocking access, so much as the reduced visibility hiding potentially dangerous animals. This was a particular problem during the first season in the foothills of the Nchindeni and the dense vegetation sometimes found near the main channel of the Luangwa River, which were relatively dangerous due to persistently high elephant numbers.

The presence of both the large and smaller game was a relatively intangible accessibility problem, and novel, when compared to the problems faced by other researchers. For example, in many European surveys such as Barton *et al.'s* (2002) in Spain, the presence of fields at various stages of crop rotation affects both accessibility and visibility, as farmers are more willing to allow survey on ploughed fields (that yield near 100% visibility of the ground surface), whereas they are unlikely to allow survey in fields with ripe crops (which completely obscure the ground) (Barton *et al.* 2002:164). At least in the case of the latter it is possible to find out in advance when the field will be uncultivated, whereas the location of wild game is virtually impossible to predict with any certainty. Nevertheless, this project had a distinct advantage over many as there are no villages, field boundaries, or potential landowner issues, as people are not allowed to live within the park, and there are few villages or dwellings close to the Luangwa on the GMA side of the river. It was necessary, though, to obtain a research permit from the Zambian Wildlife Authority (ZAWA) to undertake any work in the area. In some areas the presence of wild animals, in particular elephant, caused delays to fieldwork, or alteration of plans at the daily level.

5.4. Approach to Sampling in this Project

5.4.1. Introduction

This section outlines the theoretical basis and the framework on which the sampling strategy for this project was initially designed and then developed. The many aspects to be considered during the design stage of the sampling methodology for the survey included the limited prior knowledge of the nature of the archaeological material, the structure of the archaeological record in the valley, the logistical issues of access and safety, and the limits on manpower and time in the field. With regard to the archaeological record, experience of archaeologists who have conducted surveys elsewhere in Africa, indicated that there are often continuous distributions of artefacts in the landscape (Foley 1981:31; McIntosh & McIntosh 1993:77), due to the fact that throughout most of archaeological time artefact distribution originated from highly mobile groups (Foley 1981). Even with the more sedentary groups of the Zambian Early and Late Iron Ages, the villages would rarely have remained in one place for more than 30 years due to the tradition of relocating the village following the death of each Headman (Derricourt 1985:17). Preliminary observations, made whilst accompanying the PPE project during a field season prior to the start of this project, revealed a similar expectation for the Luangwa Valley. Lithic material is generally found in scatters, clusters, and even as continuous thick carpets of cobbles containing large numbers of stone artefacts. Iron Age pottery appeared to be less common, though clustered where present. In general, distributions appeared to be continuous with few obvious and easily defined sites (see 5.5.5 for a discussion on site definition). Discovering sites in this instance was therefore deemed to be a secondary objective, as the main objective was to determine the relations of the archaeological record to the geomorphology of the landscape, and from that point attempt to determine what parts of the landscape were favoured differentially through time. A survey design was needed that would encompass all archaeological material in the landscape, and an off-site approach was therefore taken.

5.4.2. The use of non-probabilistic reasoning and the off-site approach in this project

Surveying in archaeology arguably has its roots in the origins of the discipline in the 16th and 17th centuries (section 2.3). However, it was not until the advent of the New Archaeology in the 1960s that probabilistic sampling methods (random sampling) were explicitly used (Orton 2000:69). Their use rapidly spread in the USA (Wobst 1983) and shortly afterwards in the UK and Europe (Orton 2000:73), since when there has been considerable debate regarding the applications of sampling techniques in archaeological site and regional investigation (Cherry *et al.* 1978; Clarke 1972; Hodder & Orton 1976; Moore & Keene 1983; Mueller 1975; Orton 2000). Some researchers, including Cowgill (1975) and Orton (2000), prefer rigorous mathematical statistical techniques to be applied in archaeology wherever possible. However, this may limit what questions can be asked in archaeological research due to the descriptive nature of many statistical approaches that are designed to provide estimates of the shape and distribution of sample values, and the range of possible population values with the presumption that chance was the only factor involved in providing the sample (Wobst 1983:71). These descriptive functions can provide estimates of reliability of statistics such as means, percentages, ratios, aggregates, and comparisons of population means, but do not allow for behavioural interpretations in themselves (Asch 1975:181).

Hypothesis testing was seen as a method of incorporating behavioural interpretations into statistical work (Wobst 1983:77). However, this is not always possible as both standard statistical sampling and hypothesis testing require sample sizes to be well defined in relation to the population being sampled; this requires a large amount of prior knowledge of the archaeological record (Boismier 1991:20; Dunnell & Dancey 1983:268; Shennan 1985:13, and see section 5.4.4). As this survey was conducted in an archaeologically unknown area, it was not possible to make the requisite calculations of sample size. The methodology for this archaeological survey needed to be designed to consider the structure of the landscape, vegetation, and the practical issues discussed above. The approach to this regional investigation was therefore based on Asch's discussion on the limitations on probabilistic reasoning in archaeology, and the merits of using non-probabilistic reasoning in archaeological work (Asch 1975), Boismier's (1991) non-probabilistic methodological concept for surface survey, and the off-site approach developed by Thomas (1975), Foley (1981), and Dunnell and Dancey (1983).

Targeting sample locations to areas of known high artefact density using a non-probabilistic approach will not produce a sample useful to analysis, as there would be bias introduced by selective collection of artefacts (Boismier 1991:20) and over or under sampling of areas of low density (Foley 1981:34). Therefore, a non-probabilistic survey will require a similar methodology to a probabilistic one, and in the sample design a choice of sampling unit (5.4.3), sample size (5.4.4), and scheme (5.4.5) will need to be made (Boismier 1991:20). The essential difference is that with non-probability sampling there will be indeterminate bias (in the statistical sense where population characteristics can be estimated with little or no bias, due to the fact that the sample acts as an estimator of a population statistic) (Asch 1975:181), and the outcome is that the representativeness of any one sampling unit cannot be systematically evaluated on an objective basis for statistical reliability (Boismier 1991:20).

The off-site approach studies archaeological materials and features outside the context of the site to make inferences about human behaviour (Foley 1981:29). In North America the approach has been called non-site archaeology (Thomas 1975), and arguably was first defined and used for archaeological landscape studies by D.H Thomas. Off-site approaches are useful in many contexts, particularly in areas where artefacts occur relatively continuously, or in dispersed scatters across a landscape (section 2.3). Once taphonomic processes that have affected them have been screened out, density distributions of artefacts in the landscape should conform to the distribution and frequency of prehistoric human activities (Foley 1981:32). Artefact density analysis is one potential methodology for the off-site archaeologist (*ibid*), and is used frequently (Bintliff 2000; Gillings & Sbonias 1999; Keay & Millet 1991; and Shennan 1985). It is a simple method to use, both procedurally in the field and analytically afterwards; it can be used for other related analyses such as site differentiation (Gillings & Sbonias 1999:53), land-use (Barton *et al.* 1999:629), and examining the effects of geomorphological processes on artefact density (Barton *et al.* 2002:172).

5.4.3. Sampling unit

Following the reasoning of Asch (1975) and Boismier (1991), sampling units need to be used in a non-probabilistic survey (which is also the case for probabilistic surveys (Orton 2000:85)), as for many logistical reasons a 100% survey covering an entire region (a total survey) is not usually possible. There have been those who argue that it is not possible to conduct a regional survey by taking samples of the landscape in this manner because it would result in large areas of ignorance between sampling units (Bintliff 2000:201). Bintliff asks the question *"If we do not know the structure of the surface data, what grounds do we have for putting any reliance on the representativity of thin transect samples or spot samples across the landscape? (Bintliff 2000:201)"*. While there is both logical and statistical truth in this, particularly with regard to regional studies (see discussion in 5.4.4 below), it is not possible for most archaeological studies to perform such a survey due to constraints on time, manpower, access, safety (particularly in this survey), and funding limits. Therefore, the use of a sampling strategy is needed. The use of a non-probabilistic approach in this survey requires mapping of artefact densities based on dispersed survey units and subsequent interpolation of values between units. This interpolation in turn relies on the assumption that the individual units are representative of the area surrounding them (Boismier 1991:20). This assumption places limitations on what questions can be asked of the data; in the final design of this survey, for example, survey units were often a kilometre or more apart. As a result there would be little information that would be useful in delineating sites from background off-site scatters (Bintliff 2000; Gillings & Sbonias 1999), detailed modelling of land-use intensity (Barton *et al.*2002), or studying site hierarchies in the landscape (Favory & Raynaud 2000). The non-probabilistic approach adopted here is useful as it allows an understanding of the relationship of archaeological material to the geomorphological processes of the area. This can then be used to determine what information remains in each of these areas regarding past human activity in the target landscape. Depending on the extent and types of geomorphological disturbances, a preliminary reconstruction of land-use preferences, and possibly intensity, can be reconstructed.

Decisions regarding the choice of sampling unit are concerned with the shape and size of the units to be used for data collection (Boismier 1991:20), with some issues related to the overall sample size (discussed here and section 5.4.4). Theoretically, as long as the units are sensitive to the phenomena observed in the sample design, in this case artefacts in the landscape, the units can be squares, rectangles, hexagons, triangles, or any other shape, such as field boundaries, that seems useful (Orton 2000:86; Wobst 1983:71). Formally, the only requirement is that they do not overlap and that a sample unit could potentially sample any part of the region (Orton 2000:85). A balance has to be sought between having a large number of small sampling units and a small number of large units (Boismier 1991:20). The former would create a larger sample size than the latter (Asch 1975; Banning 2002:86; Wobst 1983:71). However, there would be an increase in cost in the form of time used when surveying a large number of small units, as more time is spent travelling between many units, thereby reducing the amount of time available for survey (Banning 2002:86; Derricourt 1985:16). For this project 100m x 500m rectangular transects were eventually chosen as the field data source and these were designed so that in the analysis stage of the project they could be split into 100m x 100m units if necessary. Other shapes and sizes were considered, and these are discussed below in relation to the first season's research (5.5.4).

5.4.4. Sample size and fraction

A key issue in probabilistic statistical sampling concerns obtaining a sample size (number of observations made, i.e. number transects (Banning 2002:124)) that permits standard errors to be produced for estimates of population parameters, which then allow confidence tests to be applied to these estimates (Orton 2000:92-4). This in turn allows statistical inferences to be drawn from the data using techniques favoured by researchers like Cowgill (1975) and Orton (2000). As noted above, such sampling is not feasible for many research projects, including this one, due to the off-site nature of much of the archaeological record and logistical limitations placed on sampling. It is also necessary to know the basic parameters of the archaeological record being sampled to determine a statistically useful sample size (Asch 1975:181; Dunnell & Dancey 1983:268; Orton 2000:92; Wobst 1983:72), which in this case was not possible without conducting an extensive initial survey. Therefore, in line with the non-probabilistic approach outlined above (5.4.2), formal statistical techniques were not used to calculate sample size. It was necessary to calculate sample size based on what was pragmatically achievable in the survey time allowed (see 5.5 and 5.7).

Related to the sample size is the sampling fraction. This is the proportion of the population that the sample includes; it mainly affects precision, which is a figure that can be used to reduce sampling errors (Orton 2000:23). Sampling in landscape surveys can rarely achieve a sampling fraction of any great size. For example, Foley's survey in Amboseli had a sampling fraction of 0.05% (ratio of transect size to the landscape). Furthermore, in a landscape context when individual artefacts are the operational unit, as in an off-site approach, rather than sites of a minimum specified size that ought to be discovered (as in Orton 2000:84), the relationship between the landscape and the artefact is key, and is approximately $2{\cdot}5^{-10}$ %, a figure more in common with astronomical sampling than with the life sciences (Foley 1978:59). The main point stressed by Foley (*ibid.*) is that the relationship between the size of unit (transect in this case) and data component (artefact) is of importance when designing a survey strategy. This relationship constrains the sampling procedure with regard to statistical tests (Foley 1981:39); this further highlighted the necessity for a non-probabilistic approach for this project.

The determining factors in this survey regarding the number of transects walked, and therefore sample size and fraction, was the amount of time that could be devoted to the fieldwork, the number of people surveying, and most importantly here, the safety of the surveyors. The exact sample size was determined after the first season's work (see 5.5 and 5.7), as it was only after this that the logistics of working in the region were fully appreciated. The transects in this project were designed and surveyed in a manner that allowed artefacts to be collected in their landscape context rather than being targeted at discovering 'sites' in the traditional, if vague, archaeological sense. Nonetheless, sites were encountered particularly from the Iron Ages, and a discussion on how they were defined and recognised follows later in the chapter (5.5.5).

5.4.5. Sampling scheme

The sampling scheme is the particular method by which sampling units are selected (Boismeir 1991:20), and thus located within the survey area. This project's concept is based on determining geomorphological controls on the archaeological record and to understand how the two are related. A sampling technique was therefore needed that would allow comparisons of archaeological material in different geomorphological zones. A stratified sampling technique was chosen. This technique divides a survey area into sub-strata, or sub-groups, the guiding principle being to minimise variability within strata, and to maximise it between them (Orton 2000:30). In this case the geomorphological zones formed the sub-groups, and the archaeological survey was designed to sample each sub-group, thus allowing the archaeological and geomorphological records to be related. Purposive sampling of archaeologically rich areas would be incompatible with trying to understand the distribution of artefacts (Boismier 1991:19; Foley 1978:52, 1981:34). Simple random sampling was an option but was not without risks as it could produce a representative imbalance between areas of interest (Orton 2000:30), a survey unit straddling two zones with differing archaeological properties (see Foley 1981:37 for an example), or could omit certain areas of the map that may be relevant to the research questions posed (Banning 2002:115).

5.4.6. Summary

The next sections outline the field methodology for the archaeological and geomorphological survey work. The archaeological survey was based on the concepts outlined above, taking into account the limited prior knowledge of the area and the logistical constraints. Limited knowledge of the nature of the surface archaeological material, and the use of the off-site approach that used artefacts rather than sites as the main focus of investigation, ruled out the possibility of designing a survey that would yield a statistically derived sample. The non-probabilistic reasoning used to design the survey required that similar methodological issues were addressed to produce an un-biased sample. For this survey rectangular transects were used, and the region was surveyed using a stratified sampling technique employing separate geomorphological zones to form the survey strata. The number of transects surveyed in the region could not be determined statistically, and the primary control over the number of transects surveyed was the time allowed in the field.

5.5. Field Methodology – First Season

5.5.1. Survey design: sample scheme and unit

The survey design was based upon the concepts outlined above, and on successful surveys undertaken elsewhere (reviewed in chapter 2). In the planning stages the exact size of the survey region was not firmly determined, and was initially defined as the geographical area of the SLNP and the surrounding GMA. The lack of a precise outline for the survey area was due to uncertainties regarding the limitations to fieldwork outlined above which precluded accurate calculations of the length of time available for survey. Within the initial plans it was hoped to gain a representative sample of archaeological material from various disparate geographic locations within the park, creating a sample that would allow comparisons of material across the width of the valley from the Nchindeni Hills to the Muchinga Escarpment. For the first six-week season of fieldwork three survey areas were chosen (figure 5.1) as this was perceived to be an achievable aim in the time allowed. In the second season, more zones could be surveyed in other areas of the park. Two of the initial areas were situated either side of the Luangwa River, and the third was deeper inside the park. It was thought that this zoning would be sufficient to compare areas near to and either side of the Luangwa River, and centrally in the park. The survey area on the south bank is representative of the area near the Nchindeni margin of the valley, the one on the north bank and one further into the park being representative of the centre of the valley.

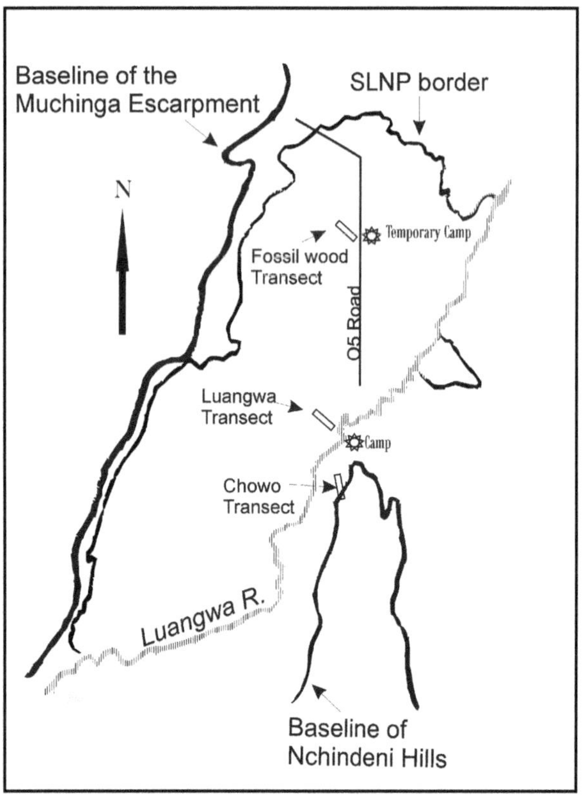

Figure 5.1: A map illustrating the positions of the two ranges of hills that comprise the valley sides, the boundary of the National Park (its southern boundary is primarily formed by the Luangwa River), and the location of the first season's transects (not to scale).

Transects of 0.5 to 1km in width are frequently used in landscape surveys, usually due to considerations regarding the structure of the archaeological record (Cherry 1982:19, Shennan 1985:13). One-kilometre wide survey units with a maximum length of 5km (figure 5.2) were planned, and it was hoped that this would yield a broad understanding of the varying density of archaeological material in the landscape. These large survey units could then be further subdivided into 100m^2 units in the analysis stage to permit the construction density maps in GIS that would provide information about spatial variability in each area. In the planning stage for the first season simple calculations revealed that in a six-week field season surveying these large units appeared feasible, even with only one surveyor (see appendix 5.1). In retrospect these large survey units seem too large, both from a logistical and an analytical point of view, as despite the calculations presented in appendix 1 and the potential to split the transects into smaller blocks, the fieldwork and subsequent data processing would have been extremely time consuming (see 5.5.3). The survey conducted was a surface survey, though sub-surface samples obtained as part of the main Luangwa project were useful in developing an understanding of the geomorphological features and the archaeological sequences contained within them (see appendix 6.1). There is the consideration that the surface remains may not be truly representative of the number of archaeological entities within the region, however most regional surveys are surface surveys (Bintliff 2000:201-203; Boismier 1991; Cherry 1982; Foley 1981; Shennan 1985). Sub-surface sampling would be a possible solution to test the representativity of surface material, but, within the spatial and logistical context of this study sub-surface sampling was not deemed appropriate due to its tendency to be resource intensive. It is also better suited to surveys sampling sites, or as a sub-sampling technique for regional surveys (Orton 2000:90).

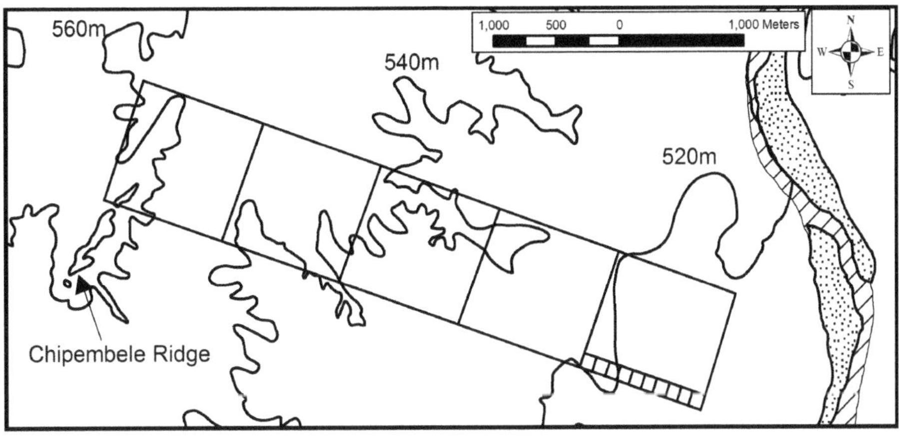

Figure 5.2: Close up of the planned transect on the north bank (the Luangwa transect in figure 5.1). The smaller squares are the 100m square blocks that the data could be split into after the fieldwork. The Luangwa River and sandbanks are shown in hatching and stippling, and the contour labels represent metres above sea level.

5.5.2. Sampling unit procedure

The first season's sampling units (transects) were plotted by hand onto 1:50,000 scale Zambian Survey maps; they were then navigated by plotting the corner point of a transect onto a Garmin Venture GPS unit with programmable waypoints. Guiding points and the other corners of the transects were then projected using the inbuilt functions of the GPS unit. This was quite time consuming and involved a lot of downtime at camp to create the necessary points, at least one every 100m, to provide enough points to navigate the transect reliably. Before starting the fieldwork, and before it was known that there were going to be restrictions on transport and scout availability, it had been planned that the transects should be walked at 30m intervals, arranged in straight parallel lines within each transect. Previous landscape studies in Africa have walked transects using intervals of 10m (Pwiti 1996:45), 30m (Bisson 1992; Dickson *et al.* 2004), 50m (McIntosh & McIntosh 1993:77), and up to 1km intervals (Robertshaw 1994:112). In Europe the spacing between field walkers is regularly in the 10 – 15m range (e.g. Barton *et al.* 1999:614; Bell *et al.* 2002:170; Bevan and Conolly 2004:130; Bintliff 1985:201; Lock *et al.* 1999:56), and generally narrower gaps are presumably due to the relative ease of working in European contexts, allowing more people to be used to gain higher resolution data. The conditions frequently experienced in African fieldwork have made it necessary to have field walkers relatively widely spaced, and in this case the decision to use 30m gaps when using only one surveyor seemed sensible based on knowledge of the working environment and high density of material.

The transects were walked and the ground visually searched for archaeological remains; individual artefacts and clusters, when found, were put into collection bags. A reference number unique to the transect and find was given, and the GPS co-ordinate taken. The recording forms used in the survey were designed so that a GPS position could be recorded for each observation or to note the boundaries of geomorphological features. Before the field work was begun it was believed that by surveying at this intensity all artefacts discovered on the transect could be recovered and assigned either individual coordinates or at least assigned to individual 100m by 100m blocks within the survey. The archaeological remains needed to be identified and typologically dated in the field or back at camp. Relative dating was based on the typological features of the artefacts and comparisons to the Stone Age technologies used throughout south central Africa and Zambia, whilst the pottery styles manufactured in the surrounding areas of Zambia were used to date the Iron Age material (see section 2.2).

5.5.3. Field adaptation of the survey design

There were problems encountered that required the methodology outlined above to be radically adapted in the field. The prospect of having only one vehicle available for the season was planned for and would have allowed one or two surveyors to be dropped off at points in the landscape while the main team drove to excavation sites. This required the anticipated services of two scouts, one for excavation, and one for survey. However, with only one scout available, the survey had to be conducted over six days (instead of six weeks), to allow the rest of the PPE project to achieve its aims. The plan to use one or two individuals to conduct the survey at 30m intervals therefore had to be abandoned. The entire team of 8 - 10 volunteers working with the PPE project had to be used and, on the grounds of safety, a maximum spread of 100m was possible when considered acceptable by the scout; volunteers were therefore spread out at 10-15m intervals (this figure varied due to the differing availability of volunteers). Instead of the transect sizes initially devised, it was necessary to use smaller units of 100 metres by 500 metres. In the time allowed it was possible to survey thirty 100m x 500m units. Some were surveyed end to end, allowing that each of the three areas initially outlined to be surveyed to distance of at least two kilometres. The GPS unit was used to guide the left hand edge of the transect whilst the volunteers maintained their distance from each other and kept in line; though not the most accurate way of walking a transect it proved the simplest and quickest under trying conditions. Occasional checks revealed that the distances between the walkers remained relatively constant with an error of around 10 – 15m to the left or right of the transect over a 500m distance, hence the transects from the first season, when plotted into GIS, do not appear exactly rectangular (figure 5.3).

It was possible in the time available to conduct a test to determine what the likely effect would be of using these smaller 100m x 500m survey units on the information gathered with regard to spatial variation of material and its likely distribution. An experimental block 500m x 500m was surveyed in two of the three areas to try to determine this. The same spacing of volunteers at 10m intervals was used, and the return transect was placed directly next to the preceding one so that there was no gap in coverage. This was intended to be repeated five times to build a 500m square block of surveyed area. These were orientated at a perpendicular angle to the other transects in order to determine what the effect of using rectangular transects would be on the results, however in reality not enough data was retrieved to determine what it might have been in this area. An example from the results of the first season is illustrated (figure 5.3), and this is the transect from the northern bank of the Luangwa. It clearly shows that 1km blocks would not have been sensitive to the archaeological scatters, as there is extensive variability in the artefact density visible in the 100m x 500m survey blocks that would have been missed by the larger units. Furthermore, the boundaries of the geomorphological zones are often sinuous and would cross the edges of the 1km blocks many times, hindering interpretation of the archaeological record in reference to the geomorphology. The option of subsequently subdividing the 1km blocks into smaller

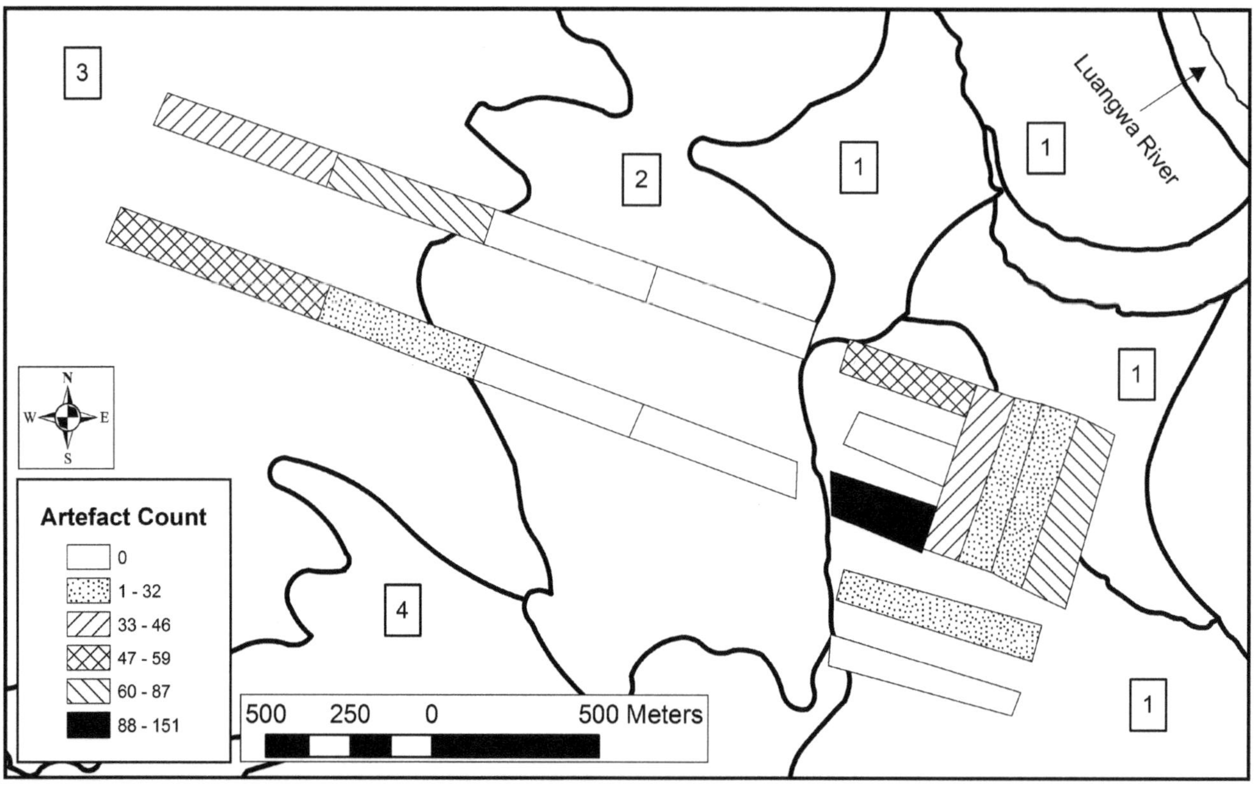

Figure 5.3: An example of the results from the first season (artefact count classes automatically selected by GIS). These results are examined in detail in section 7.9. A smaller area was surveyed than initially planned (figure 5.2 illustrates a slightly larger area with the intended transect). These transects were designed to survey one geomorphological area, and based on observations in the field and of the aerial photographs not available to the first season's survey, it is obvious that there is a great deal more variation in the geomorphology of the area than first anticipated. The numbers refer to these different geomorphological zones as they were mapped during and after the first field season (see section 5.6 below for detail). 1: Recent floodplain soils closest to the river. 2: Wooded sand bank and possible river terrace. 3: Undulating surface covered with intermittent gravels and cobbles. 4: Older floodplain soils.

100m x 100m squares within these large blocks would have been very time consuming to process after the fieldwork, though entirely feasible for the smaller 100x500m units. The difficulty in using the GPS unit is also evident in figure 5.3, as it was partially the cause for the irregularity of the edges of the transects; furthermore, the difficulty in reading the visual display of the GPS, combined with the stressful working conditions and my initial inexperience, meant that only four of the proposed five transects for the 500 x 500m block were walked!

5.5.4 Problems with the survey design

In addition to the discovery that the 1km wide transects were too wide to be surveyed or useful to analysis, the stratification used to divide the survey area was found to be too coarse, as the approach of simply dividing the area near the Luangwa River into the north and south bank proved to be too simplistic, and did not take into account the variable nature of the geomorphology within the area. It transpired that the areas in which the three transects were surveyed contained enough variation to be divided into a total of six zones. On the transect on the south bank there were two distinct geomorphological areas; one appeared to have floodplain soils of sand and silt supporting mixed riverine woodland, whilst further away, towards the foothills of the Nchindeni, the ground became very rocky with darker soils and sparser vegetation. On the transect on the north bank, it was apparent that the area could be split into three zones over a relatively short distance of three kilometres (figure 5.3): one zone formed the relatively open areas of the floodplain, the second was in an area of mixed woodland on sand, and the third in a zone with cobble capped ridges. The transect deeper in the park, towards the Muchinga Escarpment side of the valley, was a single zone, which was composed of miombo woodland on sandy soil with cobble layers within it. It was clear after the initial analysis that many of these areas had very different types and densities of archaeological material, and that an improved survey design as well as a better understanding of the boundaries of the geomorphological zones was needed.

On returning to the UK, the recording form proved to be difficult to interpret. It was not easy to visualise the distribution of features as the form only provided the capability to list grid references, which would then need to be individually plotted into GIS and the features joined up where necessary to map geomorphological features and locations of artefact collections. There were further problems with the recording methodology in relation to both the archaeological material and the geomorphological record. Despite reducing the transects from 1km to 100m widths, there were frequently densities of artefacts that were so high that total collection was impossible. If all artefacts had been collected from certain 500m x 100m blocks, the sample yielded could literally have weighed hundreds of kilos. Therefore the edges of larger scatters were recorded and estimates, based on the number of artefacts visible in a set area, were made of artefact density. The important aspect of recording in this study was the recording of the size and density of archaeological features in relation to geomorphological features on the landscape.

In the first season decisions regarding the collection of material were made on a case-by-case basis with a unified aim of determining the density of material (number of artefacts per metre square). The geomorphological work undertaken in the first season was relatively simple due to the paucity of knowledge of the landscape; observations were made on the features seen and field sketches and notes could be taken on informative geological localities. It was simple to record the boundaries of the different geomorphological areas seen, and this could be done whilst surveying, and also from the vehicle whilst driving to sites and localities in and around the park. A key problem with the geomorphological survey was that accurate data was not retrieved in the different areas regarding the type of stone material present (lithology) or the size and condition of this material (clast size and degree of rounding).

5.5.5. 'Site' definition

The observation made prior to the commencement of the project, that the majority of the material would be evident as scatters, was borne out by experience in the first season. Overall, site definition was not critical to this study as an off-site approach was being used, which entailed that all artefacts found were recorded, whether or not they appeared to derive from a site, so that an accurate picture of the variability of the density of artefacts in a survey unit could be gained. Nevertheless, 'sites' were encountered and a method of defining them was important so that, if possible, information regarding settlement location preferences could be retrieved; this was particularly the case with reference to the Later Stone Age and both the Iron Ages, where material appeared to be more likely to survive in its original context. This is easier said than done, and a large amount of debate has been focused on the issue (Dunnell & Dancey 1983:271; Orton 2000:81-82; Plog *et al.* 1978:394; Schiffer *et al.* 1978:2; Wobst 1983:68-71). Definitions of a site can vary from the vague 'a target or place with definable properties' (Orton 2000:67), to the more precise such as "...If either hut remains or over twenty potsherds were found within a 20m square area, then the location was designated a 'site'... If less than twenty artefacts were found, then the place was designated as a 'find spot' (Bisson 1992:239)". One thing is clear, there is no single definition that is applicable in all circumstances, and it is best to consider independent criteria suited to each separate piece of archaeological research (Orton 2000:68).

Observations of the landscape prior to undertaking this research indicated that in areas where there is not a continual scatter of material, the sites are relatively small, circa 30m – 100m, and in some circumstances there may only be one artefact or spot find. Early and Late Iron Age material appeared to be primarily found in contexts that would suggest they are sites, i.e. the scatters of pottery are relatively discrete and, on occasion, associated with hut floors. However, in general the presence of pottery was relatively rare on the surface. Usually a single sherd or a small scatter of potsherds would be found and these would be recorded in the same manner as the stone artefacts. Sites were identified in the field on an individual basis, usually where the artefact counts appeared anomalously high and limited in distribution, sometimes in relation to other features such as hut floors, or fragments of wall (daga). In reality, so few definite sites were discovered, each one could be identified and defined individually in the analysis stage (chapter 7). Stone Age sites were more difficult to discern from the general background scatter, and they were only discernible in the field or during the analysis where there were significant observable changes in the density or type of material in the landscape, or where artefacts were found in relation to specific features such as rock shelters.

5.5.6. Detecting archaeological material and sites

The ability to detect a site resides with how visible artefacts are in relation to varying degrees of vegetation cover. Experience in some parts of Africa has revealed that in areas of dense vegetation surfaces may have to be combed several times before it is possible to say that all archaeological sites have been located (Robertson 2000:150). Additionally there can be a problem with how obvious a site is in the first place (Schiffer *et al.* 1978:6). For example, Derricourt (1985:17) reports in his work on the Kafue Valley in central of Zambia that when the local headman took the team to the locations of the villages of his predecessors only one of these sites had been discovered by the research team; the reason probably being the location of the village had changed with the death of each headman, and on average this would have occurred every 30 years. Therefore, at each village site there would have been little time for the vegetation to be drastically altered and surface indications such as pottery would usually be absent as the sites were not occupied long enough to accumulate discarded pottery (Derricourt 1985:17).

In the area covered by this research there are various mixtures of ground cover ranging from completely bare earth, burnt scrub, grass and earth, to areas of patchy to relatively dense woodland (particularly the riverine woodland that frequently occurs near water courses), and 3m tall elephant grasses that obscure the view ahead, making it possible to see only the ground directly in front. The fieldwork was conducted in the dry season when vegetation has naturally died down. There had also been natural and man-made bush fires that clear dry vegetation and dry leaves, further increasing visibility even in densely wooded areas. Ground visibility was recorded on the survey forms, and in general this fell mainly within the 75-100% visible range, with virtually no areas falling in the 0-25% range where the ground would have been totally obscured. Studies to test the effect of ground visibility on artefact recovery in Mediterranean surveys, where there is a broadly similar environment with regards to ground cover due to similar prolonged dry periods and bush fires, have revealed that visibility does not influence artefact recovery in any usefully predictable manner (Barton *et al.* 1999:616; Fanning & Holdaway 2004:268). Although there is a general link between dense vegetation and recovery rates, adjusting for artefact density based on visibility measures can produce erroneous distribution patterns (Bevan & Conolly 2002:127- 128). In the context of this project it was not possible to test this observation rigorously. However, archaeological material in the region is inherently visible when exposed on the surface, and even under dense vegetation the ground is usually quite visible due to lower ground cover having died away by the middle of the dry season. Therefore, if a transect intersects archaeological material on the surface, there is a very high probability that it will be detected.

With regard to sites, assuming that visibility on the ground was reasonable at any one time, which in this case it generally is, the adequacy of the survey can be determined with some relatively straightforward calculations. The probability of discovering a target is a function of its size and shape, and the spacing of the search pattern (Sundstrom 1993:92). In archaeological surveys many shapes of site may be found, however, it was discovered in the first season of research that most sites are either roughly elliptical or circular. In the case of an elliptical target, where transects are laid 10m apart, and where the sites observed had an average width of 15m and length of 30m, the probability of detection is as high as 4.7 (i.e. a site will be discovered 4.7 times, see appendix 5.2 for formulae). Where sites are circular at 15m wide the probability of detection falls to 3.0. As visibility is very high in the Luangwa Valley these figures illustrate that almost five as many, and three as many transects are needed to locate elliptical and circular scatters respectively. Clearly a wider spacing of 20m intervals would be a more efficient use of time and resources with respect to site discovery. The experience in South Luangwa was similar to Derricourt's (1985) in that there may be no indication of the presence of an Iron Age site until a grinding stone is located, whereas otherwise there may be very little pottery evident. In other locations there might be a group of trees and plants that, together, indicated to the scouts guiding us that there would have been a village site located there. However, there may be little or no evidence of human habitation, and only occasionally at these locales would a piece of pottery be found indicating the presence of Iron Age activity. Derricourt's experience differed in that where there is direct living memory of a village site there would usually be obvious signs of human occupation in the form of pottery and grinding stones.

5.5.7. Points for improvement for the second season's research

Valuable information was gained in the first season regarding the realities of working with the logistical constraints in the valley, as well as providing information on the nature of the archaeological record. With this improved understanding it was possible to design the survey for the second season. The points outlined here, based on the information above, are the main guidelines for the design for the second season's research.

- The size of the survey region should be reduced from one that covers the whole park, to one with a smaller more manageable area.
- Smaller survey units of 100 x 500m are adequate for gaining necessary data regarding archaeological material, and are an efficient size for this environment.
- The recording forms needed to be revised so that they would be easier to use in the field and interpret afterwards.
- A better map of the boundaries of the geomorphological areas was needed so that the archaeological survey could be better planned in relation to the geomorphology.
- Within each geomorphological area there should be controlled random selections of transect location to ensure that subjective decisions regarding the location of transects are completely avoided.
- Separate fieldwork was needed to develop an understanding of these geomorphological areas so that their relationship to each other could be understood within the context of their formational processes.

The next section outlines how the geomorphological map that provided the base map for the archaeological survey was created. A separate methodology for answering key questions regarding the geomorphological development of the landscape is outlined, after which the methodology for the second season is outlined.

5.6. Constructing the Geomorphological Base Map and Developing the Methodology for Understanding the Geomorphology

5.6.1. Introduction

After the first season's work it was apparent that a better map of the geomorphological zones was needed, along with a clearer strategy for surveying the main geomorphological features to gain an understanding of their formation and relationships to each other. In the period between the first and second seasons a number of techniques were used to assist in the construction of an improved map. The boundaries of the different geomorphological areas in the landscape were recorded with the details of the type of features, soils, and vegetation present.

5.6.2. Background to mapping techniques used

An enormous range of techniques is available to geomorphologists to help develop an understanding of the evolution of a landscape, and there are now a numerous textbooks outlining those of particular use to archaeological research (e.g., Goldberg & Macphail 2006; Herz & Garrison 1998). For the purposes of mapping landscapes it is quite common to split an area into zones for analytical purposes (Keesstra *et al.* 2005:195), or to use geomorphic maps already created to act as the strata for an archaeological survey (Fanning & Holdaway 2004: 258-60). These approaches for defining geomorphic zones were of particular use here as GIS was used as the analytical tool to establish relationships between archaeological materials and the different zones.

Three sources of information to aid the mapping were available for this project: Zambian Survey maps, aerial photography, and field observations.

The Zambian Survey maps provided elevation data in the form of contours, and drainage system information along with river and stream networks. This data was easily, but slowly, digitised into a form useful for analytical purposes with GIS. Aerial photography has been a well-known tool to both geomorphologists and archaeologists for over a century (Goldberg & Macphail 2006: 304-6), and is frequently used in GIS related research (see section 4.3). Aerial photographs can reveal a large amount of information regarding the geomorphic structure of the landscape (Goldberg & Macphail 2006: 304). However, their usefulness depends on the unique features of the landscape under investigation and the nature of the photographs. For example, in Smith *et al.*'s (2006) study, which mapped glacial features near Glasgow, aerial photographs were not used due to the high processing time, processing requirements, and variability in type making them too complex to be integrated into the study, especially as more recent remote sensing equipment gave higher resolution results (Smith *et al.* 2006:161). However, aerial photographs have been used in a similar manner to that used in this project, for example Keesstra *et al.* (2005) were able to map the changes to vegetation, land-use, and to the riverbed morphology of the Dragonja river in southwest Slovenia using aerial photographs from the past 60 years (Keesstra *et al.* 2005:202-203).

One of the key techniques available to any geomorphologist or geologist is the mapping and description of features in the field (Goldberg & Macphail 2006:321; Herz & Garrison 1998:17-21; Stow 2006), the main aim of which should be to gain descriptions of the landforms, surface deposits, and any active geomorphological processes to aid the interpretation of the geomorphological and geological history of the area (Herz & Garrison 1998:17). A simple method of achieving this is to seek out cross sections of the strata in the local area that have been naturally exposed, which can then be used to build a general model of the type and thickness of the deposits in the area (Herz & Garrison 1998:21). This can be achieved relatively simply by taking measurements of outcrops in the field and describing their nature with field sketches and annotated photographs, using methodologies such as those outlined by Stow (2006). In this project all grain and clast sizes are cited to a standard grain-size scale (appendix 5.3) (Stow 2006:109 & inside cover; Tucker 1991:11).

5.6.3 Mapping techniques used

The geomorphological base maps upon which the archaeological survey was designed were created using a combination of the 1960s and 1970s Zambian Survey maps, aerial photography taken in June of 1999, and on field observations made in the field seasons of 2005 and 2006. It was necessary to digitise the paper maps used in this survey to make them compatible with GIS. The maps were scanned to a digital format as JPEGs at the relatively low resolution of 300dpi using a large roller scanner able to accommodate the A1 sheets. This kept the file size small, whilst preserving enough detail for the accurate digitising of features on the map, and allowing for reasonable quality images to be captured where necessary. The raw JPEG images were then opened as raster files (and maintained as such) in the GIS ArcView 9.0. The next stage was to georeference the maps so they could be displayed on screen and be recognised by ArcView in geographic space. The datum and projection system recorded on the paper maps was applied and, for simplicity, the 14-figure Universal Transverse Mercator (UTM) grid reference system used in the maps was used throughout the survey. UTM grids are metric, giving their eastings and northings in metres, rather than in degrees, minutes, and seconds of the latitude/longitude geographic coordinate system, which then have to be converted into decimal degrees for plotting; UTM measurements are therefore visually and intuitively easier to grasp.

For referencing the maps, four corner points were used; these were where the northing and easting grid lines nearest to the corners intersected. Initially, due to an internal software bug in ArcView 9.0, the maps would not reference correctly in the southern hemisphere, and the UTM zone applied would only project in the northern hemisphere. Various attempts to solve this resulted in computer crashes, maps either being projected upside down or as a mirror image, or a combination of these. Initially, after numerous attempts over a couple of weeks, the problem could not be resolved, and the first maps, files, and layers for the database had to be constructed with the assumption that Zambia was in the northern hemisphere! However, the solution was discovered by accident 6 months later, when a digital copy of Astle's (1995) vegetation map was obtained. This had been created in ArcView 3.0, which had the correctly functioning projection system. Copying this projection system into all the files that had been created in ArcView 9.0 up to this point overcame the problem. From the referenced maps the course of the river and its sand banks, as they were in the early 1970s, could then be plotted. The contour lines on the maps were important, as once these were digitised it was possible to create a relatively detailed surface elevation model for the survey area.

The Zambian survey maps, created by the British-based Directorate of Overseas Surveys, were based on fieldwork as well as photogrammetry. Although they do not match the standard found on some European surveys, field experience showed that where features are marked, they are accurately represented. However, not all features are shown; for example Chichele Hill, which has a marked spot height of 574m above sea level, does not have any contour lines. This is not a matter of scale as the hill is two kilometres long, 800m wide, and rises 54m above the floodplain (as contours are marked at 20m intervals there should be two sets). Due to these limitations, and other errors that may exist in the computing algorithms used (see Wheatley and Gillings 2002:66-69), there may be inaccuracies in any computer generated surface elevation model. The elevation data generated might have errors of 10m or so, especially where land is flatter and contour lines are sparse (*ibid.*). In this project the archaeological data was analysed at the scale of the landscape using the 100 x 500m transects and their 100 x 100m subdivisions, so errors at the level of the computer generated elevation model should not be an issue. However, hills missing on the original maps are an issue, and luckily field observations revealed that this was a unique error in the survey area, but one that must be considered when analysing the data.

Aerial photographs were not available when planning the research in the first season, but their existence was known and copies were obtained in time for the second season. They are 1:50,000 scale and were taken in 1999 as part of an ecological survey. To be made compatible with GIS they were scanned using a flat bed scanner at a resolution of 1200dpi, which created large image files that retained considerable detail while remaining manageable for a computer of reasonable specification to handle. The same datum and projection system as the survey maps was applied so that they could be overlaid. As there is no grid system on photographs the method for georeferencing them is slightly different. They need at least six known reference points from relatively equal intervals across the photograph, and these points can be any identifiable features such as junctions in roads, or corners of buildings (Gillings & Wheatley 2002:64). There were enough visible road intersections in the photograph covering most of the survey area for eight reference points to be taken. This was achieved by taking GPS co-ordinates of these points in the field. Due to accessibility problems, and the lack of suitable intersection points on other photographs, this was only possible for the one central photograph. This photograph fortunately not only covered approximately 80% of the selected survey area, but also covered all the areas of key interest to the geomorphological aspect of the study, including the active zone of the Luangwa and the cobble ridge zone (see appendix 6.2 for definition and description of these zones). Furthermore, there is a large degree of overlap with the aerial photography allowing the photographs adjacent to be georeferenced relatively accurately. While overall accuracy was difficult to determine, GPS visibility was usually very good in the field, and an accuracy of a 7m radius around a reading was typical.

Field observations were used to check the boundaries of features mapped from aerial photographs, as well as mapping zones that could not be defined on aerial photographs. The GPS unit used in the first season was combined with a borrowed Personal Data Assistant in the second field season (figure 5.8), which allowed a quick visual check of any feature's location on the map. A note was then taken of any inaccuracies that could then be corrected on a laptop kept at camp. Field observations revealed that mapping taken from the photographs was reliable, especially with highly visible features where accuracy within 10m was not unusual. This is only just below that of the average accuracy given by the GPS unit in the field. Boundaries that were not visible on the aerial photographs could be mapped on the walks and drives to and from the archaeological transects. This also allowed for descriptions of the soils and vegetation to be made which were not possible based on aerial photographs. Figure 5.4 illustrates the map of the various geomorphological zones identified for this survey. The descriptions in appendix 6.1 provide the criteria on which the zones were delineated for the survey.

The creation of the geomorphological map using these methods was only the first stage of the geomorphological work. The creation of the map allows the archaeological deposits to be directly related to each zone in which they are found. However, it was also vital to understand how these zones were formed so that any affects they may have had on the archaeological record could be understood. Therefore, each zone's formational history needed to be understood, and the relationships between different areas determined. For this three key aims were outlined:

1) To determine the modern and recent action of the Luangwa and its affect on the geomorphology of the region.

2) To develop an understanding of the erosion processes in the Nchindeni Hills and the sediment feed from them to the modern Luangwa. This in turn will aid the interpretations required for aim three.

3) To develop an understanding of the likely environment of deposition that formed the cobble-capped hills, taking into consideration the archaeological evidence for their age.

The recent activity of the river was relatively easy to interpret from the aerial photographs (see chapter 6 below), however, the older features of the landscape were more difficult to understand. The Luangwa today is predominantly a sand bed river system; this is contrasted against a large number of features in the area that have a very high gravel to cobble content. A systematic method of analysing the different clast lithologies present in each area was therefore required, and this was achieved by using the method of clast counting.

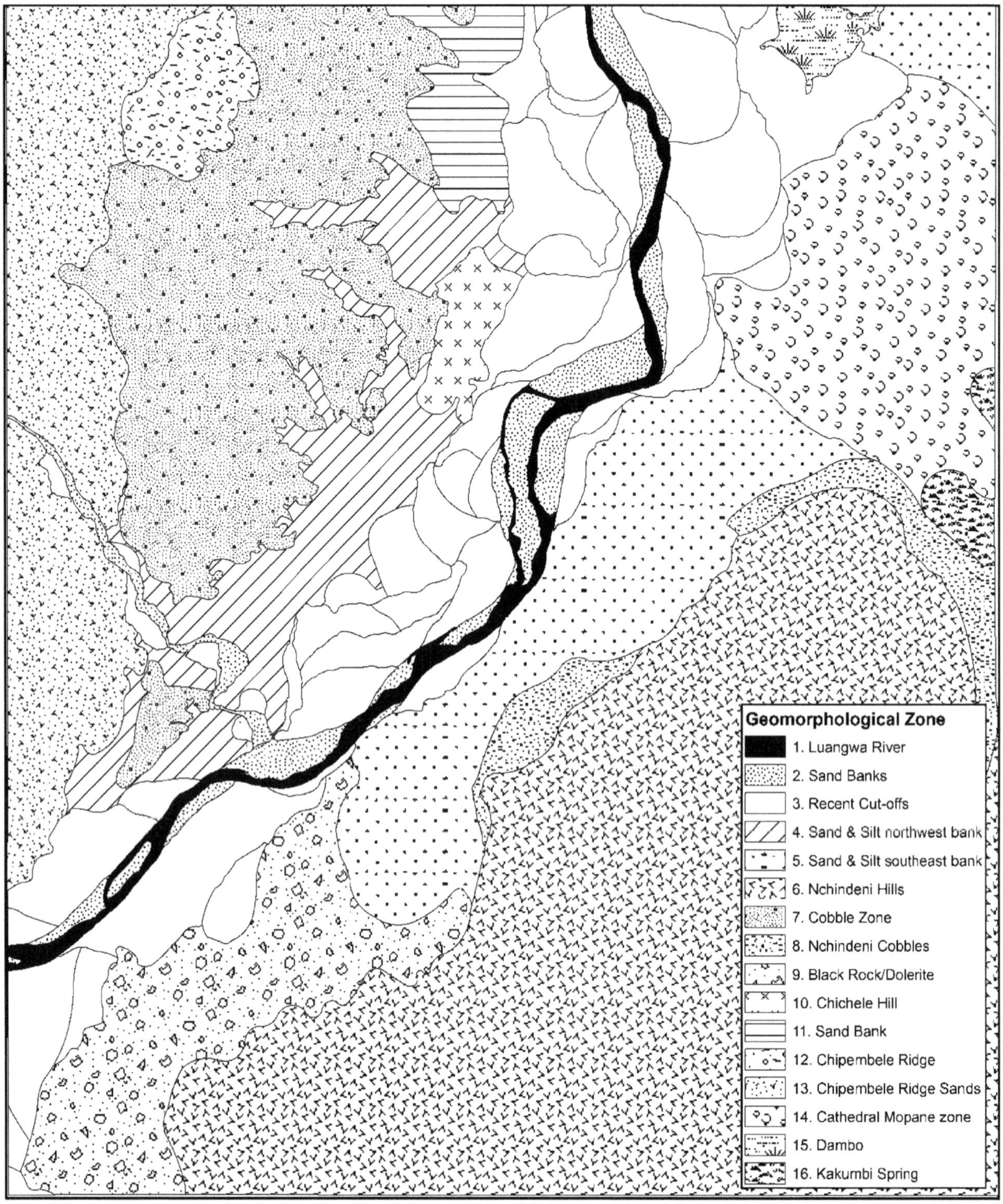

Figure 5.4: The geomorphological map created during and after the first season, labelled with key landscape features and larger seasonal tributaries. The area covered by the map matches that of the archaeological landscape sample selected in the second season and illustrated in figure 5.6.

5.6.4 Clast count analysis

Clasts are broken fragments and chunks of rock of various sizes that constitute sedimentary rocks or deposits (Prothero & Schwab 1996:3). Clast analysis at its simplest involves counting the occurrence of representative lithologies present, and determining the size and shape of the clasts either in visible geological strata, or on the surface of a recent landscape feature of interest. There are many sampling techniques for clast analysis used for both rock faces and recent surfaces, and they all have errors created by observers (Wohl *et al.*

1996:3219), as well as those introduced by varying visibility of the exposure and by localised sorting of material, which may have removed clasts of certain sizes thus skewing the results (Howard 1993). Although there have been some attempts at statistically analysing the errors associated with the datasets created by clast counting (e.g., Howard 1993; Wohl *et al.* 1996), most studies employ non-statistical methods, with virtually every study using a different methodology. Some examples are given here that have either attempted to source material lithologically, or have studied alluvial fan and debris flow deposits that appeared similar to the sedimentology of the cobble ridges in the Cobble zone which form the largest features in the Luangwa area. Thin sections can be taken from clasts and analysed under the microscope so that clasts can be compared to potential sources (e.g. Pederson *et al.* 2001:309). For this project, this would have required a large number of clasts to be brought back to the UK for sectioning and analysis, which was not possible due to their weight. Therefore lithology had to be determined using hand-specimens, with the result that the finer details of the distribution of different rock types might be missed.

Heward (1978), in an alluvial fan study in La Magdalena, northern Spain, took the average maximum clast size measured along the longest axis of the 25 largest clasts taken one to five metres either side of a section line at roughly 1m intervals (Heward 1978:455). Went (2005) used the intermediate axis (b-axis) (figure 5.5) of the 10 largest clasts of Cambro-Ordovician deposits in the Channel Islands, the results of which were, at best, semi-quantitative (Went 2005:694). The use of the b-axis in clast counts is a frequently used convention in the analysis of river sediments (Graham *et al.* 2005:12), and is also used regularly in analysis of alluvial fans (e.g., Went 2005). Steel *et al.* (1977) measured the 10 largest clasts (after omission of any outsized clasts) from an area several meters either side of the point of bed thickness measurement (Steel *et al.* 1977:1126). Adhikari and Koshimizu (2005) studied a recent 1996 debris flow in Larcha, central Nepal to determine the origin of the debris flow. Three hundred and ninety clasts were measured at one-metre intervals to determine the average size of the clasts. Additionally, to develop an understanding of the petrographic composition of the deposit to determine its origin, five clast counts were taken every three metres, with the result that a total of 650 counts were made from 130 points. (Adhikari & Koshimizu 2005:414).

In light of the many techniques employed, and the lack of standardisation for measuring clast size and counting representative lithologies, it is necessary to state how the measurements are made in any one piece of research, but it is acceptable to use any variant of the available methodologies to conduct the research (Graham 1988:18). Clast counts such as the ones outlined above are often used as a quick method for assessing the lithologies present in an area. The results can give an indication of landscape processes and potentially link different areas of the landscape. It is also possible in some circumstances to determine likely sources for material, as well as how they have been transported in the landscape. The methodology employed to analyse the clasts in the field was designed to determine the relationship of some of the different geomorphological zones and features in the Luangwa survey area (figure 5.4). For this, clast counts were taken from metre square sample locations where 100 clasts were lithologically identified, measured using the b-axis (figure 5.5), and assigned a roundness class (an estimate of sphericity based on a standardised classificatory scheme, in this case, Stow 2005:111, and Tucker 1991:16). The sites for clast counts were chosen at random in the field where there was no count already taken nearby, and wherever there was a good coverage of clasts, or a section wall existed with visible clasts.

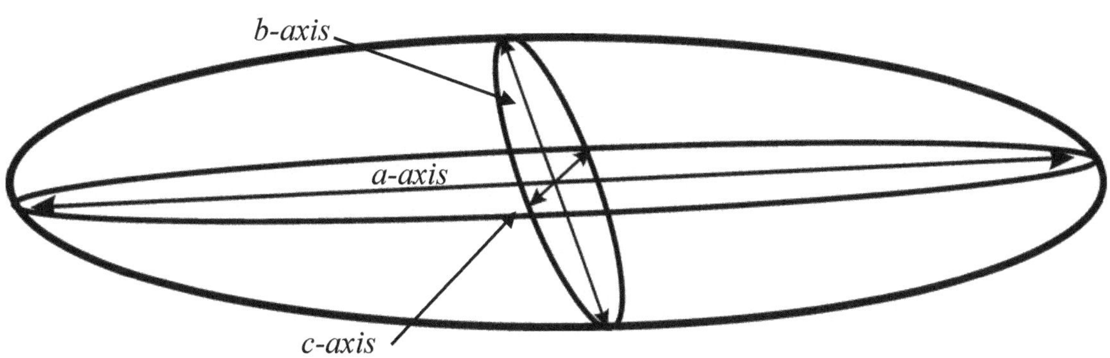

Figure 5.5: Schematic diagram of an idealised pebble illustrating the three main dimensional axes: the length or a-axis, the breadth or b-axis (used in this study), and the thickness or c-axis (after Stow 2005).

5.7. Archaeological Survey Plan for the Second Season

5.7.1. Survey design

After the first season six key points for improvement of the survey were outlined (section 5.5.7). Essentially the survey region needed to be smaller and better defined, with a separate methodology for interpreting the geomorphology. In the off-season between the first and second seasons a smaller survey area was delineated, and a 12 x 14km (168 km^2) rectangular sample of the landscape was selected (figure 5.6). Issues of accessibility are a vital logistical consideration in planning research in the Luangwa Valley (section 5.3), so this area was chosen for its accessibility from camp, and the presence of an extensive network of navigable tracks cut for game

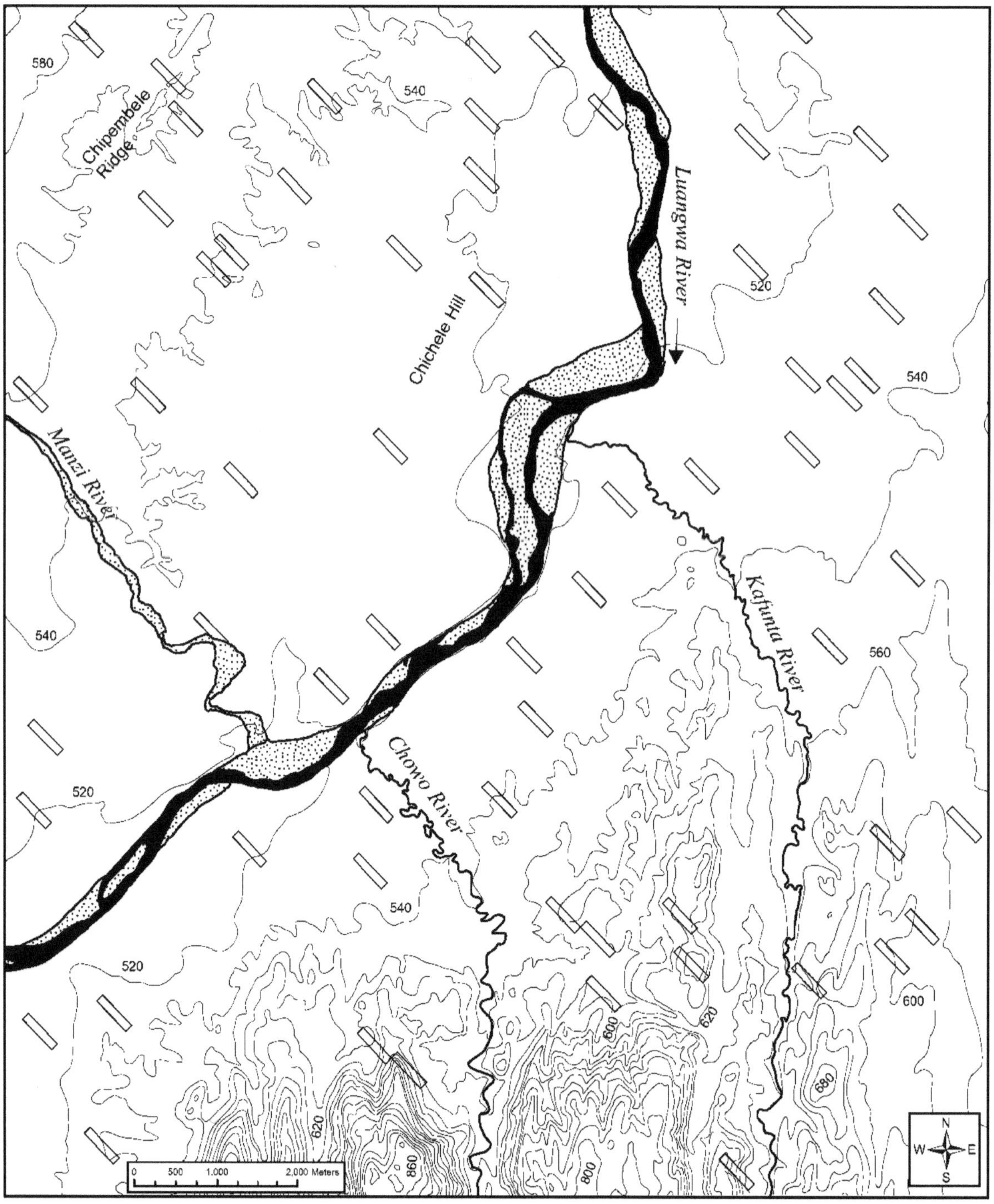

Figure 5.6: The 12 x 14km survey area selected for the second survey season. The rectangles represent the 100 x 500m survey units selected. Contour lines represent 20m intervals above sea level, digitised from Zambian survey maps (map numbers 1331B1 and 1331B2).

viewing drives. The area selected also covers a large range of geomorphological zones identified in the first season. Although these zones were not entirely understood at the time, most of them were defined and mapped in the survey area (section 5.6.5 above).

5.7.2. Sample unit and size

The experimental transects of the first season demonstrated that 100 x 500m units were achievable and useful to use, though the survey intensity of one walker every 10m was too high (see sections 5.5.3 & 5.5.5). Using one surveyor negated the hazards faced in the first season of having surveyors placed at intervals, and made it possible to use the original plan of surveying units at 20m intervals. Surveying at this intensity resulted in each survey unit taking 3.1km to walk, as there were six runs of 500m, and five 20m sections at the end of the transect to move across for the return runs. Depending on driving distance from camp, walking distance from the vehicle, and the number of finds on each transect, it could be possible to survey one or two units per day, and on rare occasions three. For the three-month season it was estimated that there would be time to survey 60 transects, which also allowed for a sample that included each geomorphological zone in the survey area. The length of time in the field was therefore the determining factor for sample size and fraction (section 5.4.4). Three weeks were allocated for work on improving the geomorphological map, cataloguing sections, and the clast counting necessary to develop an understanding of the zones.

5.7.3. Sampling scheme

To aid the planning for the second season, the paper 1:50,000 scale Zambian survey maps covering the park were scanned into ArcView and georeferenced (see section 5.6.3 above). Using ArcView the survey area was overlain with a grid of 100 x 500m units, rotated at a 45° angle, with the long axes trending northwest to south-east. This rotation was chosen as other projects have found it suitable when looking at artefacts in a landscape context to have survey units aligned perpendicularly to the prevailing trend of the units of interest (Foley 1981:34-5). In this case the general trend of the geomorphological zones, the Luangwa Valley and Luangwa River, is north-east to south-west. The grid overlay was then used to calculate the total area of each geomorphological zone; this was achieved by counting the total number of valid survey units for each zone, i.e., those that did not cross a boundary. The percentage of the survey area that each geomorphological zone occupied could then be calculated. The 60 transects were then assigned to the different geomorphological zones in proportion to the size of each zone, however, due to the large size of the Nchindeni Hills zone the final distribution of transects required weighting.

The number of transects assigned to the Nchindeni Hills zone was reduced, as the large area covered by the hills resulted in them gaining a greater proportion of transects in relation to the expected amount of data that would be retrieved. Based on observations in the first year it was known that the scatters of artefacts were discontinuous and at a low density. Whereas in some smaller zones, such as the Chichele Hill zone (zone 10) where artefacts were known to be present, a transect would not have been assigned due to its small size. The percentage of the land surface was below 1% and with 60 transects a 1% land area would only gain 0.6 of a transect. Weighting certain zones (survey strata) by the probability that strata will have archaeological material is a valid approach, as it is allows more useful information to be gathered in the time allowed (Banning 2002:116). However, caution must be exercised to avoid minimally sampling areas of low potential that could amplify any biases in understanding the archaeological record (*ibid.*). Therefore, despite reducing the number of transects in the Nchindeni Hill zone, it still retained the greatest percentage of transects in the survey. The transects removed from the Nchindeni zone were then made available to the rest of the survey, particularly smaller zones with few or no transects. One of the points of improvement noted at the end of the first season was that the location of each transect should be randomly determined. To achieve this the grid occupying each geomorphological zone was numbered, and a random number generator[5] was used to generate the numbers of transects that would be walked in each zone (fig 5.7).

---

[5] The random number generator used was downloaded from www.sunny-beach.net/random_numbers/, which is a free program for researchers that can generate random numbers from a range set by the user, e.g., generate 2 random numbers from a range of 1 to 34. The number generation process is determined by a real time number generation process, which is based on white noise from a computer sound card or microphone (http://www.sunny-beach.net/random_numbers/ accessed 11/07).

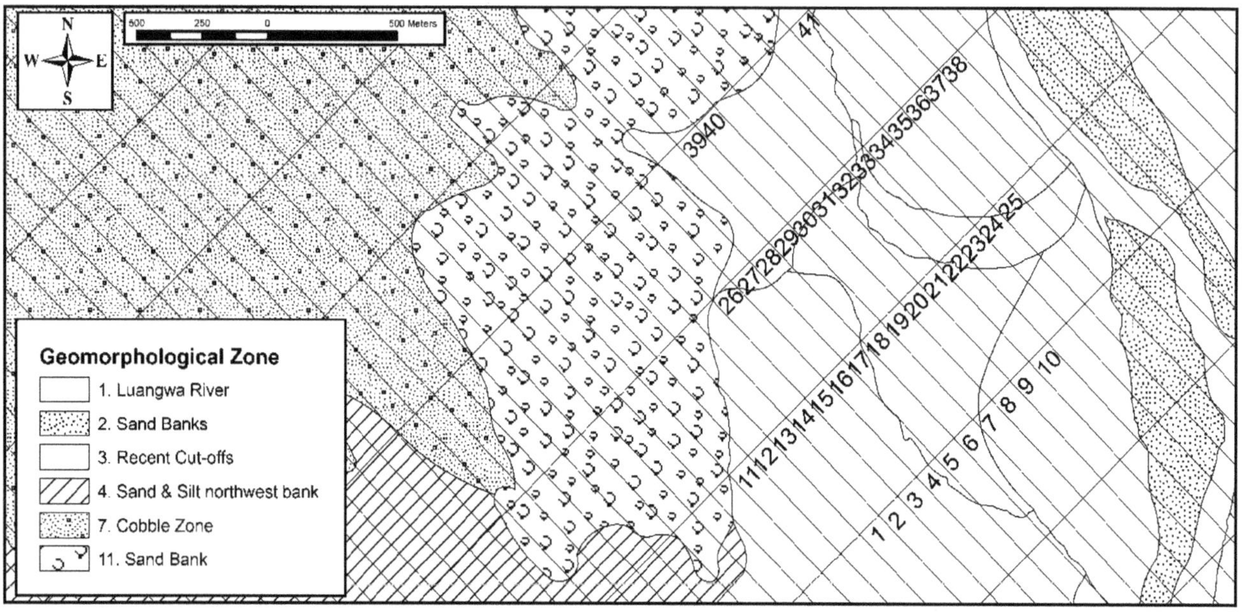

Figure 5.7: A section of the survey area illustrating the transect selection process. The 100m x 500m grid was numbered in each zone and a random number generator was used to assign the transect numbers to be surveyed. Transects crossing a zone boundary were not included in the selection process.

5.7.4. Sample unit procedure

For more consistent results a properly defined collection strategy was needed for the second season which would allow for a representative sample of archaeological material to be collected in a uniform manner, whilst avoiding collecting an unmanageable and unwieldy sample from each transect. Collecting representative samples from the field is a frequently employed strategy, however many surveys lack an explicit description of how this has been achieved (Orton 2000:83). For example, Silvester (1991) in the Fenland project of East Anglia gives no mention of how material was representatively collected from sites identified in the field (Silvester 1991:9). This makes it impossible to assess accurately artefact/site density in the survey or, therefore, the likely relative importance of the sites discovered. In other projects this problem has been accepted as a serious hindrance to the accuracy of the survey only after the survey was completed (Barker 1995:47). For example, in the Biferno Valley project in Italy, collection strategies were left to the judgement of individual teams on a daily basis, and as a result where individuals would have had to carry the samples for a long distance, they were unwilling to make adequate collections from sites (*ibid.*). For an accurate analysis of material in the landscape context, the relative density of artefact scatter, either off or on-site is needed, so in both these examples it would surely have made sense to at least outline set rules to control the sample. Admittedly, this was not fully appreciated in the first season of this survey, however, these examples are from four-year surveys (Barker 1995:14; Silvester 1991:1), and these problems faced in analysing variable samples should have been acknowledged sooner, and resolved by developing a systematic and repeatable method.

In the second season, a collection method was employed that was applied in every circumstance so that artefacts were collected from every scatter or locality in the same manner to produce consistent data. When a new scatter was discovered on a transect, a collection was made from a 2 x 2m square area around the. The transect was then resumed and a collection made from the next scatter, or when the archaeological material on the ground changed character noticeably, i.e., by technology or raw material. This strategy was designed taking into account the experience of the first season, however, it still resulted in a large amount of bulky or heavy archaeological material being collected from some transects, requiring some to be walked in two sessions. For more discrete sites, such as Iron Age huts and villages, a GPS reading was taken at the centre point and the dimensions measured. A collection was then taken from within a measured area so that artefact density could be determined. The 2m x 2m rule was not always followed with Iron Age material due to its scarcity, and diagnostic fragments were rare.

In the field, a laptop personal computer was used to manage and update the geomorphological map as fieldwork progressed. A Personal Digital Assistant (PDA) was also used, which could be updated from the laptop and linked to the GPS unit to provide easy navigation to and from transects, as well as maintaining course on them (figure 5.8). This new approach saved an appreciable amount of time in comparison with the data inputting on the GPS unit required in the first season. A paper copy of the survey area was maintained to keep track of which transects had been surveyed, as well as providing a backup in case of equipment failure. For the second season, a more visual recording form was developed (appendix 5.4), where only archaeological material, sites, and features of particular interest had grid references recorded for them. All other features such as large rock outcrops, vegetation, or soil changes could be drawn directly onto a diagram of the transect, with the PDA providing a visual output of the location. For the most part, the geomorphological fieldwork (observations of soils, some clast counts, and recording of some geologic sections) was accomplished at the same time as the archaeological field walks, or on the walks to and from transects.

Figure 5.8: A photograph illustrating the link up of the GPS (left) and the PDA (right). On the PDA it is possible to see the display of the transect, and the location on the transect (the circular symbol on the transect). The individual black lines are the 20m intervals of the transects.

5.7.5. Adaptation and development of the survey plan

During the second season, one small geomorphological feature, zone 15, was identified that was previously unknown. This area, and the transect located there, was originally classed as part of the Recent Cut-off zone (3) (figure 5.4), but was reclassified as a separate zone given its differing vegetation and soils (see appendix 6.1). The precise boundaries of some of the geomorphological zones were not known, and in the case of Chichele Hill its northern boundary with zones 4 and 7 could not be determined from the aerial photography. During the planning stages for the second season it was not possible to plot the location of a transect in the zone as the boundaries could not be determined, and the transect may have crossed the boundary into a different zone. The transect was therefore plotted whilst in the field. As it was not part of the sample originally planned before the season it meant that an extra transect was added to the survey, bringing the total to 61 transects as in the random survey.

There was the possibility in the survey plan, time allowing, to run some experimental transects that could help test any models of archaeological patterns created using GIS. Given the limited time in the field it was not possible to upload data onto GIS and conduct the modelling on location. Therefore the locations of the experimental transects were selected on the paper map using observations of the archaeological patterns in surveyed areas, and transects were planned to test the accuracy of any potential model. For a good test, a broad selection of environments in the landscape was needed, so it was decided to place transects in areas of both high and low archaeological potential.

In addition, a small number of transects were surveyed across the valley towards the Muchinga Escarpment to assess how archaeological material and geomorphological conditions varied as distance increased from the Luangwa River. This data is used in conjunction with the data from the third survey area (the O5 Road transects) of season one. The experimental transects were surveyed at intervals along the Kasweta Road (figure 7.28) which, although it zigzags across the park, is still the most direct road to the Muchinga Escarpment. There was enough time to survey experimental transects in the survey area and outside it, and in total eleven experimental transects were surveyed in areas of potential interest, six within the survey area, and five along the Kasweta Road running through the park to the Escarpment. This brought the total number of transects surveyed in the second survey season to 72.

5.8. Summary

The initial survey design used in the first season was not well suited to the environment of the Luangwa Valley. The shortcoming reflected a lack of prior knowledge regarding the variability of the geomorphological terrain over short distances, and an overly optimistic appraisal of the distances that could be travelled in the National Park. This demonstrates the value of having two or more seasons to conduct fieldwork for this type of project, as the first season can act as a test of the methodology. The redesigned methodology was aimed at specifically tackling the main aims of the project taking into consideration the working conditions of the South Luangwa area. This revised methodology was successfully implemented and demonstrates that it is possible to conduct archaeological survey in awkward environments without resorting to what can be described as the *ad-hoc* survey methods sometimes used in Africa (see chapter 2).

The use of the PDA combined with GPS provided an exceptionally efficient and easy to use method of locating and surveying transects. This combination has the potential to allow researchers to survey regions accurately and quickly, without laboriously locating transects using compasses or the limited functionality of many GPS units which, unlike the PDA, do not have sophisticated displays or uplink capabilities to mapping software, like GIS. It may not be suited to all African environments however, as it requires a laptop computer to be kept at base to manage the survey map, and easily accessible power outlets to charge up the various batteries involved (the PDA in particular required a good overnight charge most days to be assured of it functioning all day).

The geomorphological survey work was designed to gain a simple overview in a short amount of time of the type and extent of the various deposits in the survey area so that, at a minimum, a hypothetical model can be constructed of the geomorphological history of the area. The fieldwork for this aspect of the project was also completed successfully in the time allowed, and the key interpretations developed as a result of this work are considered next.

6. The Past and Present Geomorphological Processes in and Around the Survey Area

6.1. Introduction

Prior research into Pleistocene geomorphology in the Luangwa Valley has been limited to broad descriptions of the vegetation and underlying geology (e.g. Astle *et al.* 1969), and a study of recent planform change and meander development of the Luangwa River (Gilvear *et al.* 2000) (see chapter 3). The present research is the first to describe and interpret Pleistocene landscape development in the valley. Within the context of this project it is necessary to understand both present and past processes in the various geomorphological settings before the archaeological data is analysed so that natural transforms that have removed archaeological material from its primary context can be identified and screened out to reveal the remaining genuine cultural patterns in the archaeological landscape.

The methodology for the geomorphological research was designed so that an understanding of the geomorphological processes in the region could be gained. The map of the different geomorphological zones (figure 5.4) formed the strata for the stratified archaeological survey conducted in the second season. The observations made in each geomorphological zone are presented in appendix 6.1, and provide one of the key lines of evidence used here to develop interpretations regarding past geomorphological processes in the area. The clast counts form the other primary data source, the results for which are presented in appendix 6.2. Interpretations regarding the geomorphic processes of the Pleistocene are made with reference to the clast count data, photographs, aerial photographs, and orbitally derived topographic data. This is undertaken in reference to the three key aims of the geomorphic aspect of the project outlined in chapter 5 which were:

1) To determine the past and present action of the Luangwa River and its effect on the geomorphology of the region (focusing on zones 1 – 5 & 15) (section 6.2).

2) To develop an understanding of erosion processes in the Nchindeni Hills and the sediment feed from them to the modern Luangwa (focusing on zones 6 & 9) (section 6.3).

3) To develop an understanding of the likely environment of deposition of the cobble-capped hills, taking into consideration the archaeological evidence for their age (focusing on zones, 7, 8 & 14) (section 6.4).

6.2. Luangwa Planform Change and Limitations on its Recent Movement

6.2.1. Introduction

The first key aim of the geomorphological aspect of the project is to understand the movement of the Luangwa River over time. This is important as its past movement has implications for the preservation of archaeological material in the area of its current floodplain and terraces (zones 1-5) and, potentially, in other areas such as the cobble zones (7, 8, & 13). The presence of sand and silt in zones 1 to 5, and its absence on cobble zones 7 and 8, suggests that the activity of the Luangwa has affected a localised area in this part of the valley, approximately two to three kilometres across. This was examined using a combination of Gilvear *et al.*'s (2000) research, the fieldwork, and the study of the aerial photographs and satellite topography data retrieved for this project, which allowed the immediate and longer-term action of the river to be determined.

6.2.2. Planform change of the Luangwa

In their 2000 paper, Gilvear *et al.* studied the planform change of the Luangwa River since the 1950s. The lowest of the four reaches studied extended as far south as the northern part of the archaeological survey area. In this overlapping area three meanders, numbered 18-20 by Gilvear *et al.* (2000:423), exhibit straightening of the planform of the river and isolation of three meanders between 1956 and 1988. This provides a 20th century date to the land surfaces in three of the meander cut-offs in zone 3 (see figure 6.1). Where necessary this information was incorporated into the GIS to assist in creating and testing models of the location of archaeological material (chapter 7). The complexity of the geography of the cut-offs makes it difficult to suggest a sequence of dates for the surfaces of older courses abandoned before the mapping and aerial photography undertaken in the area in the 20th century. Though the date retrieved from SL10 provides a minimum age for one of the cut-off surfaces.

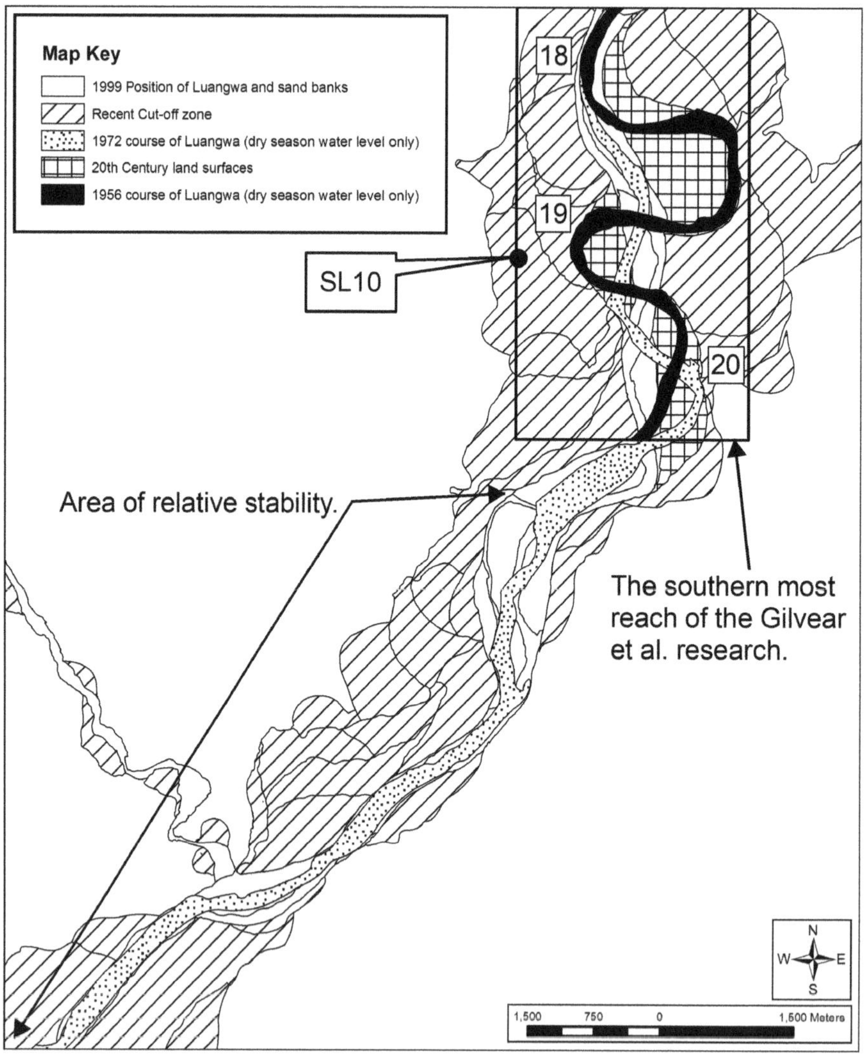

Figure 6.1: Map showing the comparison of the position of the Luangwa in 1956, 1972, and in 1999, with surfaces known to postdate 1956 delineated. The 1956 channel was mapped here by georeferencing a copy of the aerial photography from Gilvear *et al.* (2000:430) and digitising it in the database used for this project. The 1972 map provides an earlier date than 1988 for some of the channel changes identified by Gilvear *et al.*, as bends 18 through 20 had already avulsed by 1972, and have been relatively stable through the 1988 aerial photography used by Gilvear *et al.* (2000:430) until 1999. When studying the 1956 aerial photograph it became evident that the image had been accidentally inverted; it was rotated 180° to correct for the error in figure 8 of Gilvear *et al.* (2000:430), the bend numbers on this diagram are the ones illustrated in figure 1 (page 423) of Gilvear *et al.* (2000). The diagram also illustrates the relative stability between 1972 and 1988 in the south of this survey area, as well as the comparatively active area within and directly below the Gilvear *et al.* study.

For the purposes of this project, the 1972 and 1999 active channel of the Luangwa and its sand banks were digitised using the 1972 Zambian survey map and the 1999 aerial photography. This provides a comparison of the location of the river and its active sand banks between the two periods spanning 27 years. This reveals that in a similar scenario to that reported by Gilvear *et al.* (2000), some reaches of the Luangwa have experienced significant movement whilst others relatively little. It is clear that the channel that had straightened by 1988 in Gilvear *et al.*'s (2000) paper had not altered significantly by the 1999 aerial photography, with the exception of bends 19 and 20 that have moved 440m and 770m respectively, which is movement averaging at 16m/year and 29m/year. The meander to the south of bend 20, just outside of Gilvear *et al.*'s study area altered its course by 650m over the 27 years, an average of 24m/year. These figures are similar to or below Gilvear *et al.*'s figure of 33m/year. Gilvear *et al.* (2000:432) based their averaged figure of 33m/year on figures calculated for the period between 1956 and 1988 (32 years). By comparing Gilvear *et al.*'s 1956 aerial photography and the Zambian survey maps that were printed in 1972, instead of comparing the 1956 and 1988 aerial photography, this figure can be narrowed down, as meander 18 in the 1956 aerial photography had already been subject to avulsion by the printing of the 1972 map (a distance of approximately 1000m in a period of 16 years, half that of Gilvear *et al.*'s (2000) 32 years). A recalculated erosion rate based on this reveals that the maximum rate of erosion could be 60m/year.

At first glance this recalculated erosion rate suggests that the river could entirely cross the Recent Cut-off zone (3), which has an approximate average width of 2km, in 33 years. This estimate assumes that the whole river would migrate in the same direction, which is an unlikely occurrence, as migrating meanders will usually move in various directions due to differing erosive forces on each part of each meander. Movement therefore occurs both downstream (translation), and laterally (extension), as well as rotating when the meander axis changes orientation (Knighton 1998:227). It is clear from these data that the Luangwa River has the potential to erode and relocate its channel rapidly, however, this is contrasted by stability over relatively long sections of its channel (figure 6.1). Although it is possible to determine the movement over the latter half of the 20th century with the use of aerial photography, it is difficult to determine the age and the order in which older channel cuts were abandoned due to the manner in which each meander cut-off overlaps with its surrounding cut-offs. During the 2007 wet season the flood waters were exceptionally high, and an analysis of the subsequent channel position would be a valuable exercise to assess conclusions regarding movement rates.

6.2.3. The Luangwa River in the wider context of the survey area and the Luangwa Valley

Within the survey area the modern Luangwa River has a limited range of movement in a corridor between the Cobble zone (7) and the Nchindeni Hills (6). In appendix 6.1 zone 3 is interpreted as containing recent cut-off meanders of the Luangwa, and zones 4 and 5, and possibly 11, were interpreted as terraces of the Luangwa (figure A6.17a & b). Extensive field observations revealed that there are no fluvial deposits overlying the cobble zones either side of the river in zones 7 and 8 (though the type of contact between zones 7 and 11 is unknown). The cobble mounds in the Cathedral Mopane zone (14) are overlain by sand and silt. However, as no excavation has been undertaken the exact structure and relationship of the sediments is unknown. Therefore, the zone 14 cobbles may have been disturbed by fluvial action or, alternatively, have remained unaffected. The latter would suggest that the sand and silt represents overbank (flood) deposits from the main river that have buried the cobbles in the area. There are fluvial deposits between the first two or three of the cobble ridges nearest the river in zone 7 (see appendix 6.1) and these are likely to be overbank deposits that have in-filled the inter-ridge areas. Though it is possible that overbank deposits, now winnowed away, had also overlain zone 7 as they apparently to do in zone 14, it is unlikely that the channel of the Luangwa was located in zone 7. Had this been the case the cobble deposits would have been significantly disturbed or perhaps removed entirely (the cobbles themselves do not appear to be fluvial deposits, see below). This suggests that the fluvial activity of the modern channel has not been present beyond the current extent of the fluvial sand and silt zones (zones 1 – 5).

The boundaries between the older Karoo and Archean rocks and the more recent Holocene and Pleistocene deposits provide further evidence to suggest that the activity of the Luangwa has been long constrained. On the south-east bank, where Karoo bedrock is present on the boundary of the Recent Cut-off (3) and Black Rock zone (9) (localities 2 & 6), the lack of any trace of fluvial sediment or fluvial activity above the Karoo rocks suggests the river has not been present above this level in zone 9. The presence of Chichele Hill (zone 10) on the north-west bank, to the south of zone 11, appears to have constrained the movement of the Luangwa; comprising hard granites, metamorphics and silicified Karoo sandstone, it has provided an un-erodable barrier to the Luangwa. The river appears not to have been present further to the north-west of Chichele Hill, as the Cobble zone (7) extends to the hill, and has not been removed or altered by the river.

Examining the effects of some of the regional geological deformation in the Luangwa Valley in the South Luangwa National Park area provides an indication as to why the Luangwa is constrained to a relatively narrow channel in the survey area. Figure 6.2 illustrates a three-dimensional image of a stretch of the Luangwa focused to the north of the survey area. There is a significant topographic high, or bulge, in this portion of the Luangwa Valley. The bulge is approximately 120km long and extends 40km from the Muchinga Escarpment across the valley floor, with a maximum elevation of 250m above the Luangwa River. This bulge has presumably been uplifted by tectonic activity, maybe as part of the rifting associated with the formation of the valley, or the more recent Miocene movement linked to the creation of the East African Rift Valley system. Two of the main tributaries of the Luangwa in this area, the Mupamadzi and Kapamba Rivers, originate from behind this bulge and have cut paths around it to meet the Luangwa. The Mupamadzi in particular flows to the north-east for a considerable distance before it is able to cut across the valley to meet the Luangwa; this is against the direction of flow of the Luangwa, and contrary to the usual dendritic pattern of a drainage basin. The presence of this large topographic high has presented an impenetrable barrier to the action of the Luangwa, forcing the river to cut a path along the south-east margin of the valley. Further to the south, and north (not visible in figure 6.2) this barrier is no longer present and the Luangwa flows more centrally in the valley. The main body of this bulge extends to approximately 20km from the survey area; however, the uplifted area extends to Chipembele Ridge, which represents the bulge's margin in the northern part of the survey area. This bulge, along with the obstructions described within the survey area, has combined to limit the movement of the Luangwa to a small zone within which it appears to be very active.

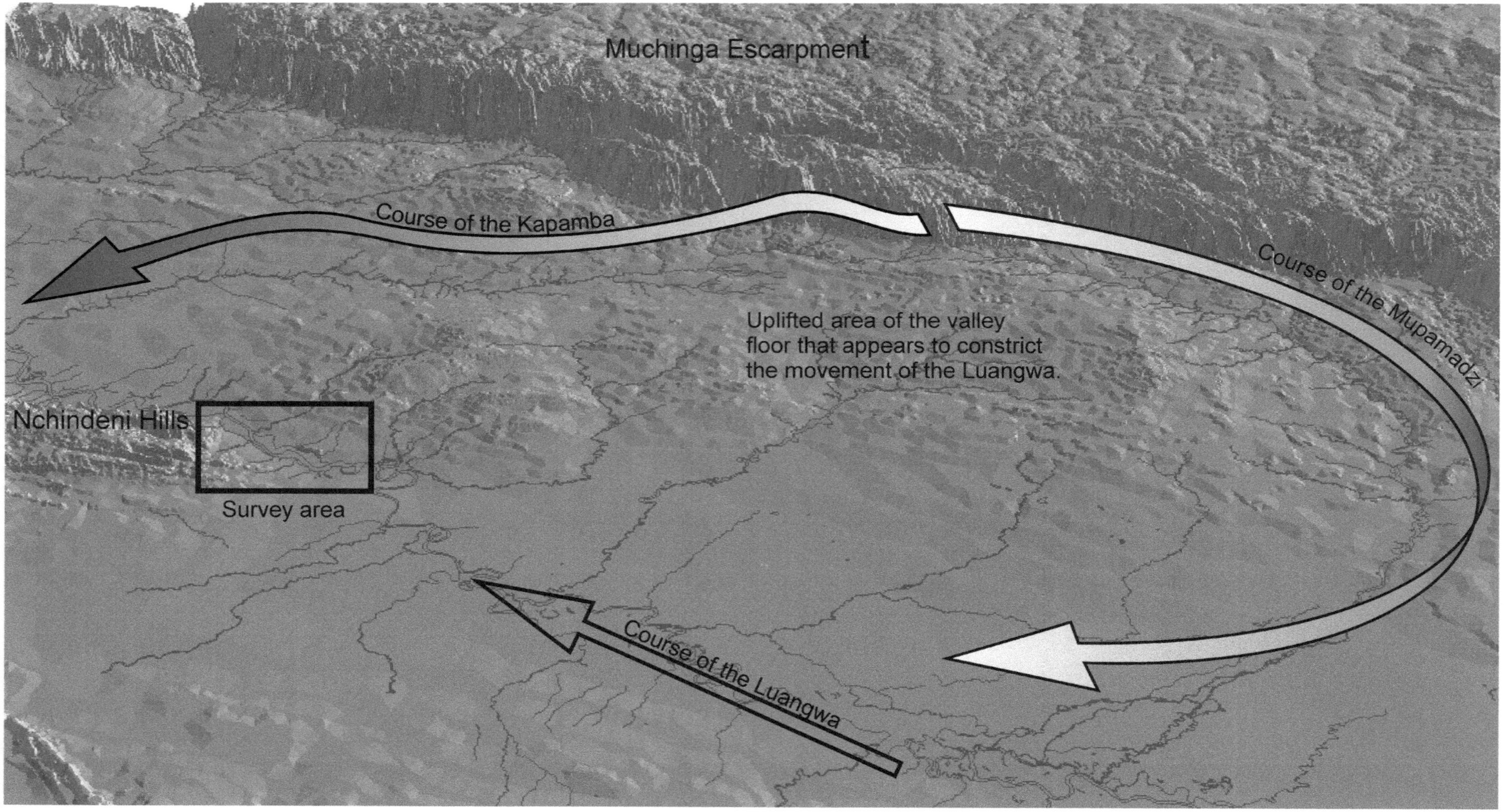

Figure 6.2: A three dimensional image of the valley floor with the bounding uplifted areas of the Muchinga Escarpment and the Nchindeni Hills (width of the image approximately 120km, though this will vary due to the perspective). Three dimensional landscape created using data from NASA shuttle SRTM 90m data, downloaded from http://seamless.usgs.gov on 10/12/06. Raw data was transformed in GIS to 20m contours that were then used to create a TIN, allowing a 3d display. River data is from Astle (1995), and is projected onto the land surface created for this figure. Vertical exaggeration is 6x. Annotation created in Corel Draw X3

6.3. Nchindeni Hills Sediment Feeds

Understanding the sediment feeds of the Nchindeni Hills (zone 9) was the second key aim of the geomorphological survey outlined in section 6.1. This is necessary as the Nchindeni Hills may have been the source for the cobble material in zones 7, 8, and 14. Clast counts undertaken in the streams in the Nchindeni Hills reveal that the majority of the material in the stream bedloads contain material characteristic of the Nchindeni Hills today. Only in streams located at relatively low altitudes is there a large quantity of quartzite (Appendix 6.2). At the higher elevation of count 19 there is little quartzite, whilst at higher elevations, counts (34 and 35), there is virtually none. The low levels of quartzite at these higher elevations suggests that there was not a significant relict cover of such material and that this was not a source for the quartzite in the cobble zones. The presence of the quartzite in the lower elevation streams on the eastern side of the Nchindeni in counts 17, 18, 19, and at count 16 in zone 14, suggests that these streams may be carrying material from deposits with quartzites similar to quartzites found on the cobble zones. This suggests there may be a source of quartzite deeper into the hills on the eastern side of the Nchindeni, or alternatively evidence for a localised relict cover of material. Their low elevation though, makes it unlikely that they could be the source for the cobble zone material (see section 6.4 below). If it were evidence for a relict cover, its localised nature raises doubts that it could have supplied enough material for the amount of quartzite found in the cobble zones.

The cobble zones, with their hyper-concentrated cobble deposits, were deposited by alluvial fans (section 6.4). Although they are the nearest potential source for quartzite cobbles, the evidence of the clast count data suggests that the Nchindeni Hills are not the source of the material. They appear to have made little contribution to the cobble zones particularly once the Luangwa has been crossed. The low counts on the north-west side are probably from local outliers of metamorphic and granite rock, such as Chichele Hill. Further towards the centre of the valley the quartzites appear in greater frequencies, with larger clast sizes, suggesting that the Muchinga Escarpment side of the valley is the more likely source for the cobble material (section 6.4).

Debris flow deposits recorded at locality 8 (figure 6.3) in the Black Rock zone (9), at the foot of the Nchindeni

Figure 6.3: Annotated photo showing the debris flow deposits recorded at locality 8. The black and white scale is in cm, with a total length is 21cm. All the clasts in the debris flow layer were weathered to a significant degree.

Hills in the south of the survey area, indicate that at some time the landscape was less stable than it currently is. The highly weathered clasts in the deposit would not have survived transport in a debris flow, as they would have rapidly been destroyed. Therefore the deposits have weathered *in-situ* and must be older than the un-weathered material recorded elsewhere in zone 9. The ground surface of zone 9 is currently stabilised by vegetation, and debris flows do not occur today, thus the debris flow must originate from a time when the landscape was less stable. An analogy for this exists in the deposits analysed in and around Chipata, where climate change from dry to wetter conditions has triggered debris flows on a denuded landscape (see chapter 3). It is not possible to suggest a date for the deposits in zone 9 as no archaeological materials were found within the debris flow. However, it is clear that the deposits have not provided a significant sediment input into other areas as the high proportion of granite in the section and the stream bed (counts 8.1 and 8.2) does not appear at any other locality. The geomorphological processes in zone 9 in this area of the Nchindeni are therefore very localised phenomena.

6.4. The Depositional Environment and Formation of the Cobble Zones

6.4.1. Introduction

While the landscape in zones 7 and 8 is more accurately described as an undulating topography of cobble capped hills, the most enigmatic features of the region are what have been described as "cobble ridges", these are the characteristic feature of zone 7, and zone 8, though they are smaller in the latter. Although there are some ridge shaped cobble hills, field measurements with a compass revealed that there is no preferred orientation to these, and the majority of the features are better described as globular in form, rather than linear. Following a description of the sedimentology and form of the cobble zones, the most probable scenarios for the environment of deposition are outlined with reference to the clast count data presented in appendix 6.2.

6.4.2. Sedimentology and form of the cobble zones

From a historical perspective, Dixey (1937a:70) reports the presence of large gravel deposits in the north Luangwa area and, though he did not undertake any fieldwork in this survey area to the south, he cites Wallace's 1897 travels in the south Luangwa Valley, and suggests that the cobbles reported by Wallace are likely to be from the same sequence (*ibid.*: 70). Wallace (1907:394) is the first to note the presence of the cobbles in the landscape, and observes them to be well-rounded pebbles of quartzite strewn on the ground. Dixey cites the presence of fossil 'dinosaur' bones in the deposits to the north as evidence for them being part of the Cretaceous sequence overlying the Karoo, specifically the Dinosaur Beds (Dixey 1937a:70). This is unlikely to be the case in the south Luangwa area, as excavations have revealed that there is an archaeological sequence, primarily consisting of Early to Later Stone Age (Modes 1, 3, and 5), contained within four excavated examples SL1, SL2, SL4, and SL8 (Barham *pers comm.*). Furthermore, a recent trip to a remote area of the North Luangwa National Park by a local enthusiast revealed that the nodules containing the dinosaur bones (more properly – early mammal-like reptiles) are found in cobble deposits similar to those in the South Luangwa area. In these there also appears to be a large number of stone tools (S. Tolan *pers comm.*). If these cobble deposits have archaeological sequences within them similar to the South Luangwa area, it would suggest that the dinosaur nodules are clasts in themselves and are contained within deposits of Pleistocene rather than Cretaceous origin. The archaeological associations in the South Luangwa National Park area, which at their earliest appear to be Early Stone Age (Barham *pers comm.*), indicate that the sequences in the cobble zones must have been deposited over the past two million years. More recently the cobble zone has been subject to erosion, leaving the cobbles isolated on small hilltops, on the geomorphological map it can be seen that wet season tributary channels have cut into the zone with the result that the sand and silt of zone 4 appears to extend into the Cobble zone at some points (appendix 6.1 and figures 5.4 and A6.1).

Field observations revealed that the cobbles in zones 7 and 8 appear not to have been deposited by fluvial activity, which was at first suspected due to their proximity to the Luangwa. If they had been fluvial deposits it would be expected that a large proportion of the clasts would show imbrication (clasts aligned and partially stacked along their long axis) indicative of fluvial transport (Prothero & Schwab 1996:73-4). In contrast imbrication is only vaguely expressed by a small number of clasts in the SL8 sections (3 loose imbrications between 7 clasts were observed in total at various points in the section). The deposits appeared to be hyper-concentrated, meaning they are poorly sorted and clast supported, as opposed to matrix-supported sediments, where finer sediment fills gaps between the larger clasts. These observations led to an early dismissal of the possibility that the cobble deposits derived from fluvial processes.

Hyper-concentrated flow deposits have a low silt and clay content (Cannon 1997: accessed 09/07), and this is seen in the cobble deposits in zones 7 and 8. Hyper-concentrated flow deposits can occur in alluvial fan settings (Leeder 1999:335), and are intermediate between debris flow and alluvial deposits (Vallance 2000:612). Alluvial fans form when a confined flow of water (usually due to steep valley sides) with a high carrying capacity (competence) for clasts becomes unconfined resulting in a reduction in competence (for example a channel opening out onto an open plain). This allows the coarser material to drop and accumulate (Leeder 1999:333; Prothero & Schwab 1996:138), whilst finer material will be transported further away to the distal part of the fan (Prothero & Schwab 1996:144). Deposits can vary with grain sizes ranging from clays to boulder grade material, and sedimentological sequences can include conglomerates, cross-bedded sandstones, debris flow, fluid flow and hyper-concentrated flow deposits, as well as channel cut and fill structures (Blair 1999:1015; Prothero & Schwab 1996:145; Wells & Harvey 1987:182). Fan development is usually controlled by tectonic factors, such as the juxtapositions of rift valley horst and grabens, and of mountains and mountain fronts with basins (Harvey 1997:236). The resulting alluvial fans tend to be wedge shaped, as the deepest sediments accumulate near the outlet from the hills, and gradually decrease in thickness further from the outlet; they also tend to be limited in lateral extent (Prothero & Schwab 1996:144). Of the numerous factors controlling the total fan area and fan slope angle, the most influential is the drainage area of the source (Leeder 1999:330). However, factors such as the source drainage basin geology can have significant effects on the type of sediments deposited and their depositional regime, whilst conceivably producing similar overall fan morphologies (Blair 1999).

The sedimentology of the cobbles in zones 7, 8, and possibly 14 indicate that they are most likely to have been deposited in an alluvial fan environment by hyper concentrated flows. Other more complex depositional features and sequences used to recognise ancient alluvial fan systems seem to be absent in the survey area, however this might be an affect of the lack of good geological sections in the area, or the result of modern streams cutting into the deposits, which is particularly noticeable in the Cobble zone (7) (appendix 6.1). In the next sections the possible sources for the alluvial fans are considered; they include the nearby Nchindeni Hills, the Muchinga Escarpment on the other side of the valley, and a source from within the valley.

6.4.3. A Nchindeni source for the alluvial fan deposits

The nearest potential source for the cobble material would be the Nchindeni Hills that protrude into the south of the survey area, and form the uplifted block on the south-east side of the valley. However, the cobble deposits are situated on the opposite northwest bank of the Luangwa River at a minimum 20 metres above its current elevation (e.g. at SL8 approx 540m), rising gently to approximately 45 metres in elevation above the Luangwa, as seen 8km north of the toe of the Nchindeni at the base of Chipembele Hill. As noted above, the thickest deposits of alluvial material will be close to the outlet of the streams and rivers draining the source area that in this scenario would be the Nchindeni Hills. The implication of this is that the thicker material would have been where the Luangwa River and its floodplain are now presently situated, furthermore these deposits would have been significantly thicker than the 45 metres of deposits now seen 8km away, which as the distal deposits would have been the thinnest and at the lowest elevation. There would therefore have been a large amount of material removed from the Luangwa's current course to allow it to be at its current elevation below both cobble zones (figure 6.4). This material would have been removed from the floodplain, the Nchindeni cobbles (8) on the south-east bank, and the area of the Cobble zone (7) on the north-west bank closest to the Luangwa, which are both now only 20 metres above the level of the river (as opposed to the required minimum of 45m). If the Luangwa were present during the formation of the fans, it would have been located either more centrally, or on the opposite side of the valley, nearer the Muchinga Escarpment. It is possible, for example, that the Luangwa occupied the channels of the modern day Mupamadzi and Kapamba Rivers. The main channels of the two rivers almost connect; their drainage basins are adjacent in the north-western margin of the valley, and the headwaters of their tributaries are only 500 metres apart. If their channels do represent a previous course of the Luangwa River, the rifting that created the bulge that now constrains the Luangwa would have interrupted this course, elevating the land in the area of the headwaters by 250m. This would have then forced the Luangwa to find a new route through the valley, resulting in it cutting through the cobble ridges and removing a large amount of material to the current elevation of the Luangwa.

Tectonic activity would have been required to disrupt the hypothesised ancient course of the Luangwa, however, the Luangwa Valley is believed to have been relatively stable since the initial uplift of the East African Rift Valley five million years ago (section 3.3). The degree of recent movement is currently unknown, but the bulge would have had to be created in the past 100,000 - 50,000 years, as this is when it is likely that the cobbles ceased accumulating (the mid to late Middle Stone Age). The lack of research into the tectonics of the area makes it difficult to determine whether or not the bulge that constrains the position of the Luangwa to the south-eastern margin of the valley is a recent phenomenon or is related to the Pliocene uplift. It is probable though that the Luangwa River was either not present or had a different form to the current river. Alluvial fans form in semi-arid conditions so it is likely that the fans in the Luangwa Valley would have formed in the drier glacial or stadial periods. The result of this would have been that the ancient Luangwa River might have been much

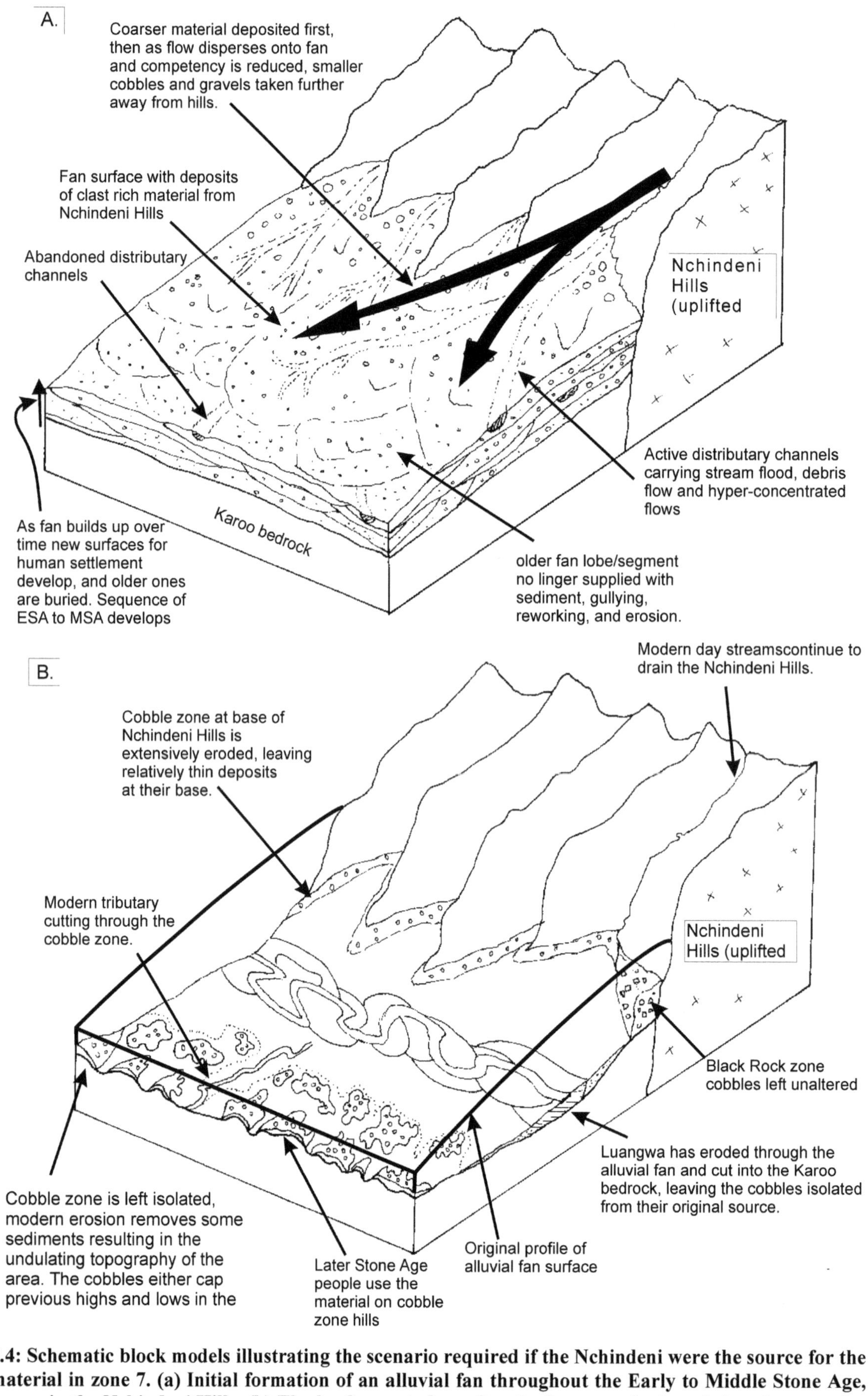

Figure 6.4: Schematic block models illustrating the scenario required if the Nchindeni were the source for the cobble material in zone 7. (a) Initial formation of an alluvial fan throughout the Early to Middle Stone Age, with a source in the Nchindeni Hills. (b) The landscape today, after the Luangwa had eroded and removed a large part the alluvial fan, leaving the cobble zone (7) isolated from its original source, with Later Stone Age settlement on the cobble surface.

reduced in size, more highly seasonal, or entirely absent. As with the tectonic activity, there is no climatic data for the Luangwa Valley directly (see section 3.4), so it is impossible to construct an accurate model for the nature of the environment during past climatic regimes. What can be said with greater certainty is that, in general, glacial periods would be characterised by drier conditions with a denuded unstable landscape. With the change to interglacial environments this denuded landscape would be prone to erosion with the potential for the Luangwa to rapidly erode unconsolidated Pleistocene deposits in the area. With no accurate chronology any interpretation based on past tectonics and climate is conjecture; however, the clast count and sedimentological observations can help resolve this issue.

For a scenario where the Nchindeni Hills are the source of all the cobble deposits there would be telltale features visible in the landscape today, and there would be evidence in cobble zones 7 and 8 of fluvial action and deposits, resulting from the river cutting across and down through the fan. Furthermore, if the material on the cobble zones originated purely from the Nchindeni the clast count data should indicate similar lithologies present both in the Nchindeni Hills themselves, and on the cobble zones either side of the Luangwa reflecting their common source (zones 7, 8, & 14). There should also be a decrease of cobble size away from the Nchindeni reflecting the reduction in the competency of the streams and hyper-concentrated flows as distance from the Nchindeni increases. However, these lines of evidence suggest that the Nchindeni are not the probable source for the cobbles on both sides of the river, thus negating the need to provide an explanation that requires the Luangwa to erode and remove a large amount of fan material.

In section 6.2.3 it was noted that there is a lack of fluvial deposits overlying the cobble deposits in zones 7 and 8, and that it appears the Luangwa has not been active outside of a relatively small 2km wide floodplain. Though it is possible that finer sands and silts of fluvial origin have been winnowed away leaving the cobbles exposed, the presence of the archaeological sequence still makes it unlikely that there was a river present as it is difficult to conceive of how the sequence would have remained undisturbed as the channel cut across and down through the fan deposits. The lithology of the material on the cobble ridges is different to that of the Nchindeni Hills today, as the cobble ridges are primarily constituted of quartzite, whilst the Nchindeni are composed of granitic and metamorphic rocks of various types. The clast count data illustrates that the cobble zone material differs from much of the material emanating from, and in, the Nchindeni Hills (appendix 6.2). In the larger tributaries emanating from the Nchindeni, quartzite similar to the cobble zones is found in the Chowo River whereas in the other large tributary, the Kafunta, the material differs. Quartzite is found in the largest quantities in the low altitude streams in the eastern portion of the Nchindeni. These data suggest that there may have been, or still is a source for the quartzite in the Nchindeni, but as noted in section 6.3 this would have been too small an area to produce all the quartzite seen in the cobble zones. Furthermore, the evidence from the clast size analysis suggests that there is no statistical difference in size between the cobbles on the north-west bank and those on south-east bank (appendix 6.2, table A6.6), whilst the cobbles nearer the centre of the valley in counts 13, 26, and 27 are statistically larger (table A6.5). Thus there is no evidence for a direction of flow from the Nchindeni, but there is for a direction of flow from the centre of the valley towards the cobbles in zone 7. The angularity data can support either a source on the south-east or north-west side of the valley (appendix 6.2), as the cobbles behind Chipembele Ridge and in the Nchindeni Cobbles zone are more angular than the cobbles in zone 7, indicating that the cobbles in zone 7 have been transported further, but potentially from either source.

There is a large amount of sand present in relation to the amount of cobbles behind Chipembele Ridge and towards the centre of the valley, and this is what would be expected further away from the fan outlet at the base of the Nchindeni, however, these sands occur at too high an elevation. At a distance of 20km they reach 760 metres, with the result that there is little difference to the Nchindeni Hills' current maximum height of 900 metres, making it unlikely that the Nchindeni could have been the source, as any alluvial fan from such a small change in elevation (140m) would be much smaller, as it would have derived from very shallow valleys. It is possible that tectonic activity could have produced some of the changes in elevation necessary to raise parts of the valley thus explaining the minimal difference in elevation between the Nchindeni and the sand and cobbles toward the centre of the valley. However, as noted above, it is difficult to assess this based on current data. The Nchindeni therefore cannot be conclusively regarded as the source for the cobbles in all the cobble zones and it is necessary to consider another hypothesis – that there are two sources for the cobble material, the Nchindeni Hills and the Muchinga Escarpment.

6.4.4. Two sources for alluvial fan material

With the hypothesis that there are two potential sources for the cobbles, the Muchinga Escarpment would supply the material for zone 7, and the Nchindeni Hills would supply zone 8. In this scenario it is possible that the cobble zones either side of the Luangwa might have differing lithological make-ups. It would also be expected that coarser material would be deposited nearer the Escarpment with smaller cobbles in zone 7. The hypothesis also requires that the cobbles in zone 7 would have had a source from the Muchinga Escarpment, a minimum of 44km away (figure 6.5). During fieldwork for the final season's research, a slight variation of this hypothesis was considered. Observations along the Kasweta road revealed that the deep sands and silts with cobble layers behind Chipembele Ridge are being eroded over large areas. The channels removing this material would have been, and, apparently still are confined channels with high flow rates, which at present drain into tributaries that meet the Manzi River before reaching the Luangwa. If in the past these tributaries had acted as feeds for alluvial fans, with occasional flash floods, hyper-concentrated flows, and debris flows, rather than seasonal streams and rivers, the channels, on meeting the Luangwa floodplain, would have become unconfined, reducing the competence of the flows which would have then deposited material in an alluvial fan environment.

For either version of this scenario to be probable it would be expected that the clasts toward the centre of the valley beyond the Cobble zone (7) would be larger and more angular. The data demonstrates that clasts are smaller and less angular in zone 7 than the counts towards the centre of the valley (appendix 6.2), although ascertaining the nature of the clasts nearer the base of the Muchinga Escarpment was not possible due to its inaccessibility. The fact that the clast counts on the southeast bank all contain large proportions of metamorphics and granites in addition to quartzite, whilst the majority of clast counts in zone 7 and to the centre of the valley only contain quartzite, supports the suggestion that there might be two sources for the material in the cobble zones.

It is necessary to address the issue of whether or not an alluvial fan could extend from the Muchinga Escarpment to the cobbles on the north-west bank (zone 7). If the Muchinga Escarpment is the source of the cobbles in zone 7 it would be necessary for them to have been transported a minimum of 44km from the Escarpment. A comparison of the scale of some alluvial fans suggests that this is not unusual. For example the Kosi megafan in India has an area of $16{,}000km^2$ and a length of 150km; while the Gandak fan has an area of $32{,}000km^2$, though notably these fans have very large catchments; $50{,}000km^2$ in the case of the Kosi megafan (Gupta 1997:13). In general, there is a positive relationship between alluvial fan area, and drainage basin size (Guzzetti *et al.* 1997:132; Leeder 1999:330), though when considered in detail other factors also have an effect such as basin slope, sediment yield (Oguchi & Ohmori 1994:405), length of the mainstream, and drainage density (Church & Mark 1980:343). In this study it is only possible to estimate the current drainage basin size, but this can be done for the rivers flowing from both sides of the valley. Using the tables from Guzzetti *et al.* (1997:132) and Leeder (1999:332) it is possible to estimate the sizes of fans created by the drainage systems operating on either side of the valley (table 6.1). This was achieved by taking the central point of any outflow as the apex of the fan, and then estimating the maximum size of the fan by counting the grid squares that would have been covered by the fan. This method revealed that it would require all the streams of the Nchindeni to be combined in order to form large enough deposits to include both cobble zones 7 and 8. However, if only the streams in the eastern part of the Nchindeni that have quartzite similar to the cobble zones are included for the potential source of the fan, then a fan is unlikely to have extended to both zones.

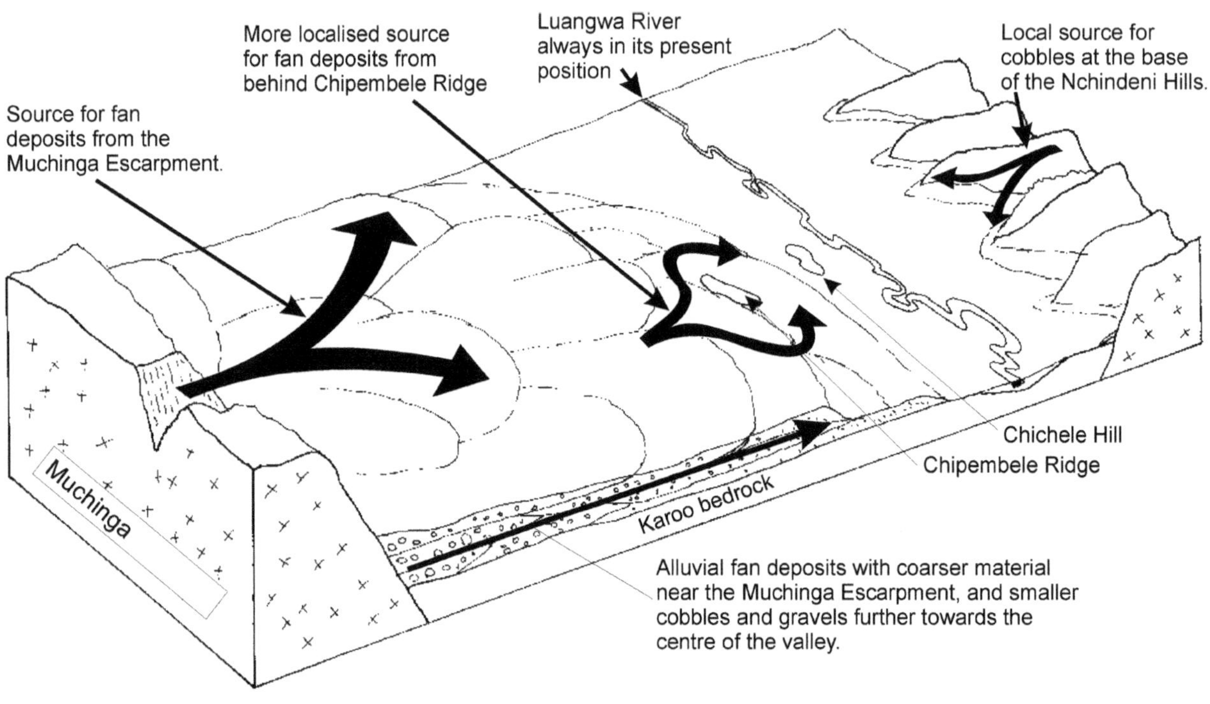

Figure 6.5: Block model illustrating a scenario where there are two sources for the cobble material; the cobbles at the base of the Nchindeni Hills (zone 8) originated locally from the Nchindeni, and the cobbles in zone 7 either originated from the Muchinga Escarpment, or more locally from behind Chipembele Ridge. These two suggestions are compatible with each other, and material could have originated from the Muchinga Escarpment before being reworked and moved from behind Chipembele Ridge to the Cobble zone (7).

Rivers and Streams included in analysis.	Drainage basin area of streams, before they reach the valley floor.	Probable range of fan size.
All streams leaving the Nchindeni Hills.	33.5km^2	19 – 104km^2
Streams with quartzite bedload in the eastern part of the Nchindeni.	13km^2	2 – 95km^2
Mupamadzi River from the Escarpment.	1,500km^2 (conservative estimate based on available maps).	102 – 500km^2
Kapamba River from the Escarpment.	900km^2	80 – 300km^2

Table 6.1: The drainage basin areas of rivers and streams in or near the research area, and the likely range of fan sizes that may be developed (after Guzzetti *et al.* 1997:132, and Leeder 1999:332). The drainage basin areas have either been directly determined from the 1:50,000 scale Zambian Survey maps, or are an estimation based on the available maps.

From the Muchinga Escarpment side, if the two drainage systems of the Kapamba and the Mupamadzi were operating as sources for alluvial fans, neither is likely to have produced fans that could have spanned the distance to cobble zone 7. The Kapamba drainage system however, may have created a fan large enough to contribute material behind Chichele Ridge, which would explain the high sand content behind the ridge where cobbles would only be transported in the event of larger flash floods. This data was extrapolated using the general relationship between fan size and basin size while ignoring the other factors noted above; it also ignores the possibility of drainage basin size changes over time that due to basin capture may have reduced or increased basin sizes it the area. Therefore this data can only act as a guide to the potential size of fans in the area; and examination of other parameters would depend on further extensive research.

Despite the limitations of the data presented above, the Muchinga Escarpment appears unlikely to have provided a direct source for the cobbles on the cobble ridges, as a fan is not likely to have extended as far, nor would relatively large clasts have been transported as far as the Cobble zone (7) on the north-west bank. The second alternative outlined above is that the cobbles derive more locally from within the valley. This suggestion is supported by observations made in the field that seasonal rains are eroding the deep sands over large areas behind Chipembele Ridge (approximately 10km away). Their unconsolidated nature having left them susceptible to erosion resulting in deep gullying (figure 6.6). The indications in the field are that this could be a relatively rapid process with powerful streams moving large cobbles of quartzite. This is further attested by the nature of the deep channels cut into the terrain that would act as very powerful confined flows during wet season floods, draining the area along the Kasweta road from the centre of the valley towards the Luangwa floodplain (figure 6.7a & b). At present these channels are tributaries of the Manzi River and ultimately the Luangwa. Hypothetically, during semi-arid glacial episodes, when alluvial fan activity was more probable, the ephemeral rivers that feed the Luangwa may have not been present and conceivably the confined nature of the channels could have fed the alluvial deposits in zone 7. In this scenario the cobbles in the sand and silt layers behind Chipembele Ridge are being reworked within the valley, and the coarser material is deposited on the proximal area of the alluvial fans in zone 7, while the sand and silt is transported to the distal part of the fan in the area now occupied by the Luangwa's floodplain. The ancient Luangwa, if and when it was present, would then erode and remove the sand and silt material deposited on its floodplain. A key issue though is that the high elevation sands behind Chipembele Ridge would need to have been the source for the cobble zone sediments for approximately 1.5 million years (to allow for the archaeological sequence). It is difficult to imagine, considering the apparent fast rate of erosion in the area, that this is likely, as it would seem probable that the material would have been exhausted some time ago. A mechanism is therefore needed where material is re-supplied to the area behind Chipembele Ridge. This is possible as alluvial fans operating from the Muchinga Escarpment could supply the cobbles and sands, and these would be reworked from the centre of the valley to the Cobble zone (7). It is conceivable that this system would be periodically abandoned and reactivated with the changes in the climate driven by the glacial to interglacial cycles throughout the duration of the mid-late Pleistocene spanning the Early and Middle Stone Age. This mechanism provides an explanation for why the artefacts in any one level at sites such as SL8 can exhibit various stages of abrasion (fresh to heavily worn) (Barham *pers comm.*), indicating that some artefacts were transported a great distance from elsewhere in the valley, whilst others were discarded and then included in the fan deposits locally, as the alluvial fans themselves would have been an attractive source of raw material for people living in the vicinity of the fans.

Figure 6.6 (above): An example of one of the gullies cut into the unconsolidated sand and silt along the Kasweta Road towards the Muchinga Escarpment (photo taken on transect K5, see figure 7.29 for location).

Figure 6.7a (right): A 3m deep channel vertically incised into the Karoo bedrock forming a confined channel with high flow rates in the rainy season.

Figure 6.7b (below): A 6m deep channel section on a meander of the same stream (figures un-scaled due to difficulty descending into the gully).

The model presented above of material being reworked from the central area of the valley is not as simple as it might first appear. It is complicated by the fact that in zone 13, which extends away from the survey area behind Chipembele Ridge, there is an archaeological site on transect number K2 that may date to the Early Stone Age. It has large >10cm quartzite flakes present and these are not only sharp, but five of them, though they did not directly refit, appear to come from the same piece of quartzite (see chapter 7 below). This suggests that this surface has remained relatively undisturbed since the Early Stone Age. Given that the process that formed the Cobble zones is known to have operated throughout archaeological time due to the sequence contained within them, a mechanism is needed to explain why some areas of the landscape were preserved whilst others were eroded to provide material for the Cobble zone (7). The nature of such a mechanism is difficult to ascertain without further detailed research into the geomorphology of the area, though it might relate to underlying geology providing some form of protection, or some other environmental or geographic factor.

The alluvial fan model is further complicated by a sequence in the channel wall of the Manzi River (SL6), near its confluence with the Luangwa River. The sequence comprises a fluvial system with occasional cobble layers that contain archaeological material (figure 6.8), which at the base of the sequence is mode one (Oldowan) dated to 0.9ma (Barham *pers comm.*; Phillips *et al.* 2005). It is clear therefore that a fluvial system was in operation at some point in the Early Stone Age close to the current course of the Luangwa. It is possible that the sequence at SL6 was deposited by the ancient Luangwa or Manzi River during a wetter interglacial phase, whilst the material on the cobble ridges was deposited by an alluvial system operating in a dry glacial phase. As noted above it is likely that a number of cycles such as this would have occurred, and this could entail that the various deposits date from different glacial and interglacial episodes, hence the fact that hyper-concentrated cobbles comprise the top layer in the SL6 section. Within the survey area, and much of the area visited close to the Luangwa around it, the Manzi section at SL6 is unique as below the armouring of cobbles at the top of the SL6 sequence there is approximately two metres of sediment above the Karoo. Whereas in sections excavated or areas visited elsewhere the rocks underlying the cobbles are either Karoo or Archean, representing an unconformable boundary spanning at least 180 million years. The Karoo at this point is at a similar elevation to that at SL8, so is higher than the modern Luangwa's channel and floodplain. Though the Karoo clays are soft and prone to erosion they are still slightly harder than unconsolidated fluvial sediments; it is suggested here that the slightly higher elevation of the Karoo has afforded the sediments in this area protection from subsequent interglacial fluvial episodes – until this one.

The cobble armouring that is present on Chipembele Ridge (zone 12) also suggests that there are two sources for the cobble zone material. The ridge rises 40 metres from the Cobble zone (7), and 20 metres from the Chipembele Ridge Sands (zone 13). For the cobbles to be isolated on top of the ridge there must have been alluvial fan activity at that elevation. If the sediment feed had originated in the Nchindeni Hills a large amount of material would subsequently have been removed from the course of the Luangwa River where the main body of the fan would have been (section 6.4.3). If the deposits originated from a fan from the Muchinga Escarpment side of the valley that has now been partially eroded, relatively little sediment would have had to be removed to isolate the cobbles on the ridge.

6.4.5. Cobble zone formation summary

It is unlikely that the Nchindeni Hills were the single source for the cobbles found on both sides of the Luangwa River. The lithological clast count data indicates that inclusions of clasts that appear to derive from the Nchindeni are only present in any quantity on the south-east bank of the Luangwa, and do not form a significant component of samples from the northwest bank. The size of the clasts also suggests that the Muchinga Escarpment side of the valley was the origin of the cobbles, whilst the angularity data supports a conclusion that both sides of the valley are a potential source. The calculations regarding the likely fan size produced by the drainage systems on each side of the valley can only provide a rough guide to the probable size of the fans produced, but these data indicate that fans extending from either valley margin would not have been large enough to produce fans extending to all areas of the cobbles in zone 7 on the north-west bank. The hypothesis that the alluvial fan deposits in the area ultimately derived from the two sides of the valley is presented as the most probable here, and that the cobbles in zone 7 originated proximally from the central portion of the valley. This hypothesis is supported by the available data, and furthermore, easily provides an explanation for both the alluvial deposits located on either side of the Luangwa and the fluvial deposits seen near the confluence of the Luangwa and the Manzi, both of which contain Early Stone Age material. It is likely that throughout the climatic cycles of the mid-late Pleistocene that the depositional regimes of the alluvial and fluvial systems alternated. If the Luangwa River was present in glacial episodes but much reduced in size, it could have continued to provide a water source. If this is the case Early Stone Age people would have been able to occupy the landscape during both interglacial and glacial conditions. This hypothesis is also presented as a least effort hypothesis, as it does not require any recent tectonic activity to explain the changes in elevation seen in Pleistocene deposits across the survey area and valley, nor does it require significant erosion by the Luangwa to have removed large amounts of cobble material, for which there is no evidence in the cobble zones.

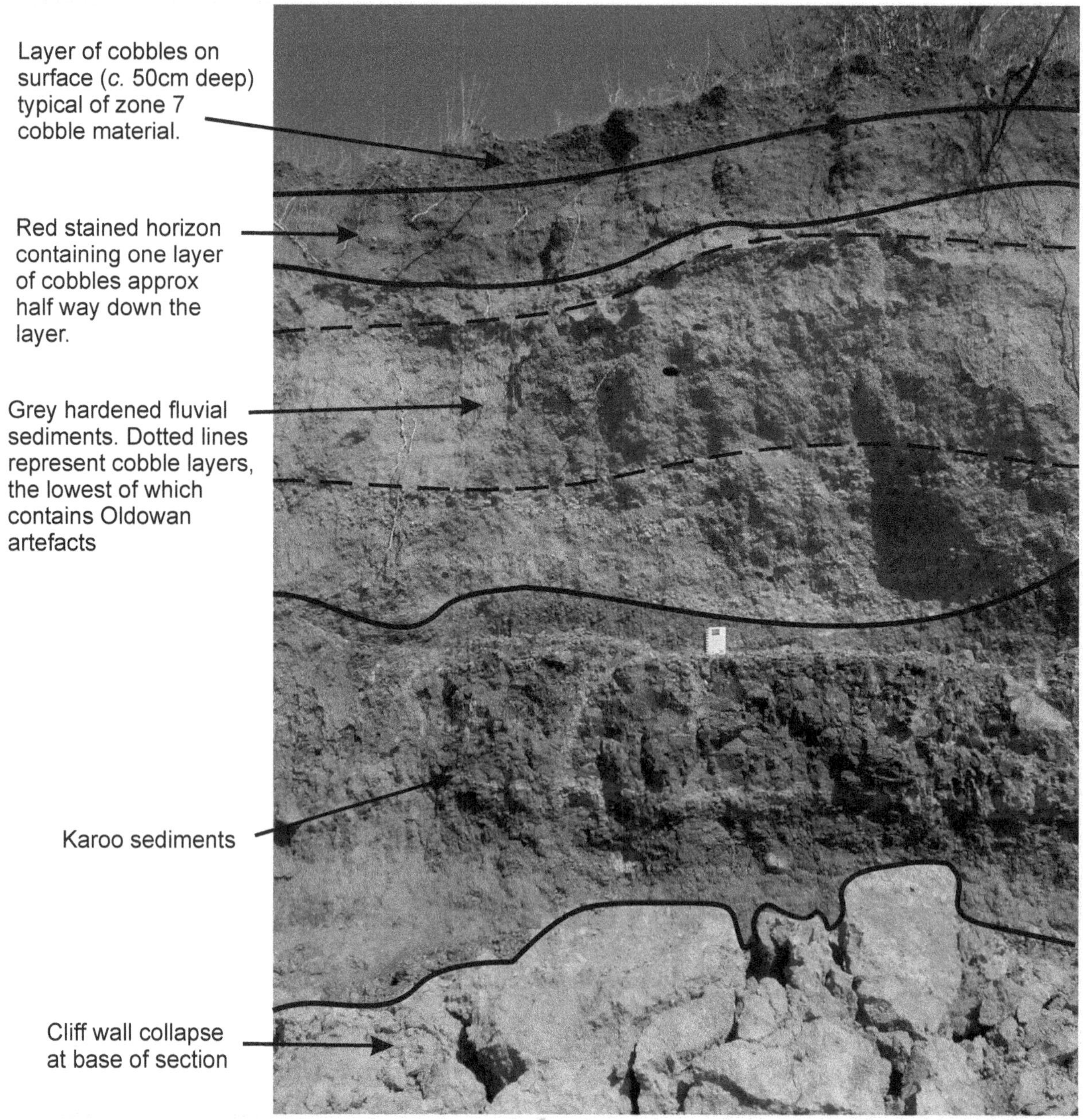

Figure 6.8: The Manzi River section in the vicinity of SL6. The SL6 excavation site itself has now been destroyed by erosion but the cobble layers identified are still present and artefacts can still be retrieved from the section. The book used for scale is 21cm high. The cobble layers in the fluvial sand and silts could be interpreted as either channel lag deposits from the base of a fluvial channel that have incorporated artefacts washed into the river, or as cobble deposits from the leeward side of a sand bar on a river that provided a raw material source.

6.5. General Summary of the Geomorphology of the Survey Area

- The present day course of the Luangwa appears to be constricted in a narrow area of the valley in the survey area. Within this narrow zone the Luangwa is a very active river with erosion/avulsion rates as high as 60m per year in certain areas.

- It is likely that this has been the case for some time as, based on current information regarding the tectonics of the area, the topographic bulge in the valley would have formed at the latest with the Miocene activity that initialised the East African rifting, though it could originate from the much earlier Karoo rifting of the valley. Throughout archaeological time it is likely that the Luangwa, when present in the valley, would have followed a similar course to that of today.

- During interglacial conditions the Luangwa would have had a similar appearance to that of today, but during glacial periods may have been much reduced or absent.

- The lack of any field evidence for fluvial sediment on the cobble zone, or at any height on the Nchindeni Hills also supports the hypothesis that the Luangwa River has occupied its current course for some time.

- The sand and silt zones (4 & 5) represent abandoned terraces of the Luangwa River comprising in-filled meander cut-offs that are either covered by overbank deposits or eroded so that their boundaries are no longer visible.

- The cobble zone material is characteristic of hyper-concentrated flows deriving from an alluvial fan depositional setting.

- The Cobble Zone (zone 7) material derives from reworked material from the Muchinga Escarpment.

- The Nchindeni Cobbles (zone 8) at the base of the Nchindeni Hills derive locally from the Nchindeni.

- The Black Rock zone (9) appears to originate from localised weathering and erosion of the Nchindeni Hills debris flows presumably from periods of landscape instability.

6.6. Implications for the Archaeological Record

6.6.1. The Nchindeni Hills (zone 6 and 9)

The Archean basement rocks that form the bedrock of the Nchindeni Hills (zone 6) and the majority of the Black Rock zone (9) have been a stable geological feature in the landscape throughout the Pleistocene and Holocene. The bedrock has undergone weathering that has produced thin soils on the hilltops and hillsides. Their development will have occurred throughout the Pleistocene whatever the climatic regime, and their erosion will have been continual. This factor will make it unlikely that any archaeological material deposited on the hillsides and hilltops in the Early and Middle Stone Ages will have survived without being transported from its place of discard. Under glacial conditions, when the alluvial fan deposits in the Nchindeni cobbles (zone 8) were fed by material from the hills, the denuded landscape will have been more unstable and prone to erosion than under interglacial conditions. The incised valleys retain deep two to four metre sediments that might in some areas cover and preserve earlier sites; however, if present they would be difficult to discover as they would be buried by sediment deposited by overbank deposits from streams which will also have removed and reworked material. Recent archaeological material from the Later Stone Age and Iron Age will have remained relatively undisturbed and should be visible on the surface. However, some hill slopes are less stable and the bases of the hills may be covered by soils washed down from above, burying even recent archaeological material. It is not possible to determine the age of the debris flows that form part of the Black Rock zone (9), but in general similar natural processes will have affected archaeological remains as in the rest of the Nchindeni.

Generally, in the Nchindeni it would be expected that only material from later periods (Later Stone Age and Iron Ages) would have any chance of remaining undisturbed. The presence of vein quartz in the hills, which can reach 30cm widths, could have been a useful raw material source for stone working, and areas of exposure may have associated sites. The Nchindeni Hills may therefore have Later Stone Age sites deep into the hills, whereas Early and Middle Stone Age sites may not have been preserved. The nature of the terrain may not have been suitable for Iron Age farming practices so the Nchindeni Hills are likely be devoid of Iron Age material, except on the lowest slopes (or possibly areas of haematite exposures quarried for iron processing).

6.6.2 The Luangwa River, its floodplain, and terraces (zones 1 to 5 and 10)

The Luangwa floodplain area, comprising zones 2 to 5, has been highly altered in recent history by the action of the Luangwa River and this alteration is ongoing. The most recent abandoned channels and sand bars will be devoid of any archaeological material, with the exception of any redeposited artefacts. As a result of the highly active nature of the Luangwa River it is difficult to determine the order in which the channels were abandoned in the Recent Cut-off zone (3), and thus determine a chronology for them. Older abandoned channels in zone 3 will have been stable land surfaces suitable for occupation by Later Stone Age and Iron Age peoples, and some are already known to retain intact archaeological sequences of some depth (e.g. SL7 on the Chowo). Patterns of landscape use may well be evident in these zones and questions regarding settlement location preferences can be addressed. It is highly unlikely that materials from the Early and Middle Stone Ages remain in their original contexts due to the highly active and concentrated movement of the Luangwa.

6.6.3 Zones containing cobbles (7,8, and 14)

The cobble zones contain deposits from alluvial fan that were active throughout the Pleistocene, encompassing the Early and Middle Stone Age. These areas would have been attractive to early humans as there is a large amount of raw material present in the area for stone tool manufacture. Any behavioural evidence in the form of knapping sites and any location preferences in the landscape will in most circumstances be disturbed as a consequence of the shifting distributary channels moving material on the fan surfaces. These channels would move a large amount of cobble material, and any artefacts present, during the floods that formed the hyper-concentrated deposits. Within this environmental setting artefacts may have been moved a great distance or may be buried locally to the area of manufacture or discard, surviving in a fresh unrolled condition. It is conceivable that in the cobble zones there may also be rare areas where an entire activity site from the Early or Middle Stone Age, such as a knapping floor, could remain undisturbed. However, there may be no way of identifying these sites on the ground surface, unless either a proportion was exposed without disturbance, or a large number of refitting artefacts were discovered from a scatter. The fan system was inactive by the very late Pleistocene and Holocene and, as a result, there is a greater chance that material from the Later Stone Age onwards could survive without alteration by the alluvial fan processes. The cobble surface would still have been an attractive area because of its stone raw material for tool manufacture, and in the Later Stone Age patterns of land-use may be evident. However, it possibly would have been less attractive to Iron Age farming peoples who were less dependent on stone for tool use. Since the abandonment of the alluvial system by its sediment feed, it has been dissected by streams resulting in the isolation of the higher points of the terrain. This erosional process will have had the dual effect of destroying earlier parts of the archaeological record and damaging some of the later deposits, as well as creating high points in the terrain that could have been attractive both as knapping and as lookout areas for observing game.

The Cathedral Mopane zone (14) has cobble areas that are very similar to the cobble zones, and these have probably had a similar depositional history that will be reflected in the archaeological material contained within them. In this area there are stable sandy soils covering much of the landscape which, as well as having buried the cobble mounds, have also provided a good substrate for the development of the mature woodland. There is the possibility that material from the Later Stone Age and Iron Ages will be intact and be visible on the surface. However, the stability of the soils may mean that artefacts from all periods are not exposed as frequently as they are in zones 7 and 8.

6.6.4 The Chipembele Ridge Sands (zone 13)

The Chipembele Ridge Sands (zone 13) display varying conditions today, from stable land surfaces to rapidly eroding gullies cutting back into the stable sands. Before the second season's research this was unknown, and it was only after the survey work had been undertaken in the area that its general geomorphology was outlined. On the stable surfaces, material from the Early Stone Age may either remain undisturbed or be covered by the sandy soil in the area, resulting in any accumulations of artefacts only being visible in the area close to the gullies where the sand is being washed away. Artefacts from later periods will, if present, also remain in their original positions of discard. Therefore any future research into the landscape archaeology of the Early and Middle Stone Age should concentrate on this area (see chapter 7).

6.7. Discussion and conclusion

This is the first research to attempt to develop an understanding of Pleistocene landscape change in the Luangwa Valley. Though it has concentrated on a relatively small area it has been possible to outline geomorphological descriptions of, and models for, the formation of many of the landscape features in the area. Within the survey area alluvial fans operated from the Nchindeni Hills and from behind Chipembele Ridge, which was fed by material from behind Chipembele Ridge that was in turn replenished by material from the Muchinga Escarpment. The Luangwa River may or may not have been present throughout the formation of the alluvial fans that deposited the cobble zone material. However, within this model it is possible to incorporate the presence of the Luangwa River, or its narrower and possibly seasonal glacial equivalent in its current floodplain throughout the formation of the fan deposits, which would have occurred in semi-arid to arid conditions not seen in the area today. These drier conditions would have occurred during glacial periods throughout the Pleistocene, whilst in the wetter interglacial periods it is likely that much of the landscape would have been stabilised with denser vegetation, and the river would have been present in a similar form to that of today. It is likely that periods of climate change from glacial to interglacial conditions have been marked by increased landscape instability, because the initial wetter conditions would have developed on a denuded unstable landscape, before vegetation had a chance to develop. This sequence would be a similar scenario to that recorded in the Chipata area by Thomas and Murray (1999:130).

More geomorphological research needs to be undertaken over the valley as a whole, but within the context of this research a basic understanding of the landscape formation processes that have operated towards the centre of the valley and the Muchinga Escarpment has been achieved. Further research would help test the likelihood that the cobble material has ultimately derived from the Escarpment, and would also assess whether this model needs to be revised or abandoned in favour of another. What is clear is that Pleistocene landscape development in the Luangwa Valley is very complex and in need of in-depth research beyond the scope of this project. This research has provided the first model for landscape development in a previously archaeologically and geomorphologically unexplored area, as well as enhancing prior research on the Luangwa River.

The archaeological record of the valley and survey area will be differentially preserved. As may be expected in any landscape context material deriving from earlier in the Pleistocene, in this case the Early and Middle Stone Ages, will have undergone more post-depositional processes than material from more recent periods. In the next chapter the archaeological material in the valley from all periods will be analysed in the landscape context with reference to the geomorphological zones mapped and the processes that have occurred in the zones.

7. Distribution of Archaeological Material in the Landscape and its Relationship to the Geomorphology

7.1. Introduction

The data collection methods and basic parameters for the artefacts in the sample are outlined in appendix 7.1 and 7.2 respectively. The analysis in appendix 7.2 reveals that the artefact sample for the Early and Middle Stone Age is relatively small, as during the analysis a classificatory system that excluded artefacts that could not positively identified as belonging to an individual Mode was used. This system was rigorously applied, and resulted in a small but reliable dataset for the Early and Middle Stone Age, and larger data sets for the Later Stone Age and Iron Age. The indeterminate categories for the artefacts act as a safeguard by removing artefacts not distinctive of a particular Mode, providing a residual sample that will definitively derive from each period. Although as a result of this practice there is obviously a loss of material from the investigation, the residual sample is more reliable, if not entirely representative.

The archaeological survey was designed to retrieve archaeological data from the landscape in a manner that would allow it to be related to the different geomorphological zones in the survey area. Descriptions and interpretations are given here for the distributions of artefacts with reference to the geomorphological processes outlined in chapter 6. Artefact patterns that are attributable to behavioural activity rather than geomorphological processes are then determined. These analyses show that, as suggested in section 6.6, geomorphological processes have altered the Early and Middle Stone Age archaeological record significantly, with the result that interpretations are limited to outlining simple behavioural attributes such as raw material selection in the landscape. The Later Stone Age and the Iron Age datasets are, by comparison to the earlier periods, both sizeable and largely unaltered, providing clearly identifiable patterns in the distribution of archaeological material in the landscape. The Later Stone Age and Iron Age distributions are examined in detail by analysing the archaeological data with reference to specific landscape characteristics and features, including the Luangwa River and its tributaries, elevation, slope angle, geomorphological setting, and soil quality. These spatial relationships and patterns are used to produce models of the archaeological landscape that provide a basic insight to the land-use strategies of the two co-existing groups, and the nature of their interaction. Data from outside the main survey area is also analysed and interpreted to provide an overview of the archaeological record elsewhere on the valley floor. These data derive from transects surveyed in the first season along the O5 road, as well as the experimental transects surveyed along the Kasweta Road in the second season. The remainder of the first season's data from transects located within the survey area is also analysed and compared with the main survey data of the second season outlined earlier in the chapter.

7.2. Description of the Overall Distribution of Artefacts in the Landscape

7.2.1. The overall distribution of archaeological material

This section provides a description of the general distribution in the landscape of the artefacts in relation to the geomorphological zones by total artefact counts, raw material, and condition. This description provides a general overview of the archaeological landscape with which individual periods discussed in later sections can be compared where necessary. In general, the descriptions of patterns in the landscape are organised geographically from south to north.

Figure 7.1 illustrates the 61 surveyed transects from the archaeological survey in relation to the geomorphological zones; the codes used to identify the transects are also displayed in this figure so that locations referred to in the text can be identified. The total number of artefacts recovered from each transect is displayed in graded classes (0, 1 – 10, 11 – 20 artefacts and so on), however, these classes are not equally graded in this figure, as more subtle patterns in the landscape were obscured by large classes of approximately 40 artefacts. As a result, the map was designed to be more sensitive to displaying transects with relatively low artefact numbers (see legend on figure 7.1). The first point to note is that the landscape is archaeologically rich. Only 12 of the 62 transects did not yield artefacts, and these were usually in zones where such a result would be anticipated due of the probable affects of geomorphological processes on the archaeological record (outlined in section 6.6). Six of the empty transects are found on the Recent Cut-off zone (3) and Dambo zone (15) near the Luangwa; both these areas contain surfaces that may date to the 20th century and are thus unlikely to have archaeological material. Artefacts recovered from the recent surfaces characteristic of zone 3 that are in their original place of discard are likely to derive from older cut-offs, but will be representative of recent archaeological time, probably either the Later Stone Age or the Iron Ages (see sections 7.5 and 7.7). The highest numbers of artefacts were recovered from the two cobble zones (7 & 8), the two sand and silt zones (4 & 5), and the Cathedral Mopane zone (14). The cobble zones are interpreted as alluvial fan material deposited throughout the Pleistocene, and the presence of artefacts from all three stone ages was expected (section 6.6). The large number of artefacts in the sand and silt zones either side of the Luangwa relates mainly to Iron Age occupation of the area, though a number of Stone Age artefacts were also recovered. Relatively few artefacts were recovered from the Nchindeni Hills (zone 6), the Sand Bank zone (11), the Chichele Hill zone (10), or the Chipembele Ridge Sands zone (13). The precise natures of the patterns in the archaeological record are explored in chronological order in section 7.3 onwards.

7.2.2. The general distribution of raw material used in the landscape

It is clear that the most common raw materials, quartz, fine, medium, and coarse quartzite (appendix 7.2), were used across most of the survey area (figure 7.2). Quartzites account for at least half of the raw material in 30 of the 49 artefact bearing transects. Due to the small sample size on some transects, care had to be taken when analysing any differences from these general patterns. With this limitation in mind it is possible to determine some other general patterns. Frequently, a large proportion of the artefacts in the sand and silt zones (4 & 5), and on Chichele Hill (10) are of quartz. The transects in the Cathedral Mopane zone (13) have a relatively high proportion of chert artefacts, as do some of the transects in the Cobble zone (7). Vein quartz is only used in the Nchindeni Hills, where it would have been locally available. The high proportion of quartz in the sand and silt zones is interesting as there are few clasts in this area, and any raw material would have been brought to the area. In the Cathedral Mopane zone (13) there is a cluster of transects with a larger proportion of chert artefacts on them than on most other transects. This might represent a localised source for chert that has been eroded and incorporated into the cobbles in this area, or a preference for this raw material and this area of the landscape, which has resulted in more of it being discarded at this locality. When these data are compared to the clast count data (appendix 6.2, figure A6.5), it is apparent that there is a little chert recorded at count 16 over a kilometre away, but none in the nearer count 31. There are also some chert outcrops at Kakumbi Spring (zone 16) (Barham *pers comm.*) The rarity of chert in most of the clast counts suggests that this small appearance in count 16 may hint at a natural localised concentration of the raw material.

7.2.3. The overall distribution of abraded and un-abraded artefacts

The majority of artefacts (67.1%) are in a fresh unrolled condition (appendix 7.2, table A7.5). It is clear however, that a larger amount of abrasion was observed on artefacts in the Cobble zone (7) and the Cathedral Mopane zone (14), both in terms of the number of abraded artefacts and the amount of abrasion observed (figure 7.3). This observation concurs with the interpretation of the geomorphological processes presented in section 6.5 that suggests alluvial fan processes would have transported a proportion of artefacts in these zones from their place of discard. Other land surfaces that have either been more stable, or are more recent, such as the Nchindeni Hills (zone 6), the Recent Cut-off zone (3) and the two sand and silt zones (4 & 5), have a greater proportion of artefacts in fresh unrolled condition. This suggests that the artefacts present derive from cultural activity, as it would be expected that if the artefacts were deposited with the sand and silt they would display a degree of abrasion.

7.2.4. Summary

It is clear that the assessment of the probable effects of the geomorphological processes on the archaeological record outlined in section 6.6 is likely to be correct, and that there is a significant proportion of abraded artefacts in areas with alluvial fan sediments such as the cobble zones (7 & 8) and the Cathedral Mopane zone (14). General taphonomic affects on the archaeological material can be screened out at this stage, as abrasion does not seem to be related to the raw material type, but to geomorphological processes (appendix 7.2, table A7.6). Initial observations and tentative interpretations of the distribution of archaeological material in the landscape have been made to provide an introduction to the distribution of artefacts before the main analysis below. Already there are hints at patterns in the landscape that might not be compatible with a purely geomorphological interpretation for their distribution, suggesting that there are behavioural causes. Examples of the latter include the concentration of chert artefacts in the Cathedral Mopane zone (14), and the quartz artefacts discovered in the sand and silt zones (4 & 5).

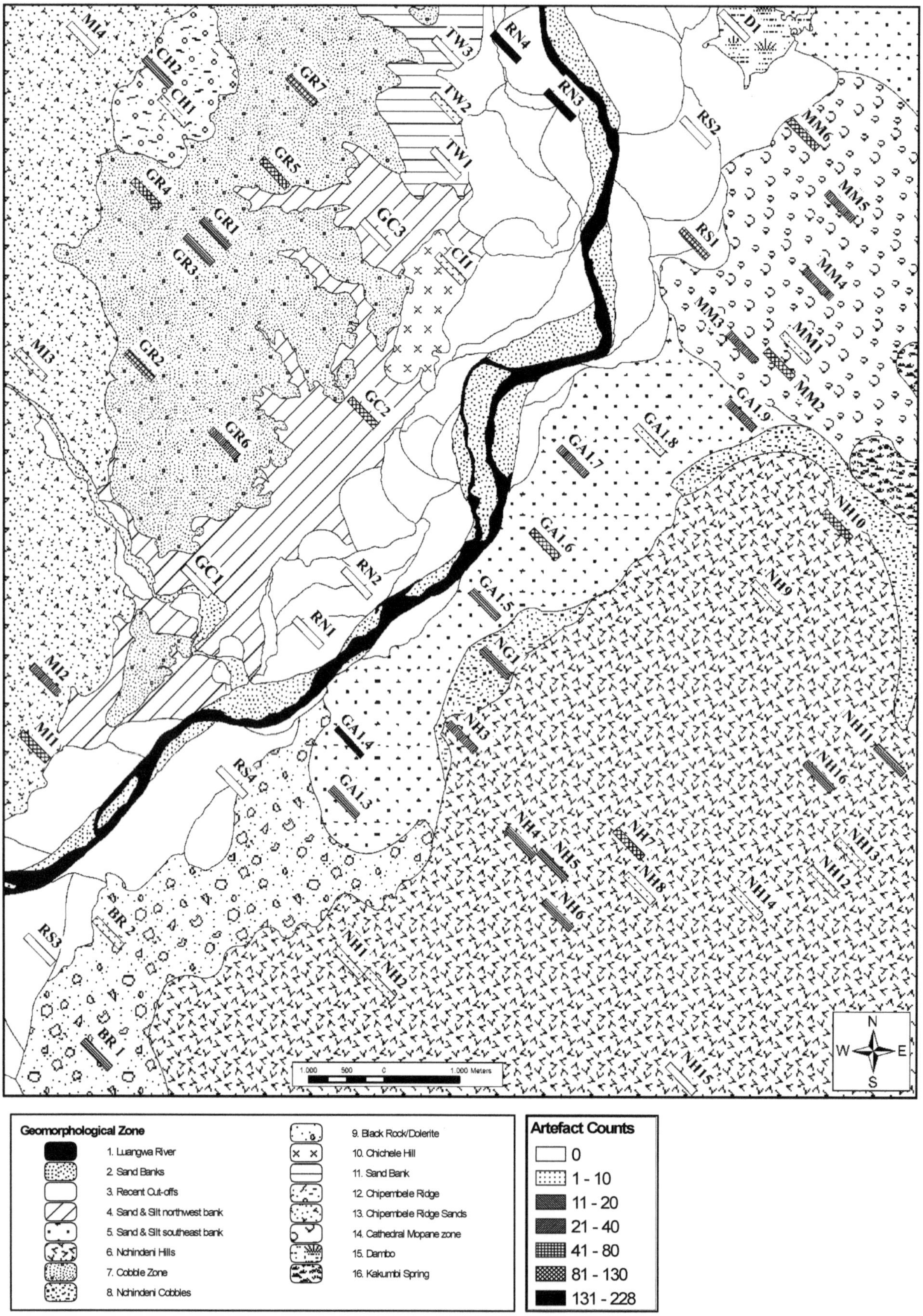

Figure 7.1: The total counts for all artefacts recovered in the 61 randomly selected survey units are displayed against the geomorphological zones that formed the strata for the stratified survey. The identifying codes for the transects are displayed so locations discussed in the text can be found. The data for the artefact counts is presented in appendix 7.3.

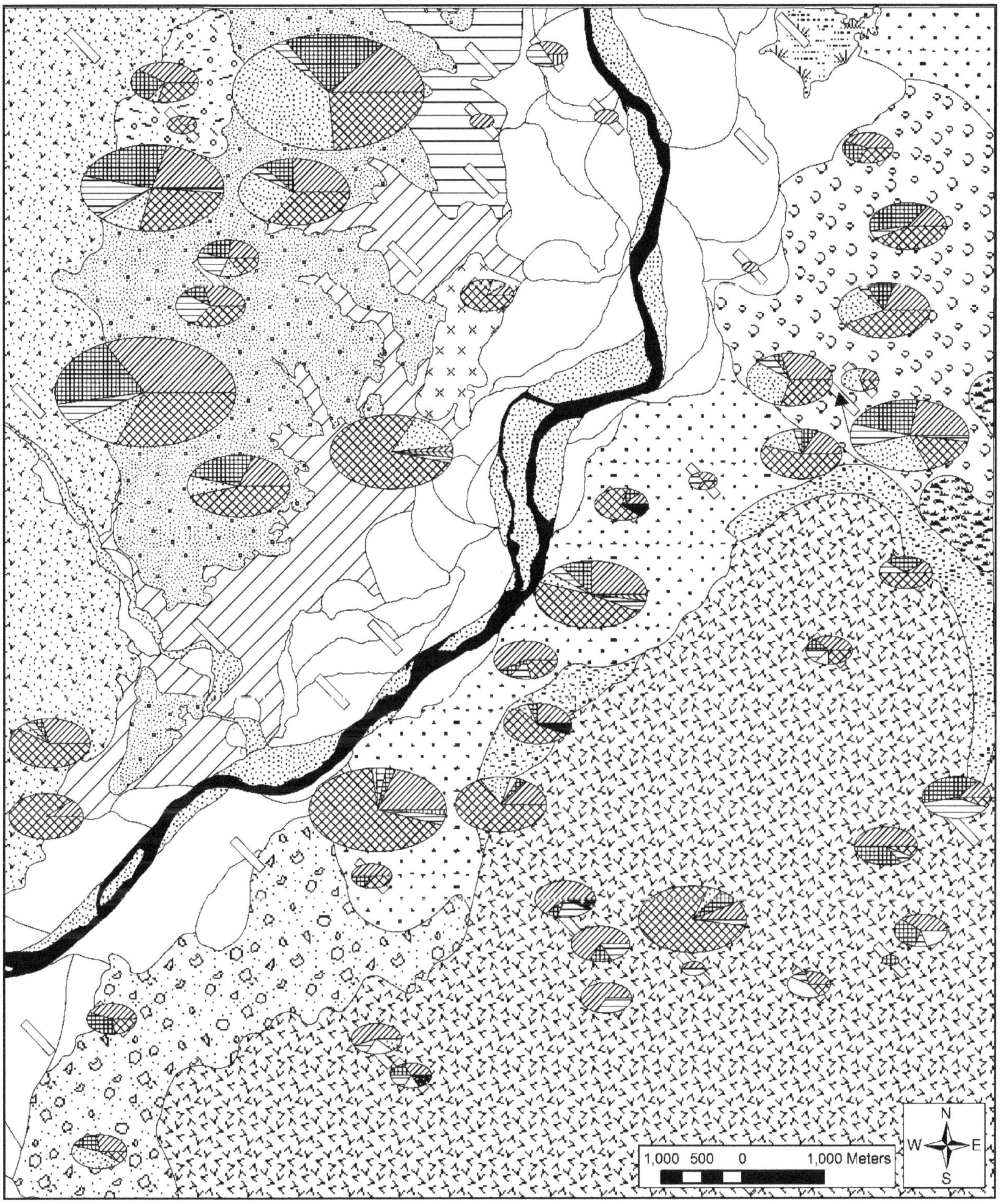

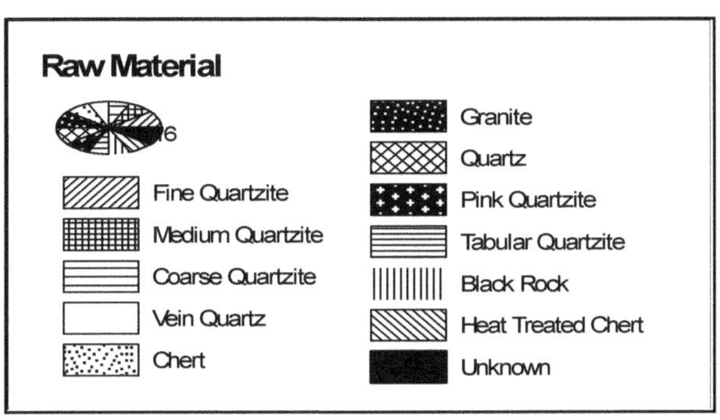

Figure 7.2: Pie charts illustrating the percentage of artefact raw materials for each transect. Each pie chart is scaled to the size of the sample collected on each transect. The data for the raw material counts is presented in appendix 7.4.

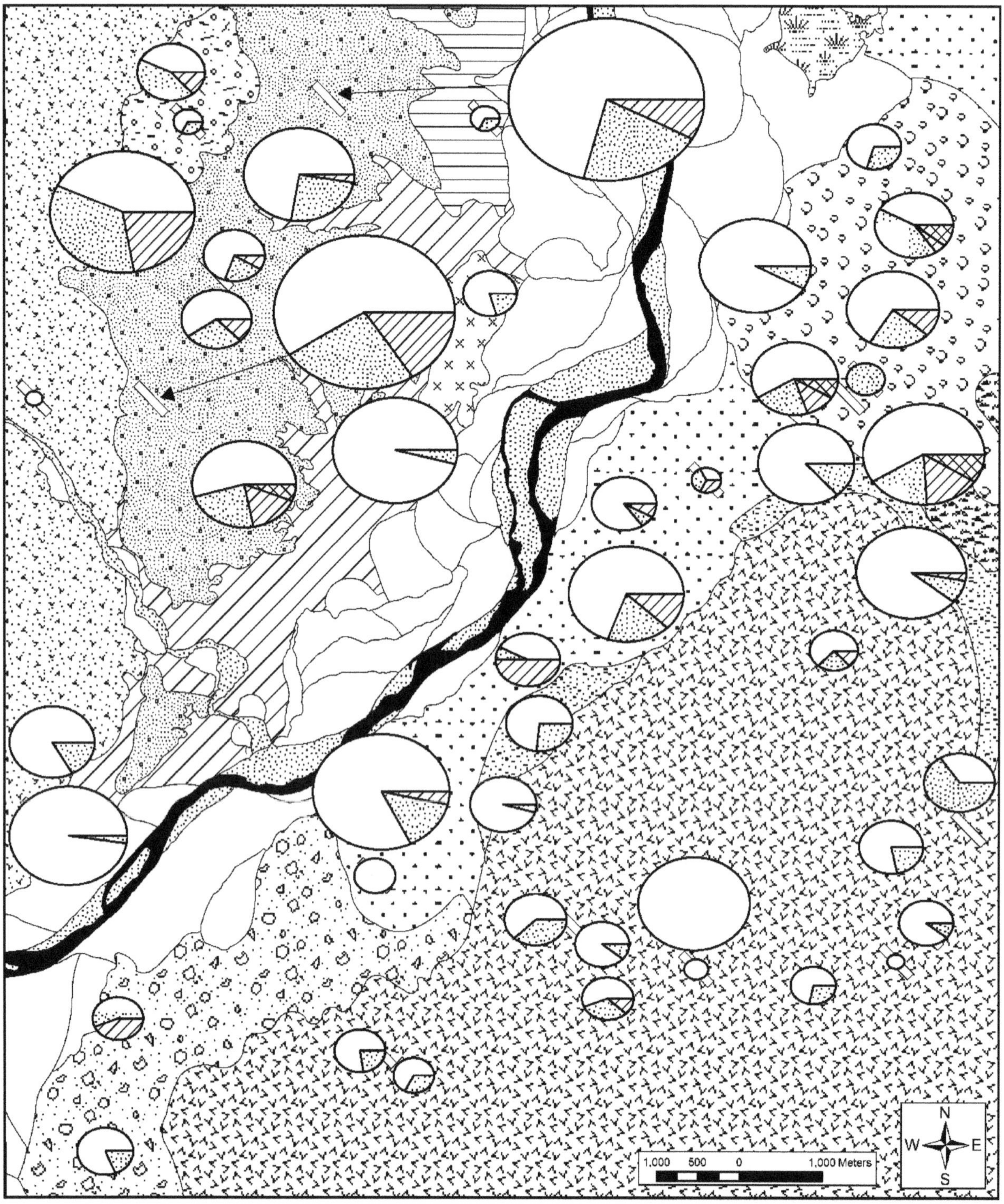

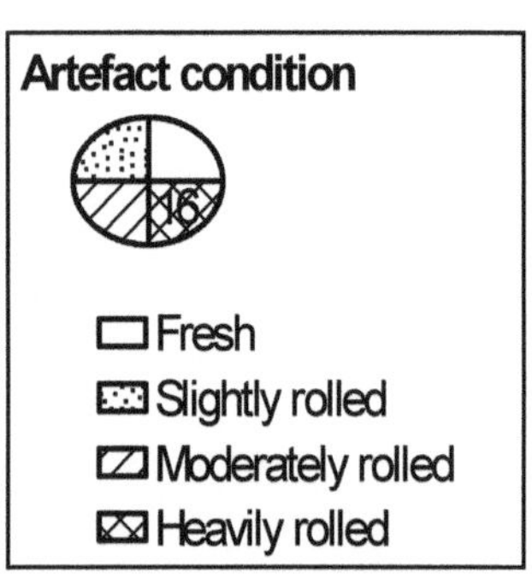

Figure 7.3: The degree of abrasion exhibited by the artefacts in the survey sample. The pie charts are scaled to the size of the sample from each transect. The data for artefact counts and percentages cross-referenced to condition and transect number is presented in appendix 7.5.

7.3. The Early Stone Age (Mode 1 and Mode 2) Archaeological Material and its Relationship to Geomorphological Processes

7.3.1. Description of the distribution of Early Stone Age material

7.3.1.1 General distribution of artefacts.

The number of artefacts that could be categorised as Early Stone Age is relatively low (appendix 7.2), so the resulting statistics are to an extent unreliable, but for methodological consistency the same sorts of analyses are applied to both the Early and Middle Stone Age as to later periods. With this proviso in consideration, a residual pattern is present in the artefact distribution in the landscape. The Nchindeni Hills (zone 6) are virtually devoid of Early Stone Age artefacts, with the exception of three counts at relatively low altitudes near the edge of the zone on transects NH9, NH11, and NH13 (figure 7.4). The highest density of artefacts occurs in the Cobble zone (7), where five of the seven transects have more than five artefacts. Five artefacts were also retrieved from CH1 on Chipembele Ridge (zone 12), which has a similar geomorphological history to the Cobble zone. Unexpectedly, the transects in the Nchindeni Cobbles (zone 8) and the Cathedral Mopane zone (14) have relatively few artefacts considering the geomorphological setting of the deposits, as it was expected that the cobbles in these zones would have similar artefact counts to that of the Cobble zone (see section 6.6). There are no Early Stone Age artefacts on the Recent Cut-off zone, the Sand and Silt zone (northwest bank), the Sand Bank zone, or the Dambo zone (3, 4, 11 & 15). On the Sand and Silt zone on the southeast bank (5) a total of twelve artefacts were retrieved from four transects (six artefacts from GA1.6 alone); this is surprisingly high as it was expected that this zone would have similarly low or negligible artefact counts to that of zone 4, which is of a comparable recent age. Three artefacts were retrieved from one of the two transects (BR2) in the Black Rock zone (9). There was only one Early Stone Age artefact discovered on the Chipembele Ridge Sands zone (13), which was located on MI2 nearer the Luangwa.

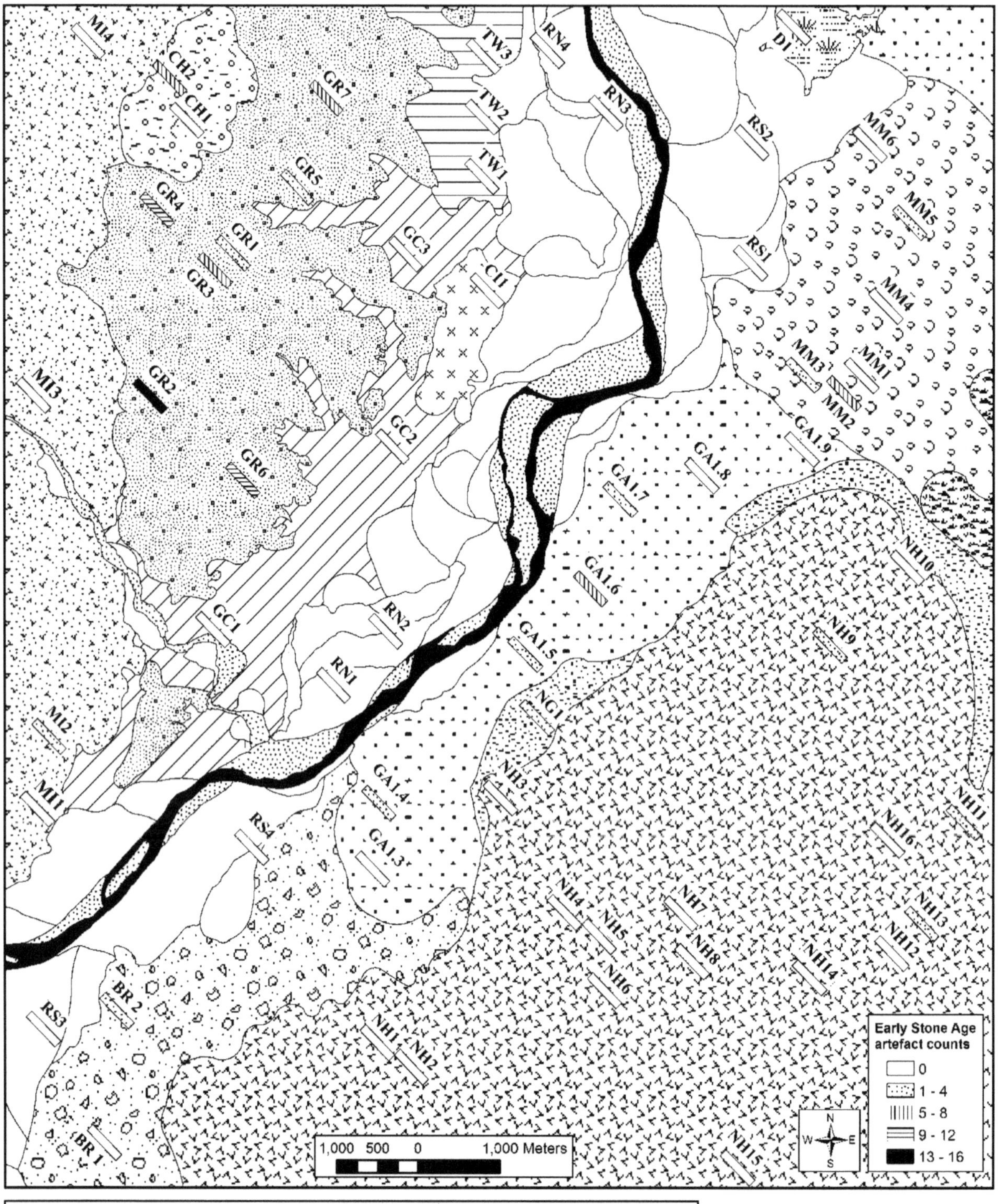

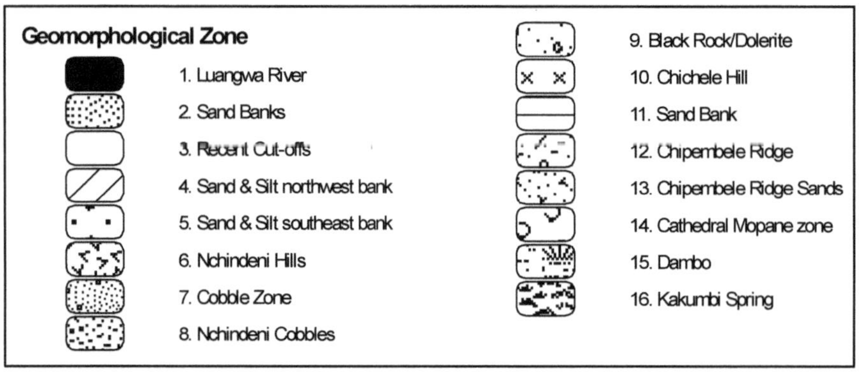

Figure 7.4: Early Stone Age artefact counts from each transect displayed with equal graded intervals. The transect codes used to identify each transect in the text are displayed here for reference purposes. The key for the geomorphological zones is also displayed. The raw data for the artefact counts for each transect is presented in appendix 7.3.

7.3.1.2 Artefact condition and raw material.

High proportions of the Early Stone Age artefacts have been abraded to some degree (figure 7.5), with the highest proportion of heavily abraded artefacts deriving from the Cathedral Mopane zone (14). It is apparent that when compared with later periods, it is the Early Stone Age that is the source for the majority of abraded artefacts in the sample (compare figure 7.5 to 7.9 and 7.14). Of the four artefacts recovered from the Nchindeni Hills zone (6) one was in fresh condition, whilst the others were slightly rolled. Artefacts found in the Cobble zone (7) frequently demonstrate abrasion to an extent, however, approximately half (24 of 51) are found in fresh condition. Almost all artefacts recovered from the relatively young surface of the Sand and Silt zone on the southeast bank (5) have been abraded, with 10 of the 12 artefacts demonstrating at least slight abrasion. The three artefacts from the Black Rock zone are also rolled demonstrating moderate abrasion.

The distribution of raw material types used in the Early Stone Age is illustrated in figure 7.6. Overall, coarse, medium, and fine textured quartzites were used in roughly equal quantities and account for the majority of the raw material used (29, 25, and 23 artefacts respectively). Only eight artefacts were not knapped using quartzite; seven were quartz, while one was of an unknown raw material. The distribution of the raw materials in the landscape does not appear to be as equally balanced, with a predominance of medium and coarse quartzite on the south-east side of the Luangwa River, whilst the majority of the fine quartzite appears in the Cobble zone and Chipembele Hill (zones 7 &12). A distinction between fine, medium, and coarse quartzite was not made in the clast count part of the geomorphological study, so direct comparisons between the types of quartzite utilized and those present in the landscape cannot be made. However, comparisons can be made with the rest of the archaeological sample, from which it is apparent that other raw materials were present and available in the landscape (appendix 7.2). These raw materials might not have been available to Early Stone Age hominins and could have been brought to the area by later geomorphic processes or anthropogenic agents. However, observations made in the exposed section at SL8 and in the field revealed that the various grades of quartzite were available in roughly equal quantities in the cobble material, with an apparent abundance of quartz, usually occurring as smaller clasts. Additionally, chert was recognised in the clast counts and was present in small quantities in the survey area, and also noted on transect GR7 as a naturally outcropping raw material source (see section 7.4.2 below). This suggests that, though present, chert and quartz were not as extensively utilised in the Early Stone Age as they were in later periods.

7.3.1.3 Artefact types from the Early Stone Age

The Early Stone Age is characterised in the sample by Mode 1 core and flake technology. Mode 2 bifaces were not found in the survey, and apart from the occasional find appear to be rare in the valley in general (Barham *pers comm.* and *pers obs*). One biface was found in each of the excavated sequences at SL8 and SL1, and single bifaces were recovered from the Chowo and Mumbwa Rivers (the Mumbwa River is situated approximately half way along the O5 road toward the fossil wood transect surveyed in the first year, figure 5.1 (Barham *pers comm.*)). It is clear more cores are present in the sample than flakes, but there is no geographical pattern to the distribution of the artefact types recovered (figure 7.7).

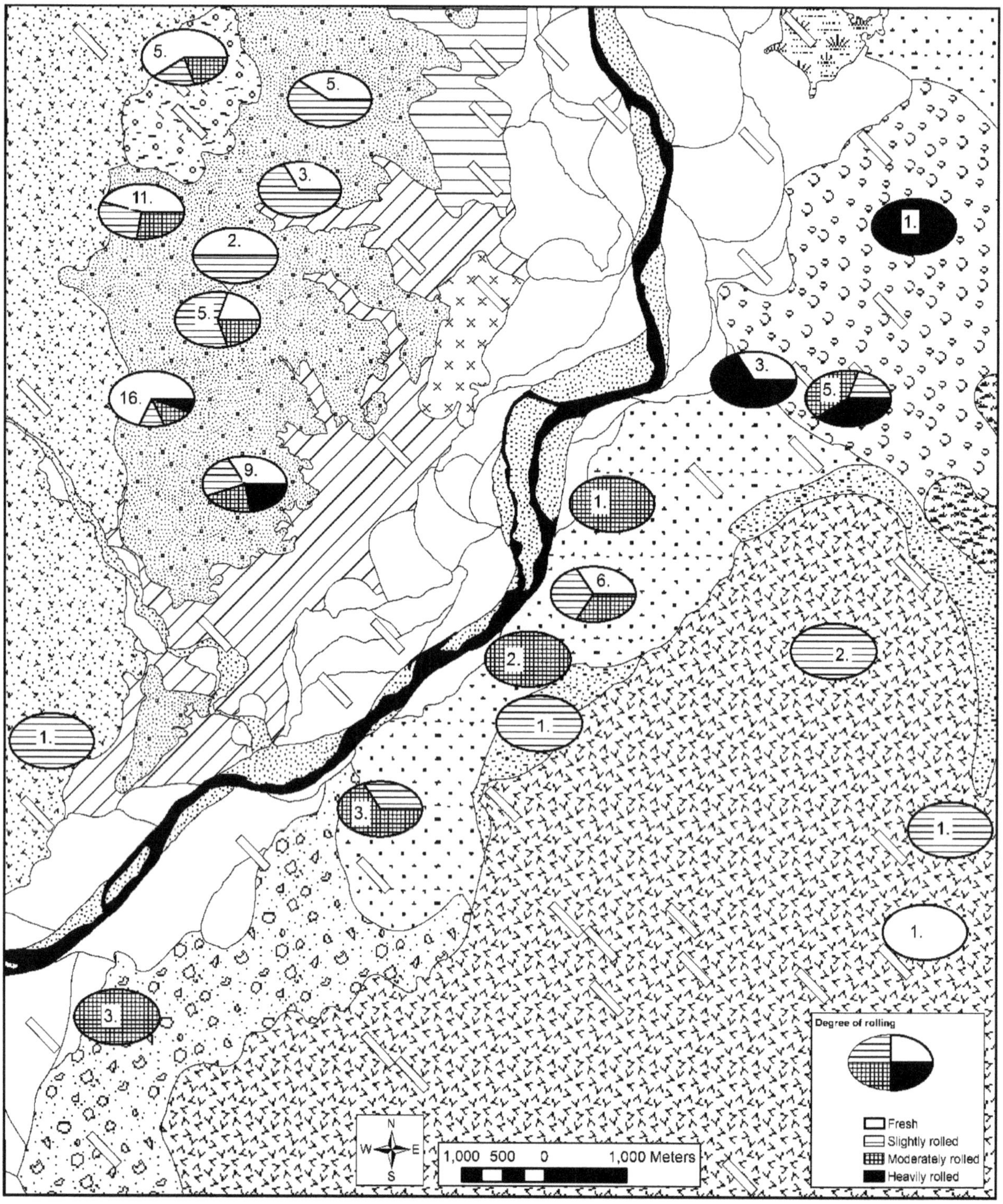

Figure 7.5: Early Stone Age artefact abrasion displayed in equally sized pie charts. The numbers on each pie chart give the sample size for each transect. The raw data for abrasion artefact counts are presented in appendix 7.6.

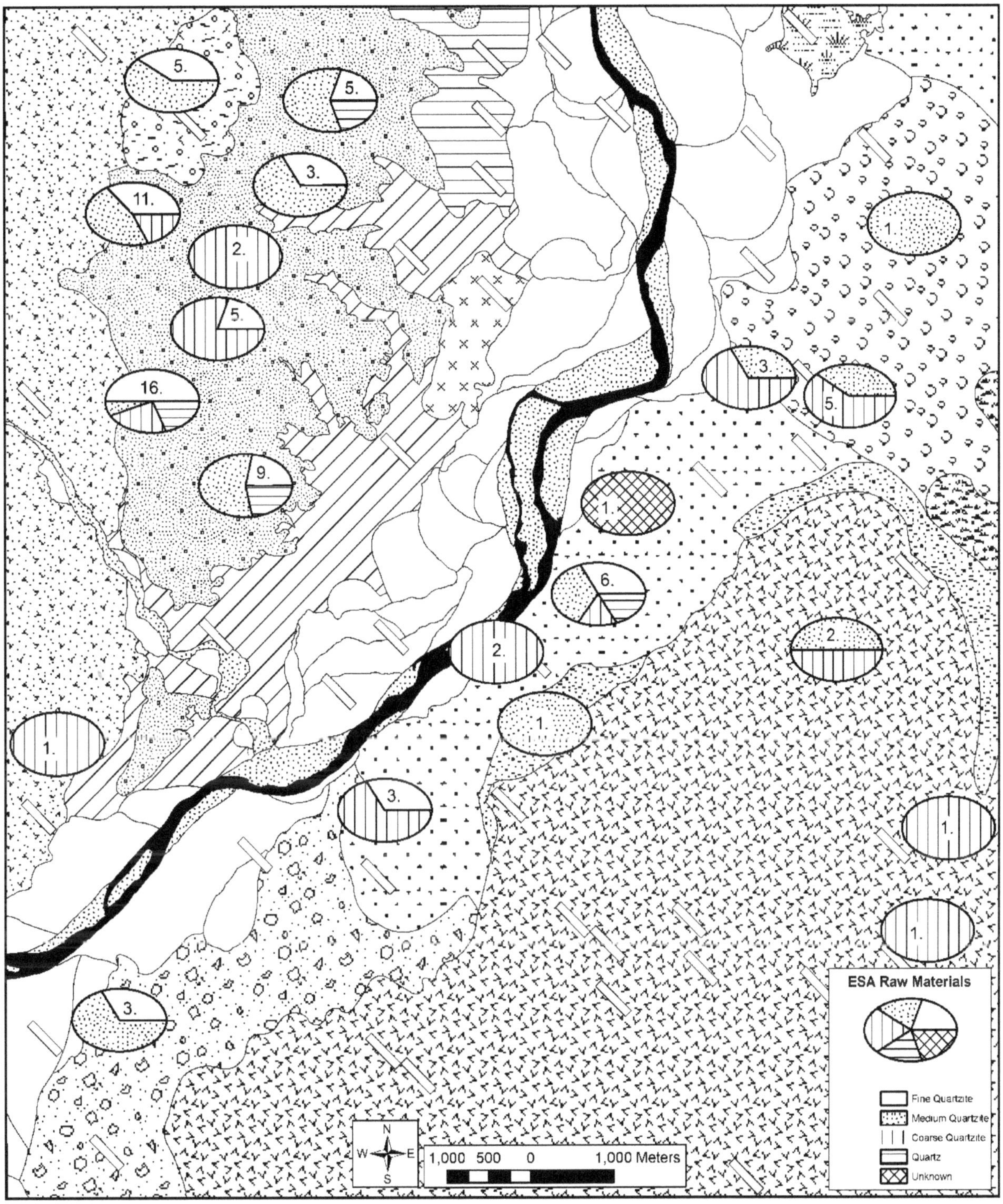

Figure 7.6: Early Stone Age raw material types displayed on equal sized pie charts. The numbers on each pie chart provide the sample size for each transect. The artefact counts for raw material are presented in appendix 7.7.

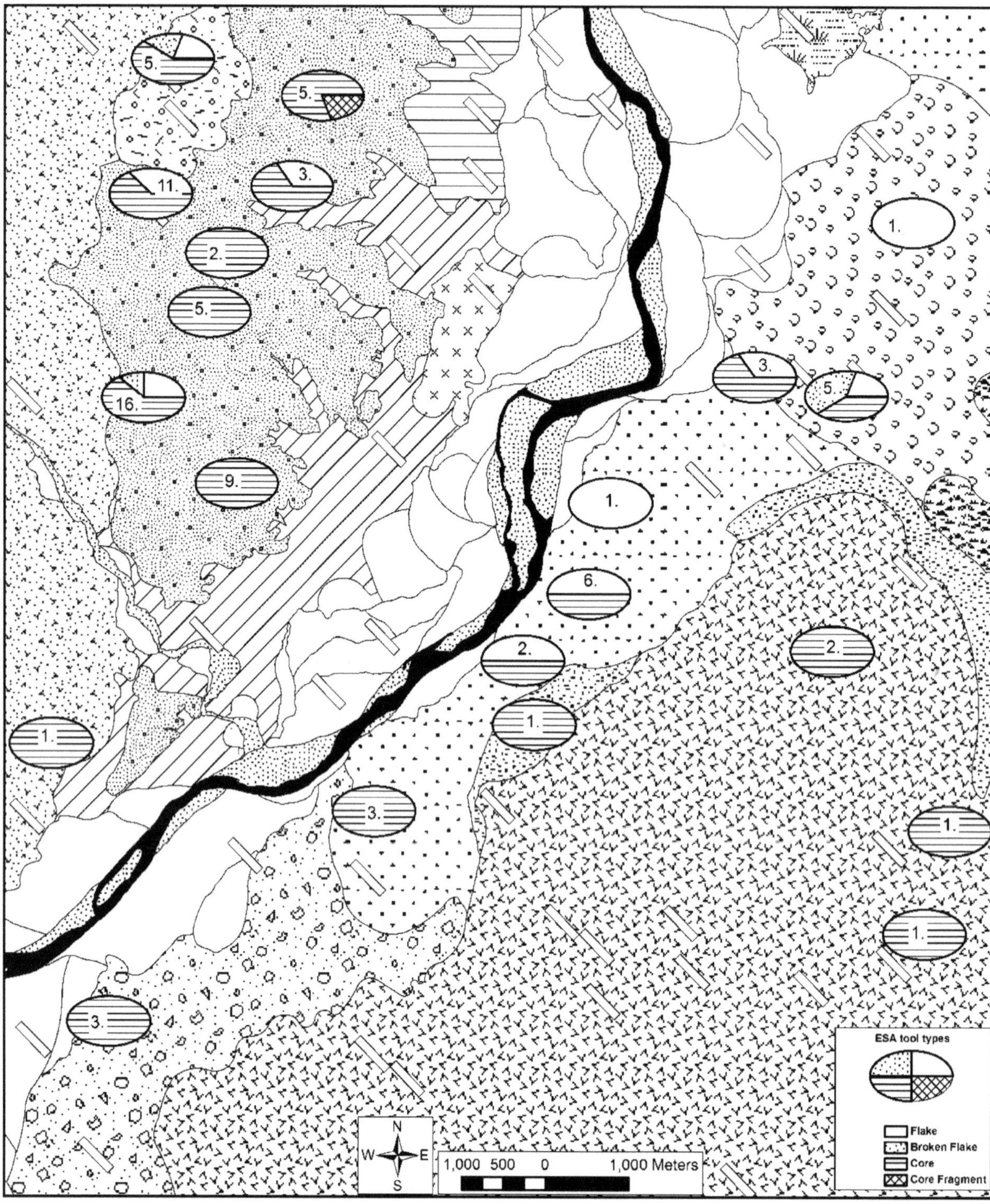

Figure 7.7: Early Stone Age artefact types displayed in equal sized pie charts. The numbers on each pie chart provide the sample size for each transect. The raw data for the tool types is recorded in appendix 7.8.

7.3.2 Interpretation of the distribution of Early Stone Age artefacts

The surfaces of the Nchindeni Hills (zone 6) were interpreted as young surfaces due to the continual erosion that would have been experienced on the higher elevations and slopes over time (section 6.6.1). Therefore, the lack of Early Stone Age material in zone 6 is not surprising as any archaeological material would have been removed by erosional processes and transported away from the hills, so it should not be seen as evidence that the area was unsuitable or undesirable for occupation by early hominins (the biface recovered from the Chowo River would have been transported from the hills). At lower elevations it was suggested that sites may be present but buried beneath material eroding from the hills (section 6.6.1), and though the presence of a single fresh Early Stone Age artefact does suggest that the lower elevations were occupied, the extent and nature of this occupation cannot be determined. The artefacts from the Black Rock zone (9) are all moderately rolled, and, considering that the material is derived from debris flows that will have been very active environments, this is not surprising.

It was expected that the artefact counts for the Early Stone Age in the Cathedral Mopane zone (14), the Nchindeni Cobbles (zone 8), and the Cobble zone (7) would be similar as the cobbles were deposited by the same depositional regime in similar climatic conditions and presumably at similar times. The fact that there are fewer Early Stone Age artefacts than expected in zones 8 and 14 on the south-east bank suggests four possible scenarios:

1) The cobbles in zones 8 and 14 were deposited after the Early Stone Age.

2) They were not accessible due to overlying sediment cover that has since been removed by erosion.

3) Active sedimentation made the surface unsuitable for artefact production and/or use.

4) The Early Stone Age material is present in these areas, but is covered by more recent sediment.

It is difficult to determine which of these scenarios is most likely with this data set; however, the final possibility is unlikely in the Nchindeni Cobbles, as the test pit undertaken in that zone revealed the depth of sediment to be only 30cm (see appendix 6.1). The high degree of abrasion present on the Early Stone Age artefacts recovered from the Cathedral Mopane zone indicates that alluvial processes might have been more active than in zone 7. This would result in the area being unsuitable for exploitation, or preservation of knapping activities. Alternatively, this could indicate that scenario one is correct, and the entire deposit is more recent and the artefacts have been transported and deposited here before the more stable geomorphological conditions of the Middle Stone Age (section 7.4 below).

The lack of Early Stone Age artefacts on the Recent Cut-off zone, Sand and Silt zone (north-west bank), Sand Bank zone, and Dambo zone (3, 4, 11, & 15 respectively) was expected, as these are interpreted as young surfaces that are unlikely to have early material on them. The relatively large amount of Early Stone Age material on the Sand and Silt zone on the south-east bank (5) had not been anticipated, as the sediments present have a similar relatively recent origin to the sand and silt on the north-west bank (zone 4). It is unlikely that the artefacts present in zone 5 derive from a preserved landscape surface from the Early Stone Age, as the unconsolidated sand and silt would not have survived erosion over the past million years, especially considering the amount of erosion and gully formation observed in the area today (see appendix 6.1). It is more likely, especially considering the abraded condition of the majority of artefacts in zone 5, that the artefacts were transported and deposited with the fluvial sand and silt from upstream of the Luangwa, or were derived more locally from cobbles in the Nchindeni Cobble zone (8) and are now being exposed by the erosion of the sand and silt. Either process is possible, and each is supported by the observations made during fieldwork (appendix 6.1) noting the presence of Early and Middle Stone Age artefacts in lag deposits at the bases of the sand and silt sequences in the area. Alternatively, the artefacts may have been misclassified, and their typological association should be with a later period. As noted above, great care was taken to avoid this, and the fact that these early artefacts are concentrated in zone 5, and not in the equivalent zone 4, suggests misclassification is unlikely. The condition of these artefacts and the nature of their deposition in zone 5 makes it unlikely that their presence represents actual behavioural choices with regard to landscape use. It is possible that Early Stone Age artefacts are also present in a secondary context in zone 4, but due to its relative stability they have not been exposed, and are only exposed in zone 5 due to the higher degree of erosion.

The mixture of abraded and fresh artefacts in the Cobble zone (7) was anticipated due both to the cobbles having been an attractive raw material source, and to the active depositional regime. Therefore artefacts could remain in their primary context (in or near their place of initial discard), or be displaced and transported by alluvial fan processes and deposited in a secondary context. The high proportion of fresh artefacts unaffected by the fan activity indicates that Early Stone Age hominins were present in this area of the landscape. One transect in particular contained a high proportion of Early Stone Age artefacts (GR2 with sixteen artefacts), and a variety of artefact and raw material types, which may indicate that parts of the Early Stone Age record remain relatively intact. A more detailed analysis of possible settlement patterns is not possible though, as there does not appear to be any discernible pattern in the data that suggests particular areas of the fans were more stable than another. Neither is there any indication of direction of transport evidenced by a proportional increase of abrasion in any direction, and furthermore there are areas where all material has been removed exposing the Karoo bedrock. It is therefore not possible with this data set to prospect for areas of undisturbed habitation in the Cobble zone (7) and, though they may indeed exist, it is likely they would be rare.

With one exception, the Chipembele Ridge Sands zone (13) is devoid of Early Stone Age artefacts. This is probably a result of the generally stable nature of the sands having not eroded to reveal any artefacts that might be present below the surface. Observations along the Kasweta road revealed that in areas that are undergoing erosion, the cobble layers containing artefacts do become exposed below these soil types (see section 7.8).

It is apparent that the most common raw materials, fine, medium, and coarse quartzite, were used in equal quantities over the survey area, which suggests that there was no particular preference for types of quartzite. When Early Stone Age raw material use (figure 7.6) is compared to the overall pattern of raw material use (figure 7.2) it is evident that other raw materials such as quartz and chert were available but rarely used. In general raw materials that are less common in the area, such as chert, were not used at all in Early Stone Age stone working. This might be a result of the general rarity of chert in the landscape, or the fact that chert clasts tend to be smaller (appendix 6.2), and thus not suitable for Mode 2 knapping strategies that produce large flakes. Occasionally chert does occur as large clasts and in blocks (table A6.7 and see 7.4.2 below), which in turn suggests that chert was not a sought after raw material, despite its superior knapping qualities and the sharper edges of the flakes produced (appendix 7.1). The primary requirement of Early Stone Age hominins in the area seems to have been the ready availability of raw materials that could be knapped to produce Mode 1 flakes, and in this case the larger quartzite cobbles that are common in the area were the preferred raw material. There is no geographic pattern to the distribution of cores and flakes in the landscape (figure 7.7), and considering that much of the material is in a secondary context, no indication of differential manufacture and discard across the landscape can be gained.

Detailed reconstruction of land-use strategies in the Early Stone Age are not possible due to the differing geomorphological processes that have operated in the area over time. Alluvial fan deposits are found in zones 7, 8, 12, and 14, and the geomorphological activity in these zones that occurred throughout much of the Pleistocene has altered the distribution of artefacts to the point that no behavioural residue remains. The more recent activity associated with the Luangwa River (zones 1 – 5 and 15) has either removed deposits that may have contained Early Stone Age material, or has deposited new material that contains artefacts that have been transported from elsewhere in the Luangwa Valley. The overall paucity of undamaged early material in these areas therefore does not support any behavioural interpretation regarding land-use, and the distribution is entirely a result of geomorphological processes partially or completely removing material. The only general behavioural observation that can be made is that Early Stone Age hominins had a preference for quartzite, and did not actively seek out rare, better quality, raw material types.

7.4. The Middle Stone Age (Mode 3) Archaeological Material and its Relationship to Geomorphological Processes

7.4.1 Description of the distribution of Middle Stone Age material

7.4.1.1 General distribution of artefacts.

The general distribution of Middle Stone Age artefacts is similar to that of the Early Stone Age artefacts, in that there are few or none on the Recent Cut-off zone, Sand and Silt zone (north-west bank), Nchindeni Hills, Black Rock zone, Sand Bank zone, and Dambo zone (zones 3, 4, 6, 9, 11, and 15) (figure 7.8). The overall density of Middle Stone Age artefacts in the landscape is very low, and a maximum of four artefacts was retrieved from any one transect. Similarly to the Early Stone Age, these data are to an extent unreliable. There was one Middle Stone Age artefact retrieved from Chichele Hill, which was not unexpected as Chichele Hill is geologically related to the Nchindeni Hills and is likely to have been present in the ancient landscape (appendix 6.1, figure A6.4). Excavations in 2003 showed the presence of Middle Stone Age material not far beneath the surface, and there are also chert deposits that would have been a raw material source for artefact manufacture (Barham *pers comm.*). Artefacts were retrieved from only four of the nine transects surveyed on the Cobble and Chipembele Ridge zone (7 & 12) (as opposed to 8 of the 9 transects in the Early Stone Age). The density of the artefacts on three transects was high relative to elsewhere, and is comparable with the Early Stone Age, in that the high densities are present on the same transects in each period (GR2, GR4, and GR7). Five of the six transects in the Cathedral Mopane zone (14) have Middle Stone Age artefacts, and in this area the concentration of artefacts is higher than the rest of the survey area, and higher in contrast to the Early Stone Age which had relatively few artefacts in zone 14 compared with the Cobble zone (7). In a similar scenario to the Early Stone Age, Middle Stone Age material is present in a small quantity and at a low density in the Sand and Silt zone on the south-east bank (5), where a total of five artefacts were recovered on four transects.

7.4.1.2 Artefact condition and raw material.

The distribution of abraded artefacts in the Middle Stone Age is similar to that of the Early Stone Age (compare figures 7.5 and 7.9), in that artefacts from the cobble zones are more likely to be abraded (zones 7, 8, and 14). In contrast to the Early Stone Age, heavily abraded artefacts do not occur, moderately abraded artefacts are rare, and the majority of artefacts were in a fresh or slightly abraded condition. The majority of Middle Stone Age artefacts have been knapped using finer grained or crystalline material, such as fine quartzite, chert and quartz (figure 7.10). The only exceptions are the two artefacts from NH10 and NH11 at the base of the Nchindeni Hills (zone 6), for which medium and coarse quartzite was used. There is a high proportion of chert artefacts in the Cathedral Mopane zone (13) on transects MM2, MM3, and also on GA1.9 in the Sand and Silt zone (5) nearby. There is also a relatively high number of chert artefacts (3 out of 4 collected) on transect GR7 in the Cobble zone (7).

7.4.1.3 Artefact types from the Middle Stone Age.

The distribution of Middle Stone Age tool types is illustrated in figure 7.11. The Mode 3 technologies, prepared cores and prepared core flakes, are clustered in the landscape on the same transects as the chert artefacts (MM2, MM3, GA1.9, and GR7). Table 7.1 provides a detailed breakdown of the artefact type, raw material and condition of Middle Stone Age artefacts. It is clear that the majority of diagnostic Mode 3 artefacts are knapped using fine grained or crystalline raw materials, and are usually unrolled (27.6% of Middle Stone Age artefacts comprise the majority of the characteristic Mode 3 technologies - prepared cores and flakes). The artefacts over the rest of the landscape are primarily Mode 1 artefacts that were identified as Middle Stone Age (see appendix 7.1). The majority of these Mode 1 artefacts were manufactured out of quartzite, and occasionally were slightly or moderately abraded (table 7.1). Two split cobbles were also classified as Middle Stone Age, though not a Mode 3 technology, their large size ruled them out as deriving from the Later Stone Age microlithic Mode 5 category.

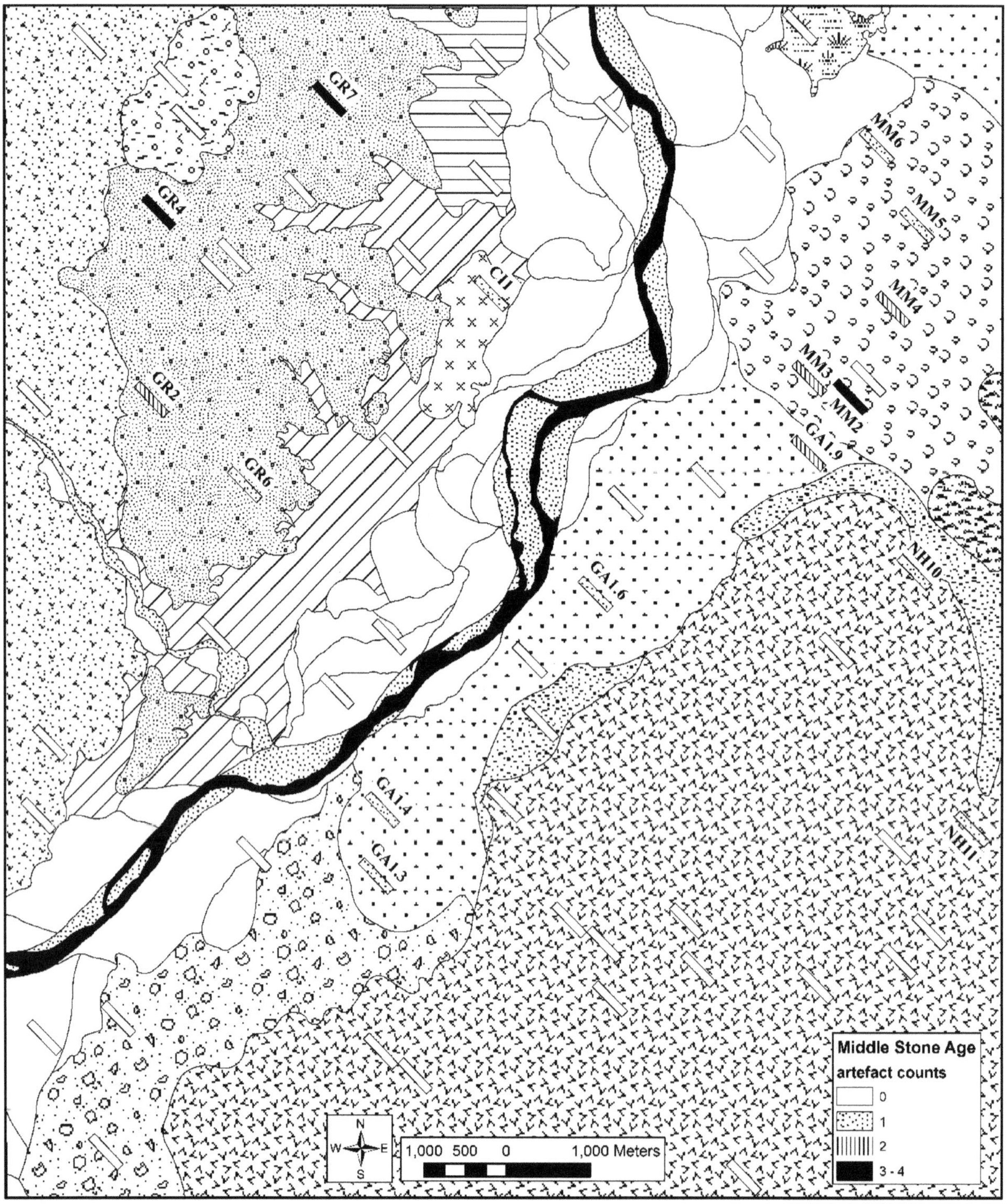

Figure 7.8: Distribution of Middle Stone Age artefacts with artefact density on each transect displayed in equally graded intervals. Only transects containing artefacts have been labelled with their code. The raw data is presented in appendix 7.3.

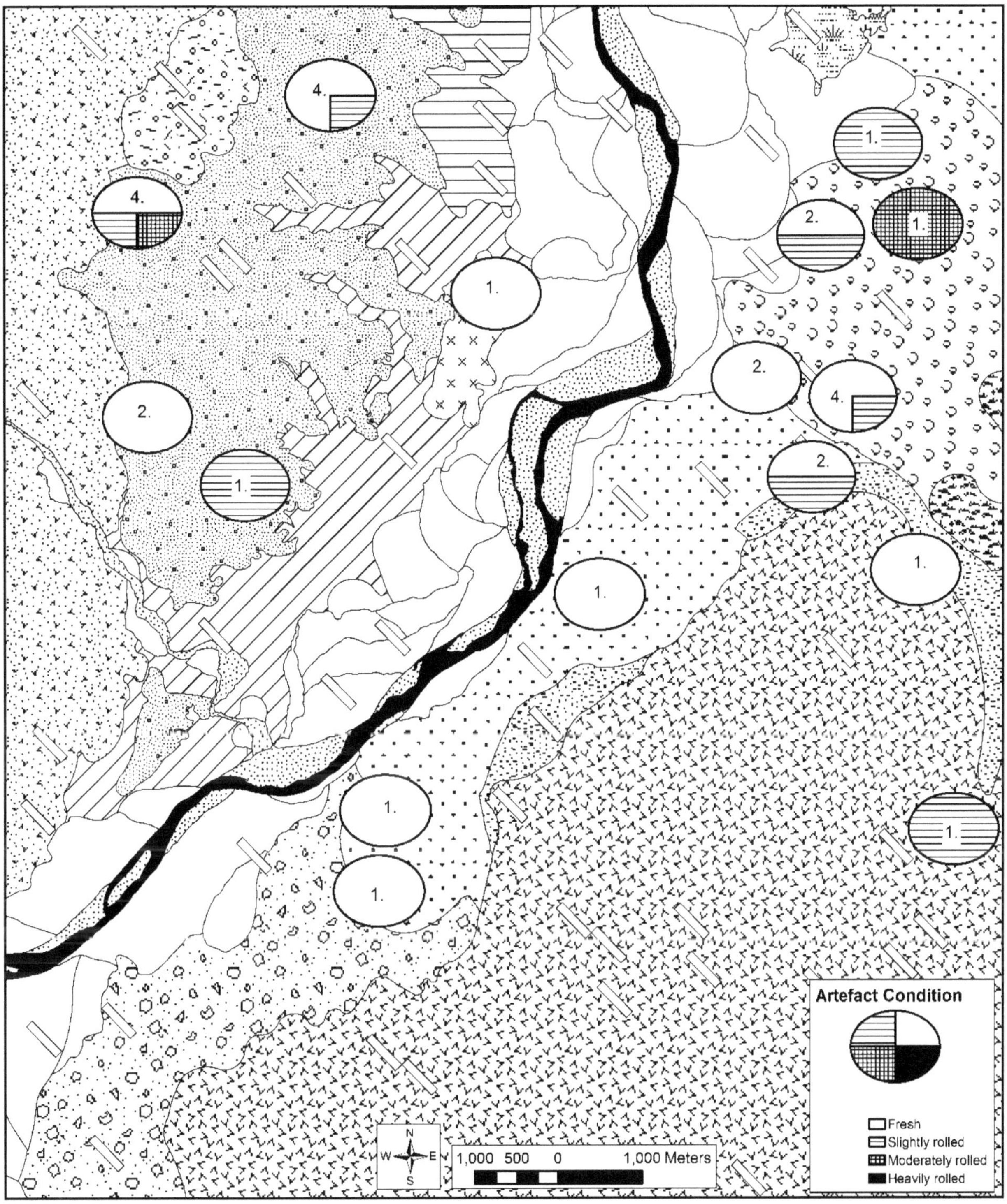

Figure 7.9: Pie charts illustrating the percentage of fresh and abraded artefacts for the Middle Stone Age, with artefact counts presented on each pie chart. The raw data is presented in appendix 7.6.

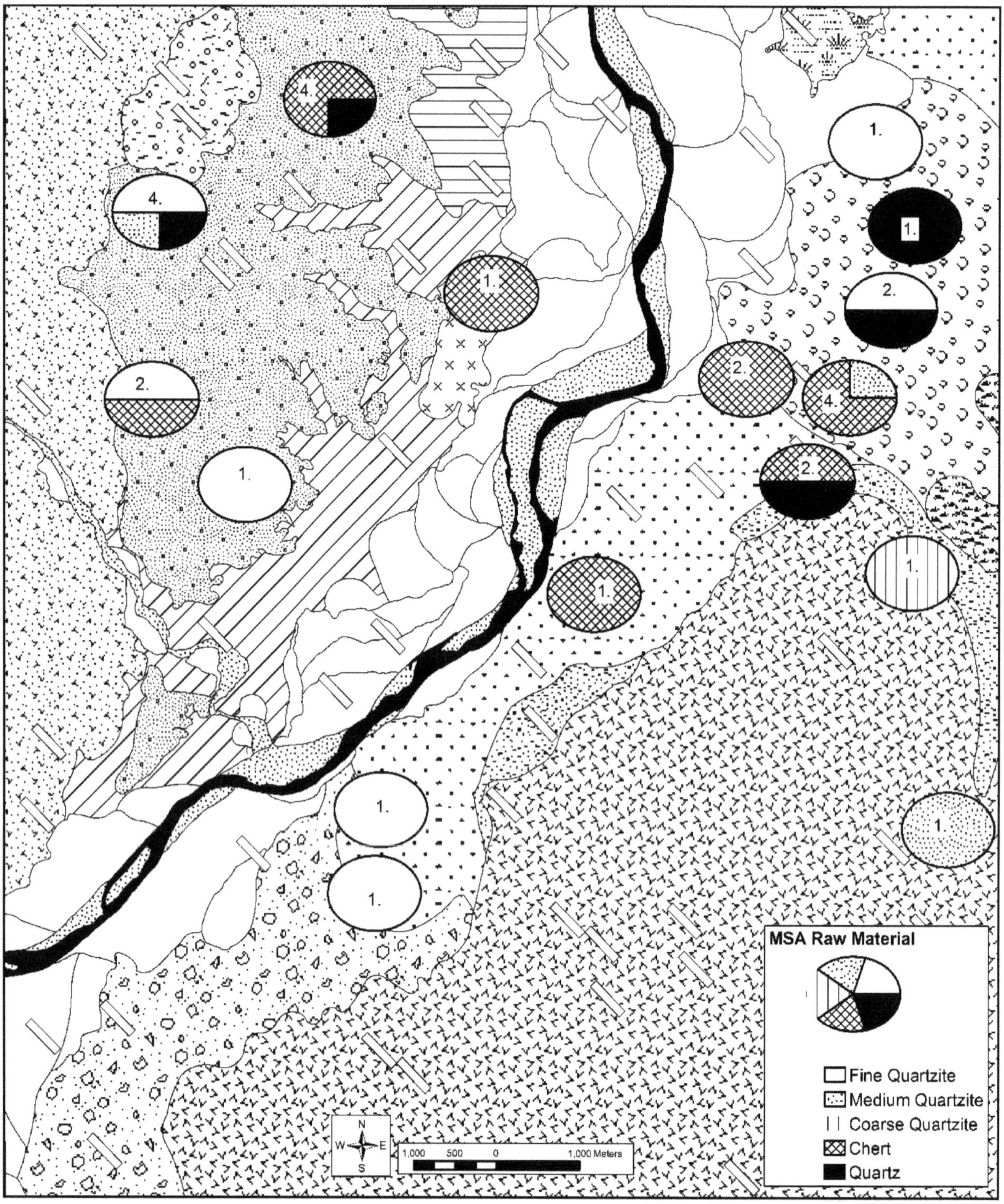

Figure 7.10: Pie charts illustrating the percentage of raw material types found on each transect. The artefact counts are displayed on each transect, and the raw data is displayed in appendix 7.7.

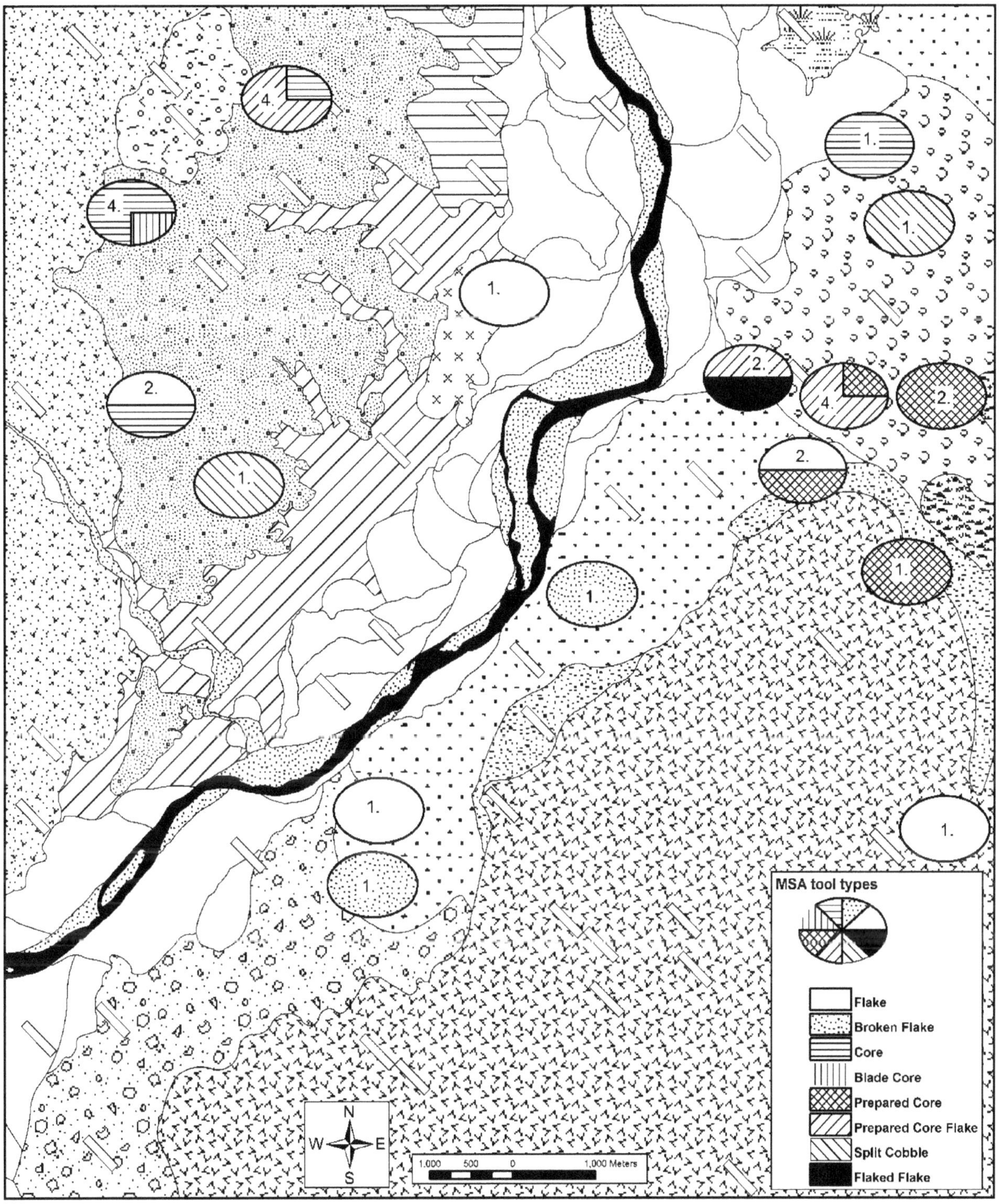

Figure 7.11: Percentages of Middle Stone Age artefact types. Artefact counts on each transect are displayed, and the raw data is presented in appendix 7.8.

Raw material and abrasion class		Artefact type															
		Flake		Broken Flake		Core		Blade Core		Prepared Core		Prepared Core Flake		Split Cobble		Flaked flake	
		Number of artefacts		Number of artefacts		Number of artefacts		Number of artefacts		Number of artefacts		Number of artefacts		Number of artefacts		Number of artefacts	
		Count	Table Sum %	Count	Table Sum %	Count	Table Sum %	Count	Table Sum %	Count	Table Sum %	Count	Table Sum %	Count	Table Sum %	Count	Table Sum %
Fine Quartzite	Fresh	1	3.4%	1	3.4%	2	6.9%	.	.0%	.	.0%	.	.0%	.	.0%	.	.0%
	Slightly Rolled	.	.0%	.	.0%	2	6.9%	.	.0%	1	3.4%	.	.0%	1	3.4%	.	.0%
Medium Quartzite	Slightly Rolled	1	3.4%	.	.0%	.	.0%	.	.0%	1	3.4%	.	.0%	.	.0%	.	.0%
	Moderately Rolled	.	.0%	.	.0%	1	3.4%	.	.0%	.	.0%	.	.0%	.	.0%	.	.0%
Coarse Quartzite	Fresh	.	.0%	.	.0%	.	.0%	.	.0%	1	3.4%	.	.0%	.	.0%	.	.0%
Chert	Fresh	2	6.9%	1	3.4%	.	.0%	.	.0%	.	.0%	6	20.7%	.	.0%	1	3.4%
	Slightly Rolled	1	3.4%	.	.0%	.	.0%	.	.0%	.	.0%	1	3.4%	.	.0%	.	.0%
Quartz	Fresh	.	.0%	.	.0%	1	3.4%	1	3.4%	2	6.9%	.	.0%	.	.0%	.	.0%
	Moderately Rolled	.	.0%	.	.0%	.	.0%	.	.0%	.	.0%	.	.0%	1	3.4%	.	.0%

Table 7.1: Middle Stone Age artefact counts cross-referenced by artefact type, raw material and condition. The percentages listed represent percentages of the total Middle Stone Age artefact count.

7.4.2 Interpretation of Middle Stone Age artefact distribution

The absence of Middle Stone Age artefacts in the Nchindeni Hills is probably a result, as with the Early Stone Age, of the gradual erosion of the Nchindeni Hills over time, and therefore cannot be taken as evidence that zone 6 was not inhabited. No Middle Stone Age artefacts were recovered from the Recent Cut-off zone or the Sand and Silt zone on the north-west bank (zones 3 and 4) and, as with the Early Stone Age, this will be a result of recent geomorphological processes, as these surfaces were deposited after the Middle Stone Age (post ~30ka). Five artefacts are present on the Sand and Silt zone on the south-east bank (zone 5), four of which are in fresh condition indicating that they are in or near their point of initial discard so, in contrast to the Early Stone Age, it is unlikely that the Luangwa River has transported them far. Their presence and fresh condition was unexpected and at first is difficult to explain, as the sediment is relatively recent. It is therefore likely that either the artefacts have been included in the deposits from a very local origin, or that recent erosion in the area, which is more pronounced than it is in zone 4 on the opposite bank (noted in appendix 6.1), is exposing fresh artefacts from a Middle Stone Age land surface that has been buried by sand and silt overbank deposits from the Luangwa River.

In general, Middle Stone Age artefacts in the Cobble zone (7) are in better condition than those of the Early Stone Age – there are two likely reasons for this: the first is that there has been less time for Middle Stone Age artefacts to suffer wear, and secondly, the alluvial fans in the area might have been less active. A reduction in activity in the alluvial fans could be a result of a number of factors, including fewer flash flood events, exhaustion or redirection of the source material for the fans, or a slightly wetter glacial phase with an associated increase of stabilising vegetation (see section 3.4). Although the excavated section at SL8 reveals that the deposits were certainly still accumulating throughout the Middle Stone Age, it does not rule out the possibility that fan activity was comparatively reduced, presumably with a slower rate of accumulation. Either factor could be likely, or they might both be contributory elements. The fresh nature of the artefacts indicates that there could be residual patterns of behaviour in the landscape. Although the sample size is small, the composition of the artefact samples and some of the geomorphological observations suggest that there is a genuine residual pattern in the archaeological record in these areas. In the Middle Stone Age there are two concentrations of artefacts to consider, both of which are in the cobble landscapes: the first is the relatively high number of artefacts on transects GR2, GR4, and

GR7 in the Cobble zone (7), and second being on transects MM2, MM3, and GA1.9 in and near the Cathedral Mopane zone (14).

The concentration of artefacts on transects GR2, GR4, and GR7 in zone 7 might be a reflection of geomorphological processes. The Cobble zone has large areas where there are low densities of cobbles lying directly on Karoo bedrock (appendix 6.1), which might represent areas where finer material from the fan sequences has been removed by recent erosion leaving lag deposits (section 6.4). Transects in these areas have relatively low numbers of artefacts from the Early Stone Age, and none from the Middle Stone Age (GR1, GR3, and GR5). It is therefore not possible to determine in this study whether these areas, devoid of cobbles and Middle Stone Age artefacts, reflect features present during the formation of the Cobble zone, or are due to later erosive events that have completely removed parts of the alluvial fans. The latter is more probable considering the erosion seen in the area today. Therefore, the higher density of artefacts on some of the Cobble zone transects cannot be assumed reflect Middle Stone Age use of the landscape in itself. However, three of the four artefacts on GR7 are Mode 3 prepared core flakes knapped in chert (appendix 7.9). There is a corresponding chert source on GR7 noted on the transect recording form, which consists of blocks of chert up to 20x10x10cm in size recorded over a $100m^2$ area. These blocks are not rounded, and are relatively large compared to the cobbles; it seems likely, therefore, that they had a local origin, possibly from chert veins eroding from the Karoo bedrock under the Cobble zone. Likewise the unrolled condition of most of the Middle Stone Age artefacts suggests that they are in a primary context. In the second concentration of artefacts (transects MM2, MM3, and GA1.9) five out of the eight artefacts are in fresh condition, signifying that the majority are in a primary context. There are also a high proportion of Mode 3 chert artefacts suggesting that, similarly to GR7 above, there may be a source of chert in the locality. Although no blocks of chert were identified on the transect recording forms for the Cathedral Mopane zone (14), and the data from the nearest clast count does not contain any chert in the sample (count 31), there is a small amount of chert in clast count 16 approximately 1.5km away, and in the adjacent Kakumbi Spring zone (16) chert does outcrop in large quantities, so chert is available relatively locally.

In these two areas the sources for chert visible today, which were presumably being incorporated into the cobble deposits during the Middle Stone Age, provided a draw to these areas of the landscape. The evidence is not as strong for a raw material source in the Cathedral Mopane zone (14) itself, and the area possibly held another attraction, not now evident, to which raw materials or artefacts were brought. Whichever is the case, these two groups of artefacts suggest that Middle Stone Age hominins had a definite preference for using chert to manufacture the more complex tools in their repertoire, and that there is a genuine residual behavioural pattern in the landscape as a result. Where chert was not readily available the more specialised Mode 3 technologies were not practised. A low sample size, and the fact that the landscape had not stabilised in either of the zones containing cobbles or those near the Luangwa River, makes it impossible to verify this pattern or determine its nature in any further detail.

7.5. The Later Stone Age (Mode 5) Material and its Relationship to Geomorphological Processes

7.5.1 Description of the distribution of Later Stone Age material

7.5.1.1 General distribution of artefacts.

The Later Stone Age material (figure 7.12), in contrast to the Early and Middle Stone Age, is present on most of the zones in the survey area, and most transects have 1 – 10 artefacts present. There are, however, some outliers to this general pattern; no Later Stone Age material is found on the Recent Cut-off, Dambo, and Sand Bank zones (3, 15, and 11), whereas four of the seven transects in the Cobble Zone (7) contain greater than 11 artefacts. The highest counts in the survey overall were from transects NH7, GC2, and GR7: 34, 44, and 34 respectively. There are five empty transects on the Nchindeni Hills that appear to be randomly distributed in the zone, however closer examination of the data reveals that factors such as elevation and slope angle explain the distribution of most of these transects (see section7.6). There were also no artefacts collected on the two transects furthest north in the Chipembele Ridge Sands zone (MI3 and MI4, zone 13).

During the survey it was noticed that much of the Later Stone Age material was either collected from discrete scatters, or from a number of small adjacent scatters. To enhance the visibility of any patterns that might be present, the 100 x 500m transects and the data was broken down into 100x100m blocks (figure 7.13). The distribution shows that there are clear clusters of artefacts on the three transects with the highest artefact counts: NH7, GC2, and GR7 (spread across two blocks on GR7). There are also some lower density clusters (11 – 20 artefacts) on three transects in the survey: GA1.4, GR6, and NH3.

7.5.1.2 Artefact condition and raw material.

The Later Stone Age artefacts are almost all fresh (figure 7.14), with some showing a slight degree of abrasion, and only one recorded with moderate wear. The majority of the abraded artefacts occur in the Cobble zone (7), or in other zones with cobbles (8 and 14). The primary raw material used in the Later Stone Age across most of the survey area is quartz (figure 7.15). The transects in the Nchindeni Hills (zone 6) are frequently the exception to this where fine quartzite forms the primary component of the sample, with the exception of NH7, where quartz remains most common. Chert is usually the second most commonly used raw material in the Later Stone Age, but overall is used proportionally less than in the Middle Stone Age. In particular there is a marked difference in the proportion of chert artefacts found on transects MM2, MM3, and GA1.9 (figure 7.15), where only three of the fourteen artefacts are chert (21%), compared with six out of eight artefacts in the Middle Stone Age (75%). Similarly, chert forms a relatively minor constituent of the raw materials used on transect GR7 in the Cobble zone (38% or 13 of 34 artefacts), where the majority of artefacts are quartz and quartzite (72%). Chert is even less frequent on GR2, GR4, GR5, and GR6 (13%, 11%, 7%, & 6%, sample sizes 22, 9, 13, & 16 respectively (appendix 7.7)).

7.5.1.3 Artefact types present in the Later Stone Age.

The distribution of artefact types found in the Later Stone Age is illustrated in figure 7.16. The proportion of flakes collected across the survey area appears to be uniform, usually representing 35-50% of the sample. Cores are also present on most transects and form around 10-45% of the sample. Split cobble cores are the third most common artefact type (appendix 7.8). Where sample sizes are larger, a small proportion of each collection is usually made up of a wider variety of artefact types that can include segments, blades, and bipolar cores, along with undiagnostic features of Later Stone Age stone working such as chunks and broken flakes. There are two transects that are notable exceptions to the broad pattern of large proportions of cores and flakes. The first of these is NH7 in the Nchindeni Hills where split cobbles form the highest proportion of artefacts, with flakes forming just over a quarter of the artefact sample, the remainder comprising a hammerstone, a crested flake, and a prepared core. Transect MI1 is the second exception where the greatest proportion of artefacts are also split cobbles, with the remainder of the sample comprising flakes, a broken flake, a chunk, and a segment. In general there appears to be no particular pattern in the raw materials used for different artefact types, with the exception of the split cobbles which are almost all made from quartz (94%).

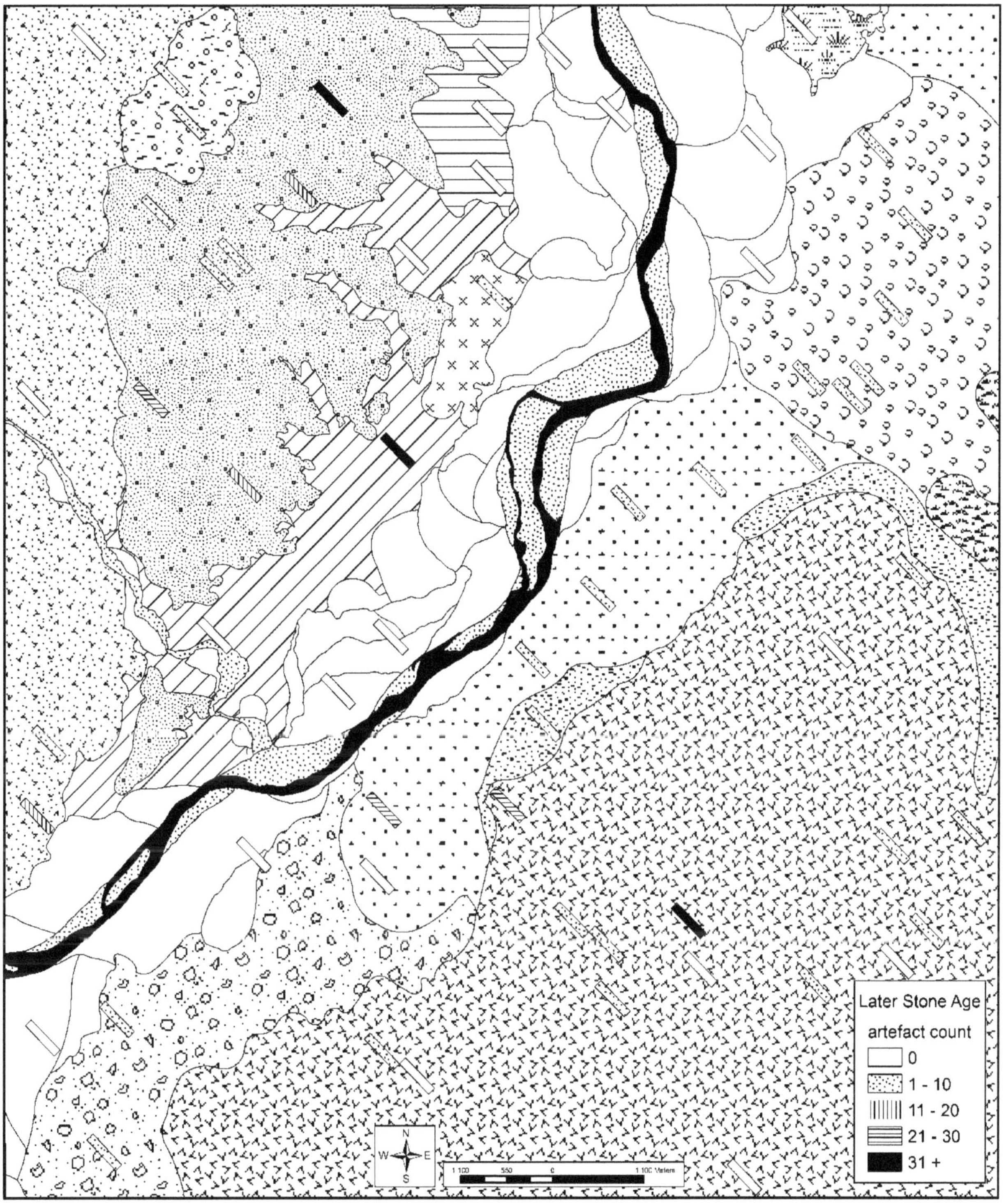

Figure 7.12: Later Stone Age artefact density illustrated by equal grades with the exception of the last category (31+), as only one count with more than 40 artefacts was present (GC2 with 44) an extra grade was not used. The raw data for artefact counts is presented in appendix 7.3.

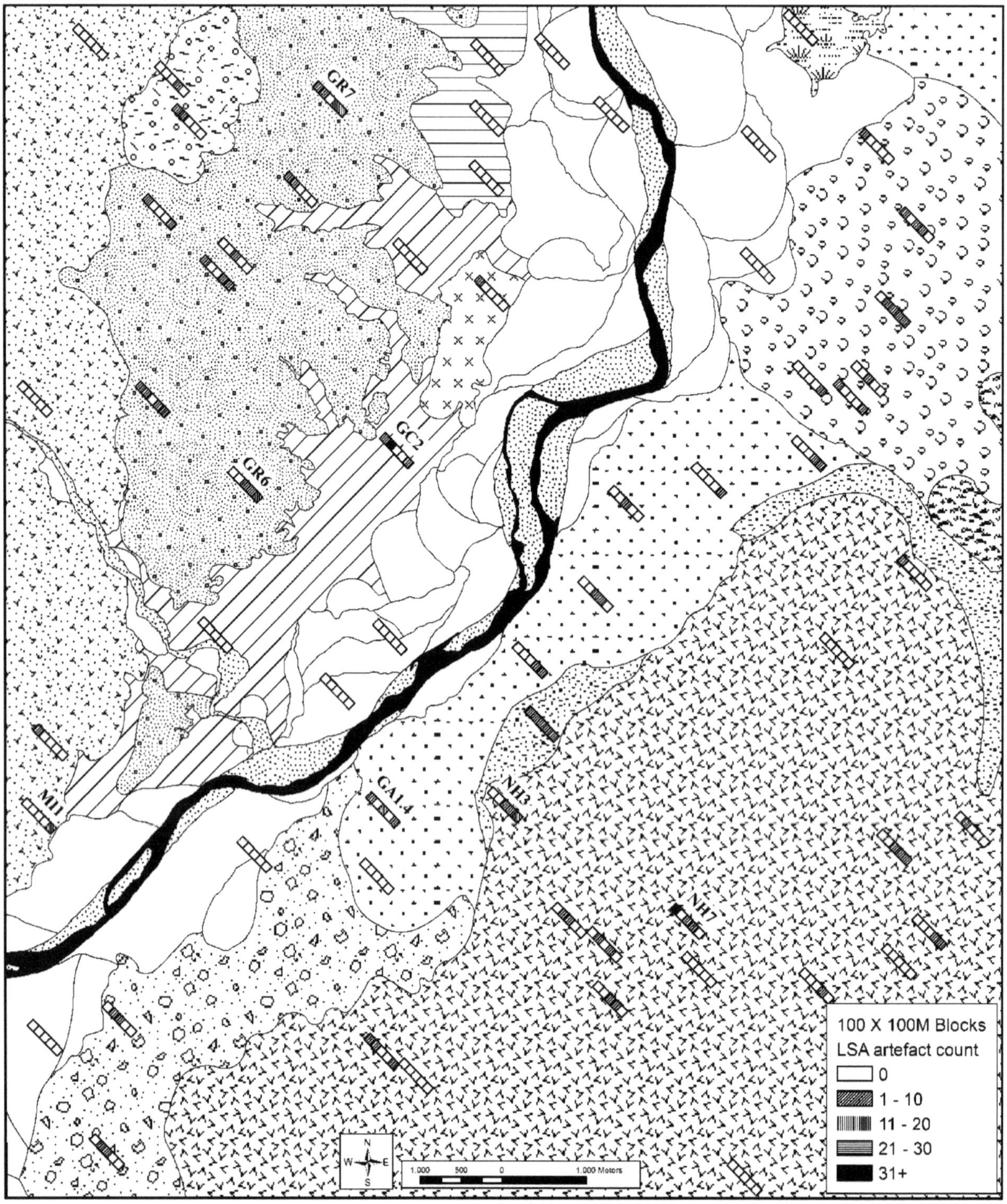

Figure 7.13: Later Stone Age artefact density split into 100 x 100m survey blocks illustrating the concentrated nature of some of the scatters. Transects discussed in the text are labelled. The data for the artefact counts in each block is given in appendix 7.10, the individual blocks were identified by an extra letter code, A to E, for each transect with A starting at the southern block of each transect.

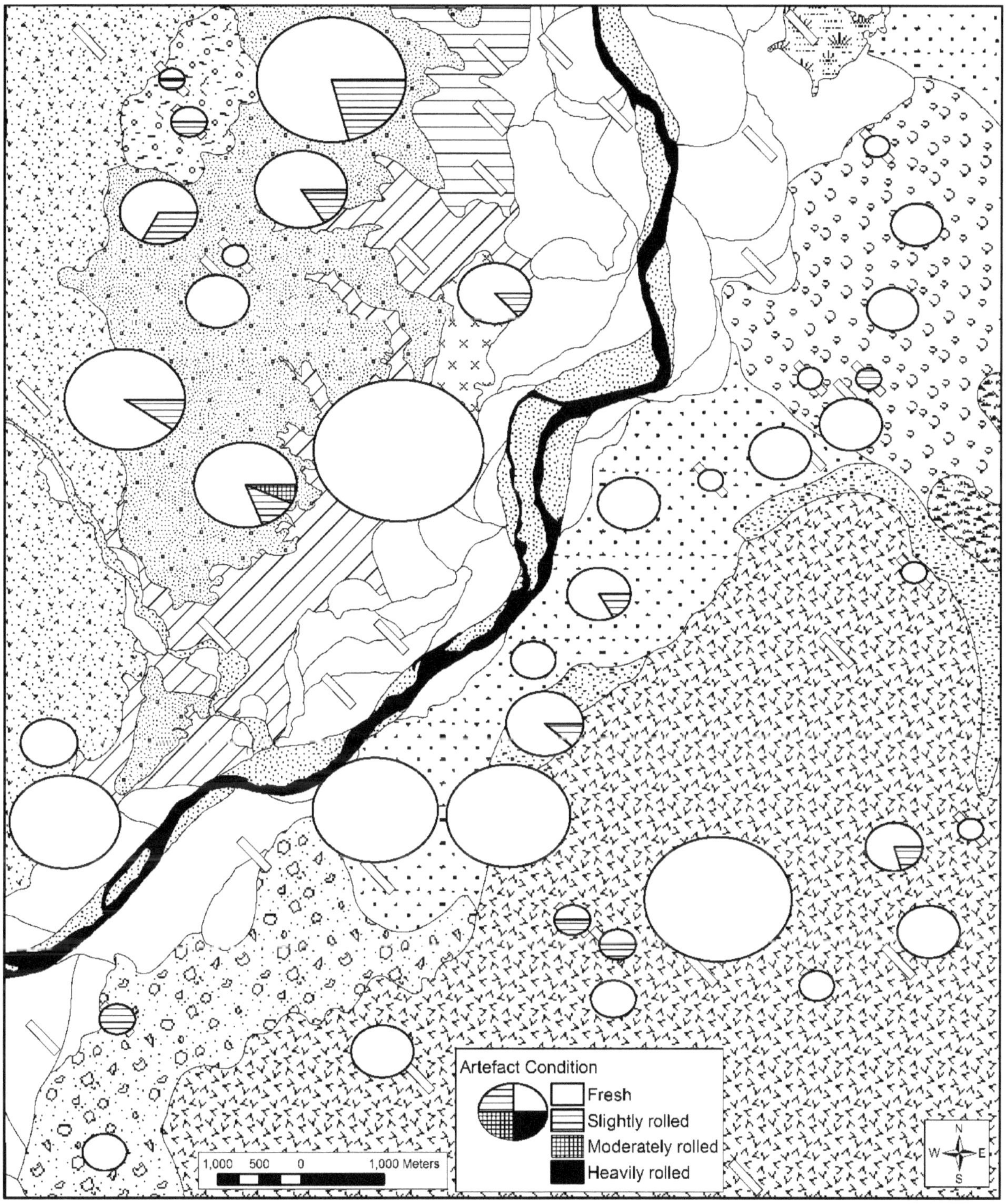

Figure 7.14: Pie charts illustrating Later Stone Age artefact condition; the pie charts are scaled by the sample size on each transect. The raw data is presented in appendix 7.6.

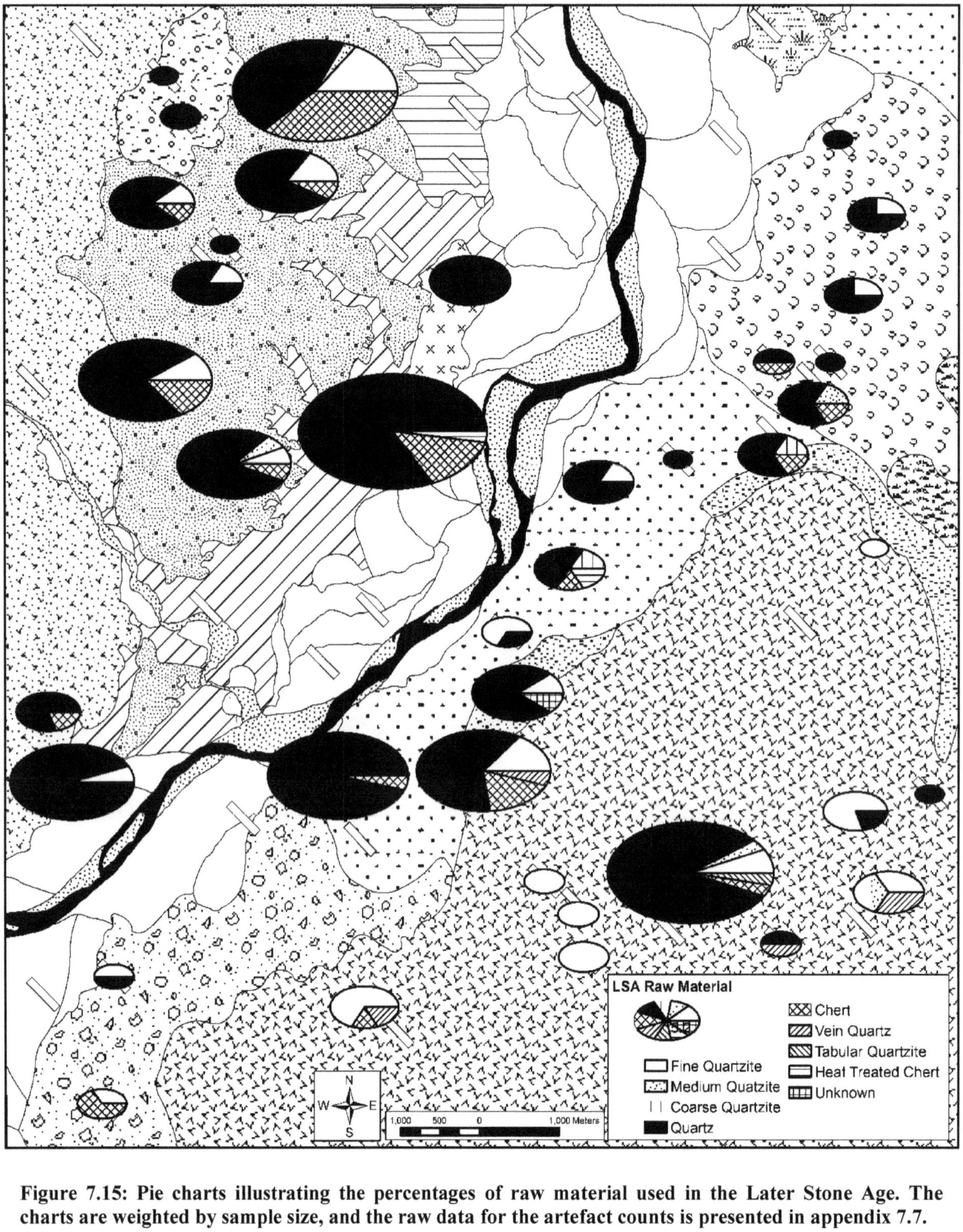

Figure 7.15: Pie charts illustrating the percentages of raw material used in the Later Stone Age. The charts are weighted by sample size, and the raw data for the artefact counts is presented in appendix 7.7.

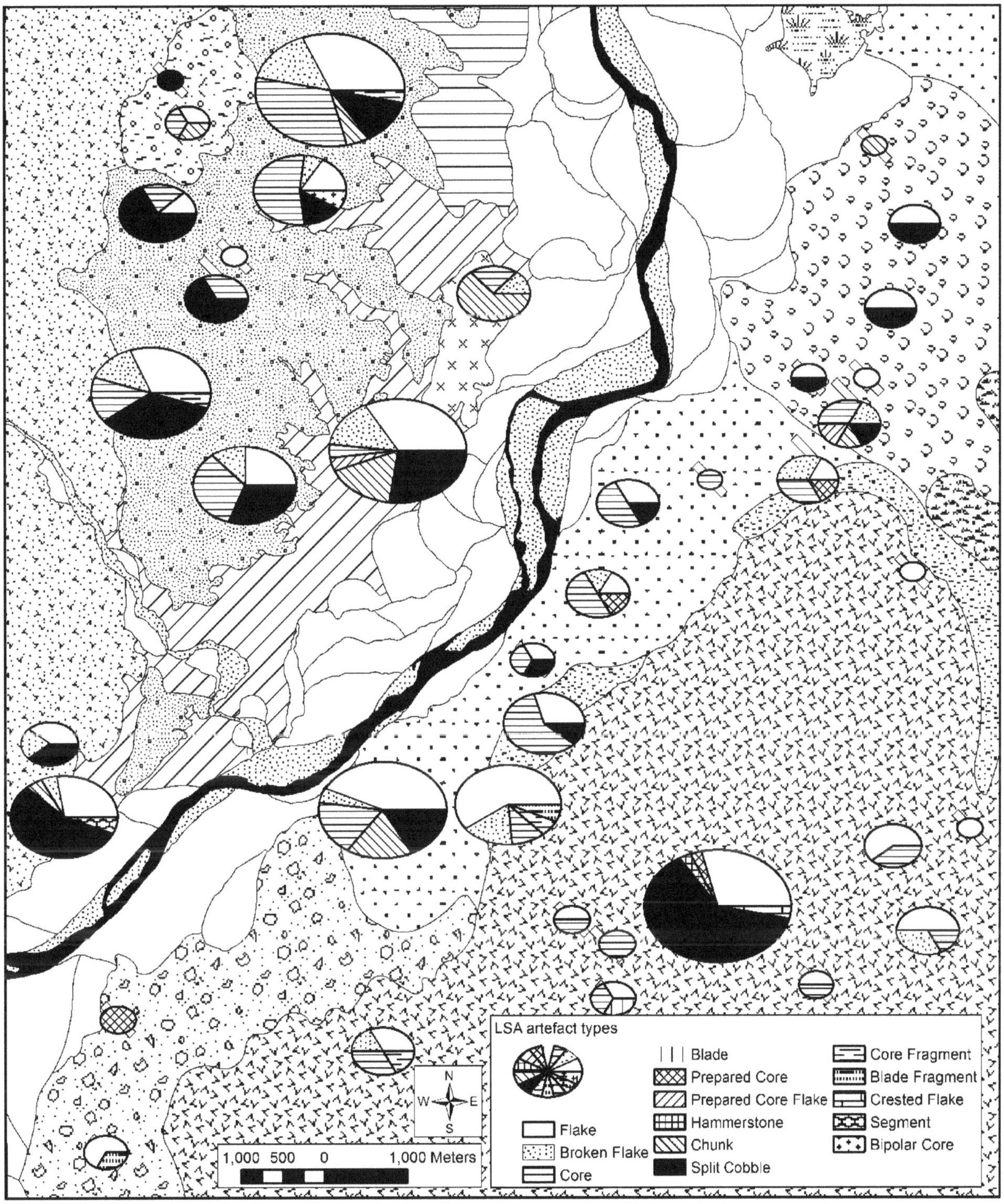

Figure 7.16: Pie charts illustrating artefact types identified in the Later Stone Age. The charts are weighted by sample size, and the raw data is presented in appendix 7.8.

7.5.2 Interpretation of Later Stone Age artefact distribution

The lack of abrasion on the artefacts combined with the wide distribution in the landscape suggests that the archaeological material was discarded across a primarily stable landscape. This is consistent with the geomorphological interpretation made in chapter 6 that states that the landscape has been relatively stable since the Last Glacial Maximum, with the possible exception of some landscape instability on hillslopes similar to that seen in Chipata. Throughout the Holocene, after this instability, it is likely that the land surface has remained virtually unchanged. It is possible, due to the persistence of the hunter-gatherer lifestyle in the region until the 19th century that the distribution in the landscape represents a palimpsest of the last two or three centuries of activity, rather than one representing the entire 20,000 years of the Later Stone Age. This is particularly likely on the Sand and Silt zone on the south-east bank (zone 5) where, in parts of the zone at least, the rapid accumulation of sediments (for example at SL7 on the Chowo River, section 7.7.2) would mean that the land surface is relatively recent. The artefacts that demonstrate abrasion may have been moved by recent stream activity or, due to their smaller, thinner sizes, may have been abraded by the trampling action of animals in the area in recent history. This is most probable in the cobble zones where the majority of abraded artefacts are found (zones 7 & 8), where in particular elephants use their trunks and feet to dig small holes up to 30cm across and 10cm deep in the cobbles (*pers obs*), presumably in search of food in the form of roots and tubers as water is not present below the cobbles. This is likely to have caused wear because the cobbles in the cobble zones are relatively large when compared to most Later Stone Age artefacts, and will act as grinding stones when the ground is disturbed in this manner.

The key exception to the general stability of the landscape is the Recent Cut-off zone (3) where the channel of the Luangwa has avulsed regularly. In the adjacent, and geomorphologically related, Dambo zone (15) there are also no Later Stone Age artefacts found. Some surfaces in zone 3 have already been identified as 20th century (see figure 6.1 and section 6.2.2), and these are thus not expected to retain any undisturbed archaeological material. The lack of any material on the transects indicates that the Later Stone Age record has been completely removed from zone 3, however, this is slightly surprising as the Later Stone Age persists alongside the Iron Age in the Luangwa Valley to the 19th century, and Iron Age material is found on parts of zone 3 (see section 7.7 below). Furthermore, Later Stone Age material was recovered from a transect in this zone in the first season (see section 7.9), and radiocarbon dates from a small excavation (SL11) revealed that some surfaces in zone 3 could be as old as 400 years (Barham *pers comm*). It is possible that the lack of Later Stone Age material in zone 3 is a result of a combination of genuine cultural patterning and geomorphic affects. Overall, zone 3 might have been less intensively used than other parts of the landscape in the Later Stone Age, and subsequently the rare scatters have either been destroyed during very recent avulsion, or missed by the survey units.

Elsewhere the relative stability of the landscape suggests that the artefact scatters are more certain to derive from c-transforms, and will therefore represent genuine behavioural residues. The greater proportion of artefacts on transects in the Cobble zone (7) is not unexpected as the large amount of raw material in this zone would have been an attractive resource to hunter-gatherers. Independent sample t-tests were used to confirm the statistical validity of the increased artefact counts that had been observed. This test compares mean counts between two groups of variables, in this case the difference between the numbers of artefacts present on transects in cobble zones 7 and 8 and the rest of the survey area. The result of this test is given in appendix 7.11, and it revealed that there is a 95% degree of certainty that there is a statistically significant increase in the number of Later Stone Age artefacts present on the cobble zones. When the Cathedral Mopane zone (14) is included in the cobble sample (appendix 7.12) the result is an insignificant increase in the number of artefacts in the cobble zones (7, 8, and 14), suggesting that as fewer cobbles exposed in zone 14 made it a correspondingly less attractive resource. Statistical tests were also undertaken to determine whether other zones had statistically higher numbers of artefacts present on them as, for example, the Nchindeni Hills zone (6) might have been an attractive location due to the presence of granite inselbergs suitable for habitation. The results of the tests are given in appendices 7.13 – 7.15 and they reveal that in the Nchindeni Hills zone (6) and the sand and silt zones (4 and 5) there are not statistically higher numbers of artefacts, indicating that, based on artefact distributions, they were equally used by Later Stone Age hunter-gatherers. Other zones were not tested as they had too few or no artefacts on them, and were clearly less used or altered by geomorphic processes (the Recent Cut-off zone (3) for example).

The concentrated scatters of material present on some transects (figure 7.13) relate to areas of repeated or more intense activity, rather than the low-density off-site material collected on most transects. The high count on NH7E can be described as a site due in part to the comparatively high concentration of material, as well as to the presence of a small rock overhang within the survey block (c. 5m away)(figure 7.17). This overhang was located on the edge of a large hill of basement granite and granite boulders, and would have provided a small shelter to the inhabitants of the area, accounting for the focus of activity. At this locality the majority of artefacts are split cobble cores for which water worn cobbles of quartz were the raw material. This material must have been brought at least a short distance to the site, as stream beds at that altitude in the Nchindeni rarely have small water worn cobbles and primarily constitute large angular cobbles and boulders that are only moved in flash flood events. There is a similar, but lower density, concentration of artefacts on transect MI1 that also appears to relate specifically to a site. It is similar to NH7 in that there are an unusually high proportion of split cobbles in the sample (figure 7.16). Another unusual factor in the scatter at MI1 is that the concentration is directly related to Iron Age finds in the same collection (see section 7.7, figure 7.24 below), and though Later Stone Age and Iron Age artefacts are often found on the same transect, they are usually from different scatters and do not appear directly related.

Figure 7.17: Rock overhang on transect NH7, below which (on the flatter ground out of shot to the right) the artefacts were collected on block NH7E. There is no scale, but three people could sit comfortably under the overhang.

The high count on GC2 is located in the Sand and Silt zone on the north-west bank (4) (figure 7.13) where there is no raw material for stone tool manufacture. MI1 is similarly located on a surface where there is no raw material. The closest raw material source is on the cobbles in zone 7, which in the case of GC2 is 300m away, and for MI1 900m away. The presence of 15 pieces of shatter indicates that knapping activities were undertaken at these localities, rather than the discard of finished artefacts that have been produced elsewhere (appendix 7.16) (although shatter is undiagnostic, its association with only Later Stone Age knapping suggests that it derives from this period). It is likely therefore that at least some raw material was brought to both the GC2 and MI1 localities for knapping, as there is no raw material present in the vicinity. On transect GR7 there is a concentration of Later Stone Age artefacts in the same locality as the chert boulder sized material noted in section 7.4.2 in relation to the Middle Stone Age chert artefacts. In the Later Stone Age, there is a higher proportion of chert artefacts on GR7, but this probably reflects opportunistic use of the chert source as overall it is used less, and is not extensively transported across the landscape. It is clear though that the finer grained and crystalline raw materials are preferred over coarser material that, judging by the knapping experiments conducted, is a result of the more predictable knapping qualities of these materials. This is the case in all the environments in the landscape that Later Stone Age hunter-gatherers occupied, and in some areas, demonstrated by the data on transects GC2, MI1, and NH7, these raw materials would have been specifically collected, and in some cases transported at least a short distance for use.

No Later Stone Age artefacts were recovered from the two transects further from the Luangwa River on the Chipembele Ridge Sands (MI3 and MI4, zone 13), and none were recovered from the Sand Bank zone (11). In the case of both zones this could either represent a genuine behavioural choice of landscape use or, alternatively, an effect of the vegetation stabilising the soils. The scarcity of raw material in the areas is an unlikely explanation, as it has not prevented Later Stone Age hunter-gatherers using other parts of the landscape that lack raw material sources (zones 4 & 5). The vegetation in these two areas is dense and continuous, and though areas of zones 3, 4, and 5 are equally densely vegetated, the cover is usually discontinuous. As noted in section 5.5.6, dense vegetation rarely causes an issue with ground visibility as most grasses have died off in the dry season, however the vegetation roots do stabilise the soil, which in other areas is often deflating. As a result, and somewhat ironically for this survey, the landscape in parts of zone 13 and zone 11 may be too stable and thus any artefacts present may not be exposed. Alternatively, the vegetation itself may be a limiting factor on the activity of Later Stone Age people, with the lack of artefacts reflecting a genuine lack of activity in these areas. This suggestion is supported by the fact that the vegetation in each area is unique to the survey area (though not to the valley), in that zone 13 contains the only miombo woodland, and zone 11 the only continuous dense thicket. Presuming that this pattern of vegetation has remained relatively unchanged during the last portions on the Later Stone Age it is possible that these vegetation types were somehow unsuitable for Later Stone Age people. Without subsurface testing it is not possible to say for certain whether this, or the stability of the soils, is the main cause for the absence of evidence for occupation.

7.6. Detailed Examination of Factors Affecting Later Stone Age Artefact Distribution and Modelling Later Stone Age Land-use

7.6.1. Introduction

It is clear that Later Stone Age artefact distribution more accurately represents the pattern of land-use of Later Stone Age hunter-gatherers than the geomorphologically altered distributions of the Early and Middle Stone Age datasets. The Later Stone Age persisted alongside the Iron Age in much of Zambia until the 19th century (Musonda 1987:156), and a model is constructed in this section so that the land-use of the two groups can be compared. To undertake this, the distribution of Later Stone Age artefacts is examined more closely with regard to elevation, hillslope angle, and distance to water sources. These factors, along with observations of the known density of artefacts on the transects, are then used to create a predictive model in GIS of Later Stone Age artefact density across the survey area, which can act as an indicator for land-use intensity. Section 4.3.3 above outlined the theoretical issues regarding predictive modelling, the aims of the predictive model produced here, and the basic procedures behind the weighted model technique. A detailed description of the weighting used to create this model is given here in sections 7.6.2and 7.6.3. The model for Iron Age land-use is developed in section 7.7 after the data for the Iron Age artefact distribution has been described and analysed.

7.6.2. Elevation, slope, and distance to streams

In the Nchindeni Hills zone (6), Chipembele Ridge Sands zone (13), and Sand Bank zone (11) a number of transects were devoid of any artefacts. Considering the clustered nature of some of the artefact distributions noted above this might be considered to be a consequence of the survey design, where transects have missed the more discrete locales of activity present in an area or, in the cases of zones 11 and 13, a consequence of stable soils. Overall, the majority of transects do detect a low-density background scatter of artefacts in the landscape, so it is unlikely that the survey transects missed all evidence of occupation. Although it cannot be determined with any certainty whether the stability of soils is the factor behind the lack of artefacts in zones 11 and 13, the absence of artefacts on transects in these zones and parts of the Nchindeni Hills is taken as evidence that these areas were not suitable or desirable for occupation.

Gradual erosion is the probable key factor behind the scarcity of Early and Middle Stone Age material in the Nchindeni Hills, and though it is possible erosion has removed material deriving from the earlier part of the Later Stone Age, it is less likely for scatters that survive from the most recent Later Stone Age. As noted above, there are empty transects that appear randomly distributed in the Nchindeni Hills (zone 6). To examine the factors that might affect the distribution of artefacts in more detail the transects are displayed in relation to the contours to illustrate elevation (figure 7.18), from which slope angle can be visually estimated. Steep slopes are clearly visible in the closely spaced contour lines in transects NH2 and NH15 that were presumably rarely traversed, and certainly would not have been suitable for habitation. The empty transect NH8 spans a hilltop, and though it is only 20m higher than the artefact rich NH7 nearby, it is bounded by steeper slopes. This suggests that the steeper hillsides made the flat hilltop less suitable an area than NH7, which has a low background scatter in addition to the concentration of artefacts at the base of the granite outcrop. The lack of any artefacts on NH9 and NH12 does not appear to be related to elevation, and is puzzling as transects in similar environments surrounding them do record scatters of material. The distribution of the random sample resulted in no transects being placed at the highest altitudes of the Nchindeni (up to 880m), with the result that NH15, as well as being on a steep slope, was the highest elevation transect in the Nchindeni Hills at 720m. It is therefore not possible to know the nature of the archaeological record in these relatively small areas of the survey area. Based on the observations of transects on steep slopes (NH2, NH8, and NH15), it is unlikely that the highest hills in the survey would have been a regularly utilised part of the landscape as they are surrounded by steep slopes. Therefore, though not tested in the field, the data suggests that slope angle and the isolated nature of the highest hills would have made these elevations undesirable at this point in the valley. In fact, because of the difficulty in climbing the highest hills, it is unlikely that a transect located on one of them could have been surveyed, due to the possibility of heat exhaustion on the climb.

It is difficult to determine visually from figure 7.18 whether or not distance to water sources is also a factor affecting Later Stone Age artefact distribution. To aid interpretation 250m and 500m buffers were applied to the streams in the entire survey area to determine whether survey blocks closer to potential water sources were more likely to contain artefacts (figure 7.19). Examination of the map suggests that distance to water does not seem to be a factor, as equal numbers of transects contain artefacts inside and outside of the buffers. To assess this observation more accurately a binomial statistical test was undertaken to ascertain whether the 100m x 100m blocks are more likely to contain artefacts if they are close to potential water sources (appendix 7.17). The test revealed that there is no significant increase in the percentage of 100 x 100m blocks containing Later Stone

Age material in the 500m buffer, though on balance this is not entirely surprising, as there are relatively few places in the landscape that are further than 500m away from streams (figure 7.19). There is, however, a significant increase in the percentage of survey blocks containing artefacts within the 250m buffer, suggesting that distance to water sources is a factor affecting Later Stone Age land-use. This also provides an explanation as to why NH9 and NH12 do not have any Later Stone Age material, as they both occur outside the 250m buffer the land surfaces they cover were less intensively used.

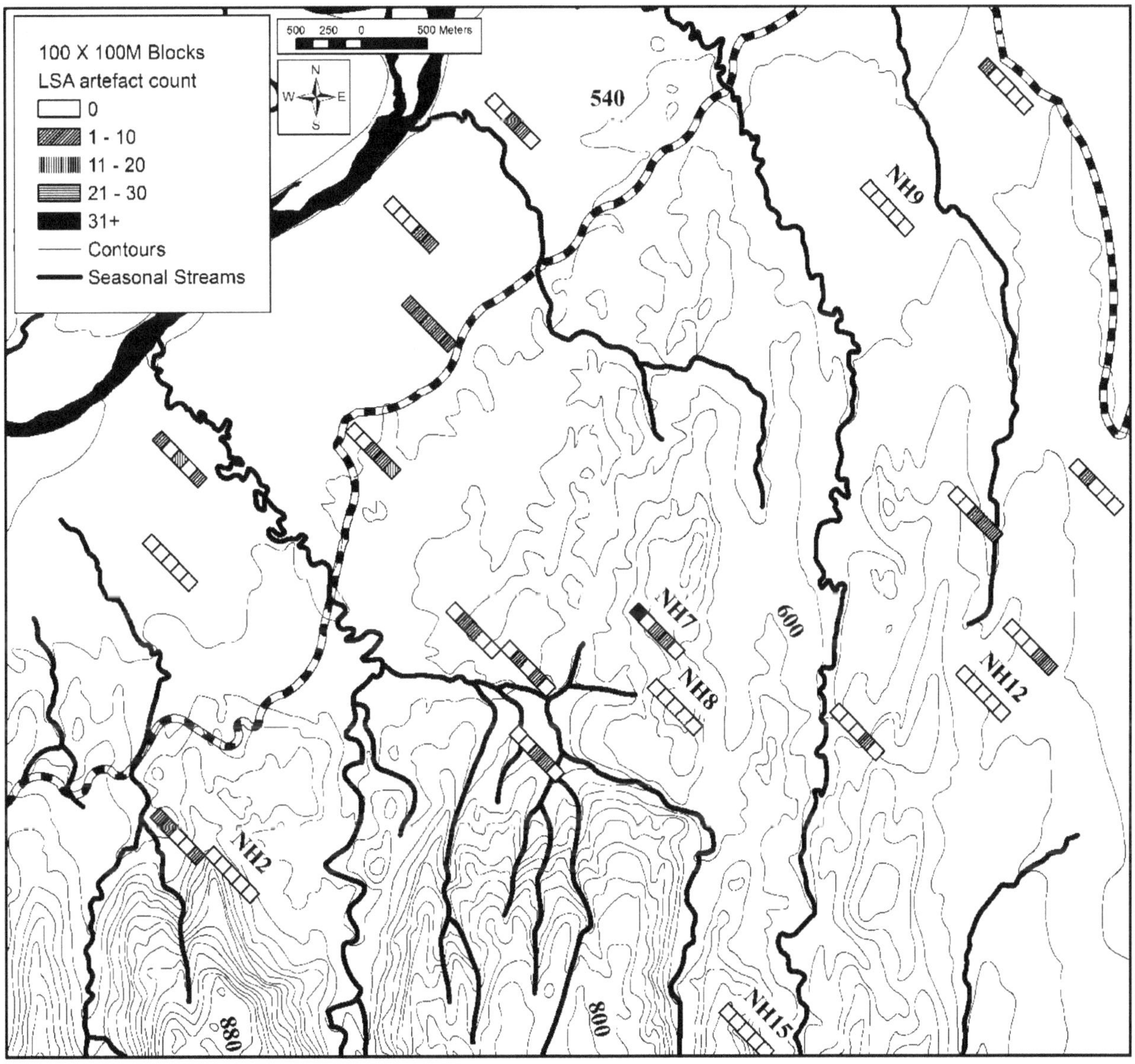

Figure 7.18: A map of the Later Stone Age distribution of artefacts in the Nchindeni Hills illustrating the relationship with elevation and the seasonal streams (contours at 20m intervals with some labelled to provide reference points). Transects discussed in text are labelled. Appendix 7.10 provides the artefact counts for each survey block (blocks numbered A – E from the southern most point of each transect). Contours digitised from Zambia Survey maps 1331B and 1331B (copyright Zambia Government 1972).

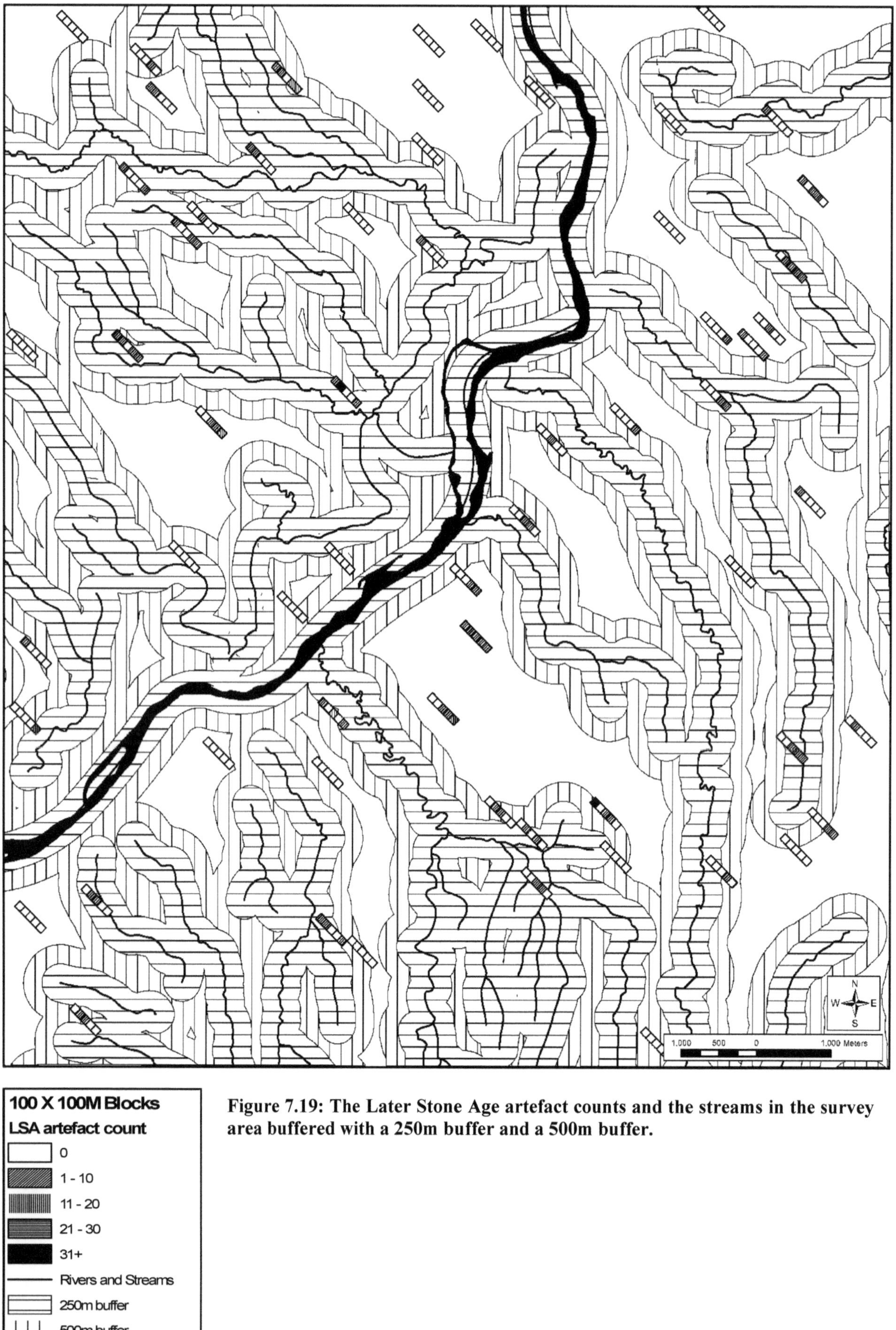

Figure 7.19: The Later Stone Age artefact counts and the streams in the survey area buffered with a 250m buffer and a 500m buffer.

7.6.3. Building the model

In this section, the selection procedure for the weighted map-layer approach is outlined. In sections 7.5 and 7.6.2, the key environmental factors that have influenced the distribution of Later Stone Age hunter-gatherer activity have been determined. These include the increase of finds related to raw material sources on the Cobble zone, and the decreased number of finds that occur with increased angle of slope, elevation, and distance from water sources. The comparatively few geomorphological factors that have affected Later Stone Age artefact distributions have also been identified, and are limited to the Recent Cut-off zone (3). These observed factors are combined in GIS to create a predictive model of the intensity of Later Stone Age activity in the survey area. The theoretical approach behind the model's construction is slightly different to that which is usually undertaken (see section 4.3.3), and the aim of this model is not to predict the location of sites, but to map the probable land-use intensity of Later Stone Age hunter-gatherers based on the observed relationship between the density of the off-site material and the landscape factors identified here. A predictive model is made by creating raster files of the necessary data, and then reclassifying, overlaying, and weighting map layers to create a new summarised layer representing the model (Brandt *et al.* 1992:271). Four factors were used to create the model: geomorphological zone, elevation, distance to streams, and slope angle. The model has to be based on a number of raster layers, and is produced as a new raster layer. These at their simplest represent the absence or presence of a feature in the landscape as a grid of equal sized tiles (Wheatley & Gillings 2002:167). This is a necessity because continuous surfaces are required by GIS to enable it to conduct the calculations necessary for producing the model.

The first task was to convert the necessary files representing the geomorphological zones and elevation into raster files, and then derive the extra datasets needed, which in this case was the slope angle and distance to streams. The geomorphological map was simply converted from a vector file (shape file) into a raster file. The elevation data had been digitised from survey map contour lines, which were therefore stored as linear features and were converted into a continuous surface for elevation using the inbuilt topography to raster function in ArcView. The tile size chosen for all the layers was 50m and this formed the tile size for the rest of the analysis; this choice seemed sensible based on the probable accuracy of the survey maps and the size of the survey blocks (100x100m). The derived layers required were slope angle and distance to streams. Distance to streams was derived using the spatial analyst function in ArcView, which provided a continuous surface representing the distance to the nearest streams and rivers that had been mapped from the survey maps and aerial photographs. Slope angle was also derived using the spatial analyst, and the input raster in this case was the continuous elevation surface derived from the contour lines.

The next step was to reclassify each raster layer to a common scale. This required each category in each map layer to be assigned a number on the common scale, which related to the positive and negative correlations of the archaeological distribution. The common scale used for this model was one to five, where one represented the least desirable and five the most desirable conditions for Later Stone Age activity (see table 7.2). The weighting in each layer was assigned subjectively in a methodology similar to Brandt *et al.* (1992:273-276), as regression analyses could not be used in this instance (see section 4.3.3). However, the statistical tests undertaken above to determine the nature of some of the archaeological relationships, could be used to provide guidelines to assigning the weighting. The raster calculator function in ArcView was then used, by which ArcView sums the weights of the layers on a cell-by-cell basis to create the model. The model is effectively a composite of the layers included into the model where highest values represent areas that are the most intensely used parts of the landscape, and the lowest values represent rarely occupied areas.

7.6.4. Later Stone Age modelling results

The results from the model are illustrated in figure 7.20. The map presented in figure 7.20 is based on the raw calculations undertaken by the inbuilt raster calculator function in ArcView, the result of which is a composite map with a scale of 10 graded colours. Considering that the inputs were to some extent derived subjectively, and that transects were widely distributed, it is likely that the automatically generated scale of 10 overemphasises the amount of variability of land-use intensity that could reliably be derived from the datasets. To simplify the model in response to this, the map was reclassified on a scale of one to five in line with the scale of the inputs (figure 7.21). In this simplified model areas of low intensity, scoring one or two, usually have no artefacts present, whereas the survey blocks with the highest counts usually occur in the areas scored as four or five. Transects that contain denser or more continuous artefact accumulations occur in areas of the map where greater intensity of land-use is predicted, and transects that contain no artefacts relate to areas of expected low land-use intensity. Based on this visual assessment the modelling exercise has proved successful. A notable outlier to the predicted pattern is NH7 (figure 7.21), which contains one of the highest artefact counts despite being located in a lighter area of the map. The NH7 locality is present due to the rock overhang, a feature that would probably have been attractive for the shelter it provided, and therefore relates to a site on NH7 rather than the off-site land-use that is primarily being modelled. This indicates that the model cannot account for smaller unique factors in the landscape that have provided attractions to certain locales. The cave site at SL3, 'Poachers' Cave', is located in a darker area of the map, which is due to the low angle of slope and the proximity of the area to the Kafunta River. In this case the model does not specifically reflect the presence of the cave site, which was in itself a centre of activity in this part of the landscape. Encapsulating these individual factors in the model would be possible, but would require a very detailed map of all the granite inselbergs in the survey area, which is information simply not available to this project.

Scale: 1 = Least suitable – 5 = Most suitable.		
Layer and justification for scale.	**Scale.**	**Category weights**
Geomorphological Zones. The weighting applied to each layer assumed that if there were no artefacts present on the zone then land-use is likely to be low and a score of 1 was given. The middle ranking of 3 was given where artefacts are present but the statistical tests revealed that there was no significant increase in the number of artefacts present. The highest score, 5, was given where the statistical tests revealed a significantly higher proportion of artefacts present.	1 2 3 4 5 6 7 8 9 10 11 12 13 14 15 16	1 1 1 3 3 3 5 5 3 3 1 3 2 3 1 1
Elevation. The weighting assumed that as artefacts were found at elevations below 640m that these areas were most suitable and were given a score of 5, the other three classes were then weighted accordingly, the least favourable class was not used as the data does not suggest a strong relation to elevation (see text)	498-575m 575-652m 653-728m 729-805m 806-881m*	5 5 4 3 2
Slope Angle. The steepest slopes have no artefacts present and were unsuitable for habitation so were given the lowest weighting, and the subsequent classes were graded equally, with the flattest areas receiving the highest score.	0 - 6° 7 - 12° 13 - 19° 20 - 25° 26° +*	5 4 3 2 1
Distance from streams. Statistical tests revealed that Later Stone Age material is more likely to be located near streams so areas closest to streams were given the highest weighting, and as distance increases the weighting was graded equally.	0 – 380m 381 – 760m 761 – 1140m 1141 – 1520m 1521 – 1900m*	5 4 3 2 1

Table 7.2: A table describing the weighting system used to create the predictive density model for the distribution of Later Stone Age artefacts.

* The classifications were set using the inbuilt functions of GIS to reclassify each layer into 5 equal intervals.

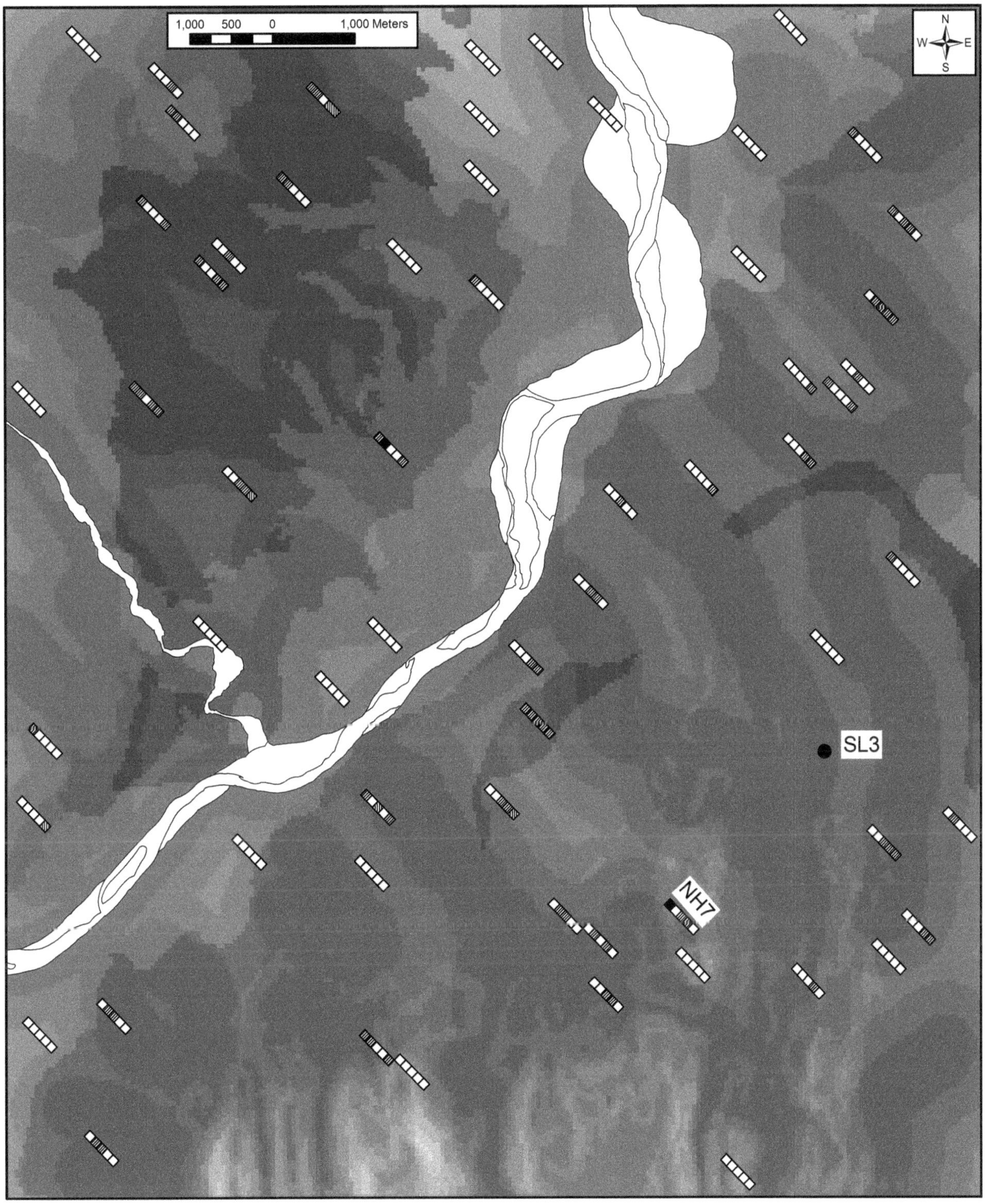

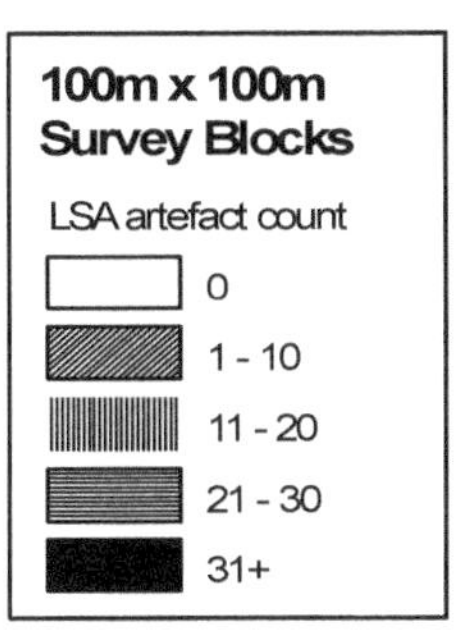

Figure 7.20: The Later Stone Age model of predicted land-use in the survey area, with the artefact counts overlaid. The dark areas represent areas of high intensity land-use, whilst the lighter areas low intensity land-use. The highest intensity land-use relates to cobble zones 7 and 8. The Luangwa River, its sand banks and the 20th century land surfaces are overlaid white to represent no expected land-use in these areas. NH7 is labelled, and Poachers' Cave is labelled by its site code SL3.

Figure 7.21: The simplified model of Later Stone Age land-use where only five colour grades are used. One represents the least used areas of the landscape, whilst five the most intensively used.

7.6.5. Testing the model

Where possible a model should be tested to determine its suitability in accurately predicting the location of archaeological material (Brandt *et al.* 1992:276; Connolly & Lake 2006:184; Warren & Asch 2000:9). In section 4.3.3 it was noted that developing a method of statistical testing was not possible for this type of model as the sample sizes in off-site landscape survey are too small in relation to the survey environment. However, it is still possible to test the model at least qualitatively, if not quantitatively by employing statistical analyses as is usually done. Although the model was constructed after the fieldwork, some of the patterns in the Later Stone Age artefact distributions that are analysed, interpreted, and modelled in this chapter were noticed during fieldwork. Extra transects were plotted and surveyed while in the field to provide a test for any models produced in GIS. In addition to this the survey data from the first season can also act as a comparative sample, though its use is limited by the different collection strategy used.

Figure 7.22 illustrates the model and the Later Stone Age data from the experimental transects, as well as the first season's transects in the survey area. In general, the experimental transects surveyed in the second season confirm the model's predictions, and in the Nchindeni Hills NH17 and NH18 are located in areas of medium/high (class 4) predicted land-use intensity and in each case they do contain artefact scatters on two of their survey blocks. NH19 was situated on a steep slope that was predicted to have low land-use intensity, and on the rocky hillside little soil and no artefacts were present. The location of NH20 was chosen for two reasons: it was situated near to NH9 because of the unexpected absence of artefacts in comparison to similar transects in the area, and it was also situated relatively close to SL3, Poachers' Cave (*c.* 100m away) to determine the nature of the off-site scatter near a known site. Surprisingly, no Later Stone Age artefacts were found on NH20, which is difficult to explain as a low number of artefacts are usually recovered from these areas, and indeed were seen nearby on the walk to the transect. NH9 and NH20 demonstrate that the off-site scatter is variable over short distances in a manner that is both difficult to explain and model. RS5, in zone 3, is located in an area of low land-use intensity (classes 2-3), and unexpectedly one block contains a relatively high number of artefacts, $n = 56$, this concentration was also related to a fragment of pottery (table 7.5). The transect was situated on a slight rise in the landscape, and this scatter would appear to relate to a site rather than an off-site scatter, which further indicates that the model is not useful for predicting locations of sites. Of the experimental transects, GR8 has the highest number of artefacts. This transect was surveyed across one of the larger cobble ridges on the edge of the cobble zone (SL8 was also located on this ridge). The high density of artefacts was predicted, but this cluster of artefacts might relate more specifically to a site than off-site activity, particularly as the cobble hill in this locality provides a good observation point for the game on the floodplain below (zones 4 & 3). This localised factor in the Cobble zone (7) is another factor, like the granite inselbergs in the Nchindeni Hills (zone 6), which cannot be included in the model at this stage without more detailed mapping of the locations of the higher cobble capped hills.

The two transects from the 2005 season are slightly more difficult to incorporate into the analysis due to the initial difficulty in designing a systematic collection strategy during the survey (see section 5.5.4). The data therefore only indicate what the coverage of material in each area is likely to be, and higher densities of artefacts must be treated with caution as they might reflect more intensive collection rather than a representative increase in artefact density. On LU1, a large number of the survey blocks contain artefacts on the Recent Cut-off zone (3), which suggests that this cut-off is old compared to other parts of zone 3. A small excavation undertaken as part of the PPE project yielded charcoal that was radiocarbon dated to 400BP, associated with microlithic quartz debitage and glass beads (Barham *pers comm.*). This is further evidence indicating that the overall absence of Later Stone Age material in zone 3 is related to recent geomorphic processes, rather than a genuine lack of occupation in this part of the landscape. The empty blocks in the centre of the transect's two arms that extend to the north-west relate to the Sand Bank zone (11) (outlined in black in figure 7.22), which is an area of predicted low intensity land-use, whilst on the cobble ridges beyond, a number of the blocks did yield Later Stone Age artefacts. There is also one block of a very high density of material on zone 3 (120+) and, despite the less systematic sampling of the first season, probably relates to a site rather than the background scatter of the area. The recorded density of the scatters in the Cobble zone (7) is probably lower than what is present in reality. This is because some figures are conservative estimates based on records made on the recording forms, necessitated by the limited time to survey this area in the first season, which resulted in fewer actual collections being made. Similarly, fewer collections were made on CHO1 than would have been in the second season, so the artefact density or number of survey blocks with artefacts cannot be taken as entirely accurate. Nevertheless four of the twenty-five survey blocks on CHO1 did contain Later Stone Age artefacts in an area predicted to have a relatively high density.

The experimental transects and the first season transects contained two concentrations of artefacts that are probably sites and are located in areas predicted to have low overall land-use intensity. This brings the total number of anomalous high-density areas to three: NH7, RS5, and LU1. In the case of LU1 and RS5 their position in an area of predicted low intensity land-use is a result of the highly active geomorphic processes in zone 3 that have removed archaeological material from large parts of the zone. The second season's transects of the main survey have by chance, due to the random selection of the units, been placed on some of the most recent surfaces. Overall the model is good at predicting land-use intensity in the survey area with the exception of the Recent Cut-off zone (3) that might have been more intensively used than the remaining evidence in the landscape indicates. The model cannot be used to accurately predict the probability of sites occurring in areas where localised attractions to hunter-gatherers would have been present that are currently unmapped. Both these issues could be resolved with further fieldwork aimed at building a chronology for the cut-offs, and improving the mapping of smaller features.

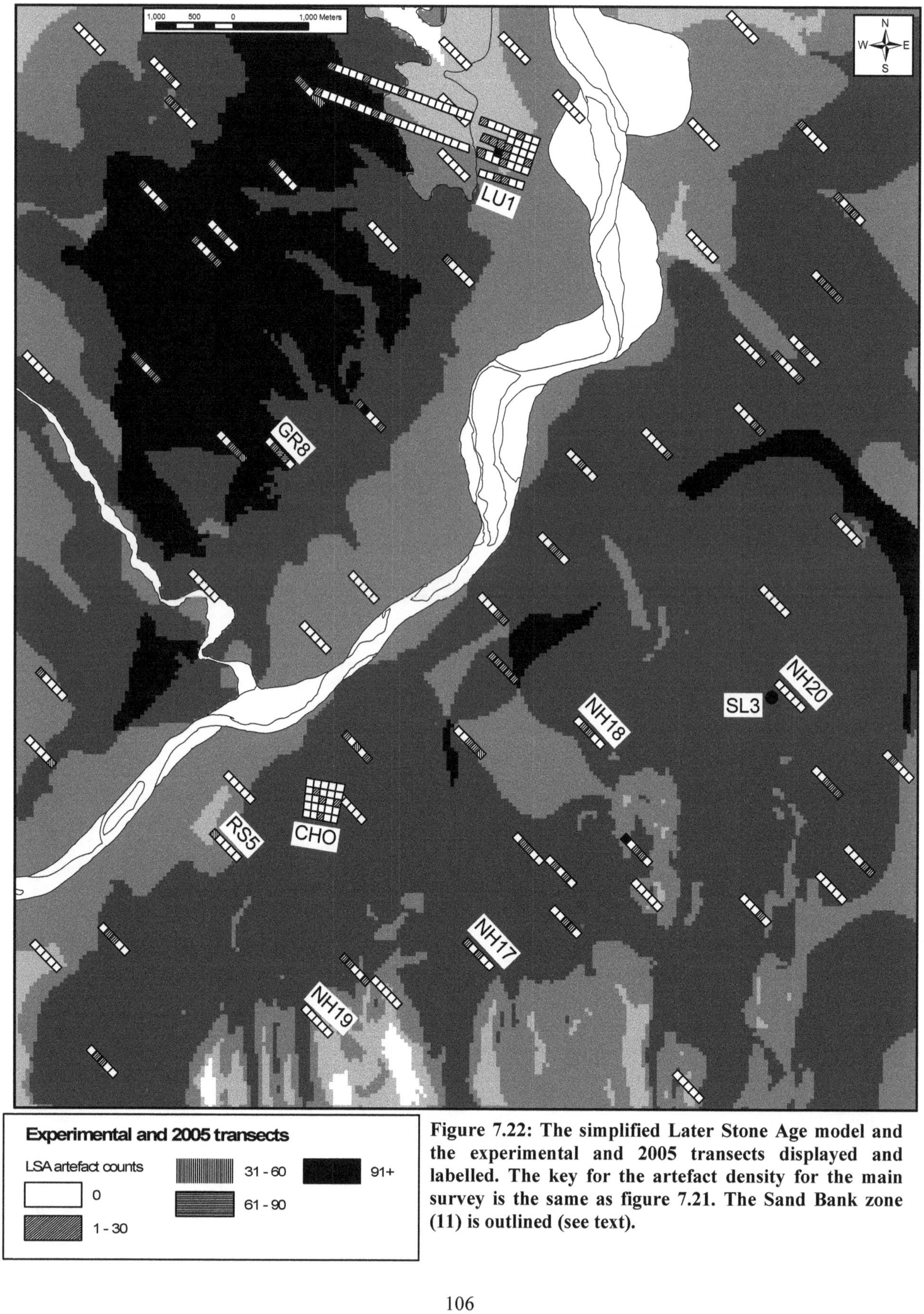

Figure 7.22: The simplified Later Stone Age model and the experimental and 2005 transects displayed and labelled. The key for the artefact density for the main survey is the same as figure 7.21. The Sand Bank zone (11) is outlined (see text).

7.7. The Iron Ages

7.7.1. Description of the distribution of Iron Age material

Data from the Early and Later Iron Ages is combined and the two periods treated as one. This decision was taken in the analysis stage as only one Early Iron Age artefact was found and identified in the survey, and diagnostic artefacts were only present on six of the fourteen transects containing Iron Age material. In general Iron Age material is not located very far from the Luangwa River (less than *c.* 2.5km) (figure 7.23), with two exceptions, NH10 and MI3 (3.5km and 4.5km from the Luangwa respectively). Transects RN3 and RN4 contained the highest density of artefacts, both of which are in the Recent Cut-off zone (3). Most of the remaining transects with Iron Age material are located in the sand and silt zones (4 and 5). No Iron Age material was found in the Cobble zone (7), Black Rock zone (9), Cathedral Mopane zone (14), or over the majority of the Nchindeni Hills (zone 6). The archaeological material encountered on transects with high artefact counts was usually concentrated in very discrete scatters. This pattern is confirmed by figure 7.24, which illustrates the breakdown of the material into 100x100m blocks for each transect. A total of seven clusters of material can be seen on the 21 survey blocks containing Iron Age material, whilst just under half (n = 10) of the survey blocks contain ten or fewer artefacts indicative of a general background scatter of material. Table 7.3 provides a breakdown of the artefact types present in the Iron Age sample, and this reveals that potsherds are the primary constituent, with the occasional pestle and/or mortar. Three fragments of $19^{th}/20^{th}$ century china were also present on GA1.4, and these could be a 19th century import to the area, or an inclusion from recent game viewing activity from one of the Safari lodges. None of the Iron Age material appeared to be rolled, though occasionally pieces appeared to be weathered due to exposure on the ground surface.

7.7.2. Interpretation of the distribution of Iron Age material

The presence of Iron Age material on some of the abandoned channel surfaces in the Recent Cut-off zone (3), on the north-west bank of the Luangwa River indicates that those individual abandonments antedate the 1960s, as this was when all residents were ordered to leave the park. The Iron Age material on the south-east bank in zone 3 has no similar minimum age as residence is permitted in the Game Management Area surrounding the park. However, there are no occupied villages in the survey area today and most residents have relocated to Mfuwe, a nearby town created to house the displaced population, so the minimum age is probably similar. The empty transects in zone 3 at first suggest that these particular cut-off surfaces are too recent to have had any occupation. While this might be the case, it is possible that the discrete nature of much of the Iron Age material in the landscape resulted in evidence of occupation being missed in the survey. Additionally, the nature of the sample meant that not every individual channel abandonment was assigned a transect, which combined with the impossibility of determining the chronological order of the abandonments from the aerial photographs (section 6.2), makes it difficult to suggest exactly which abandonments are too recent to retain evidence for occupation. The only certain evidence derives from the aerial photographs and maps that clearly demonstrate 20th century channel movement and channel abandonment (figure 6.1).

The seven dense clusters of Iron Age material arguably relate to village sites, rather than a general off-site scatter (figure 7.24), and this is confirmed by the nature of some of the finds. For example, a top and bottom grinding stone, weighing 1,503 grams and 5,000+[6] grams respectively, were recovered in close proximity to each other on GA1.4. These, having been found in association with pottery and being comparatively heavy, are unlikely to have been deposited by overbank floods. On RN3 150 pottery fragments and a pestle were recovered in association with what appeared to be hut floors (circular areas of compacted clay), and on RN4 fragments of daga (wall material) were recovered. These daga fragments are very fragile and would not have survived any fluvial transport. The four other dense clusters were primarily made up of high densities of pottery. The less dense scatters might relate either to older village sites that are less visible having been buried by overbank deposits and soil development, or off-site activity contemporaneous with one or more of the other village sites. The latter is possible as the inhabitants of the area still collect foodstuffs from the local environment as well as practising sedentary farming today (*pers obs.* and K.Kakumbi *pers comm.*). However, to determine whether these are off-site scatters or the tops of deeply stratified sites, such as SL7 where there is little evidence for its presence on the surface, would require sub-surface sampling to determine the depth of the archaeological deposits. The lack of any definitive Early Iron Age pottery is also suggestive that the surface remains are not representative of the entire Iron Age record. Furthermore, the practice of moving the location of the village with the death of each headman is followed in this part of Zambia, and as noted by Derricourt (1985) these relocations will average every 30 years and can result in little or no archaeological material visible on the surface (Derricourt

[6] This is a minimum weight, as the scales available in Zambia had a 5,000 gram maximum.

1985:17). It is likely therefore that the known distribution of the Iron Age scatters and sites in zones 4 and 5 provide a minimal indicator of the density of archaeological material in the area, and primarily relate to the most recent Iron Age activity. Despite this bias, the distribution of material will still provide a reliable indicator of preferable habitation areas as much of the landscape had stabilised by the late Holocene, so artefact scatters are more likely to represent genuine cultural patterning in the landscape.

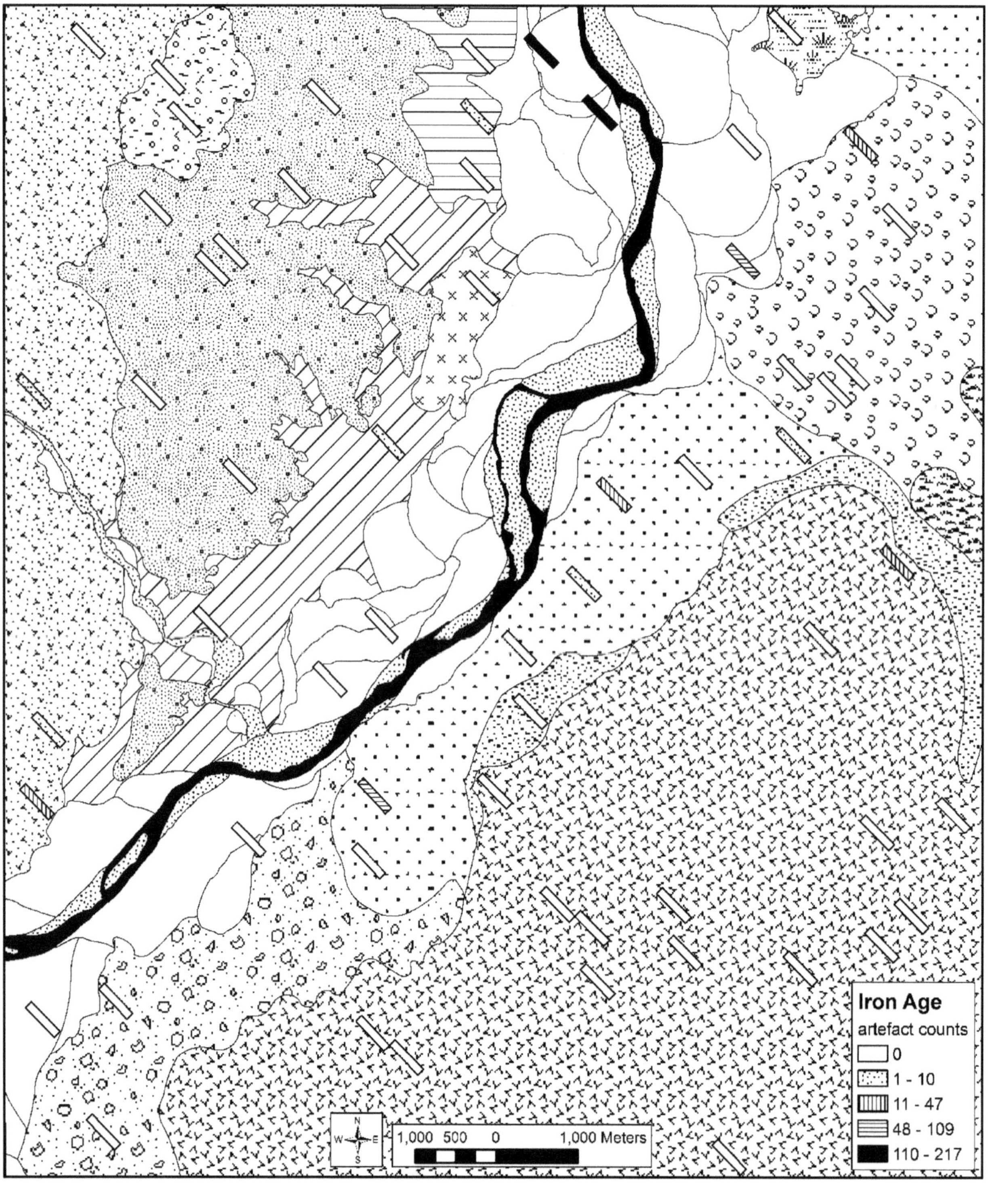

Figure 7.23: Iron Age artefact distribution. The artefact counts are provided in appendix 7.3. The grades for artefact density are not equal so that the low-density counts can be compared to the high density counts easily without using a large number of classes.

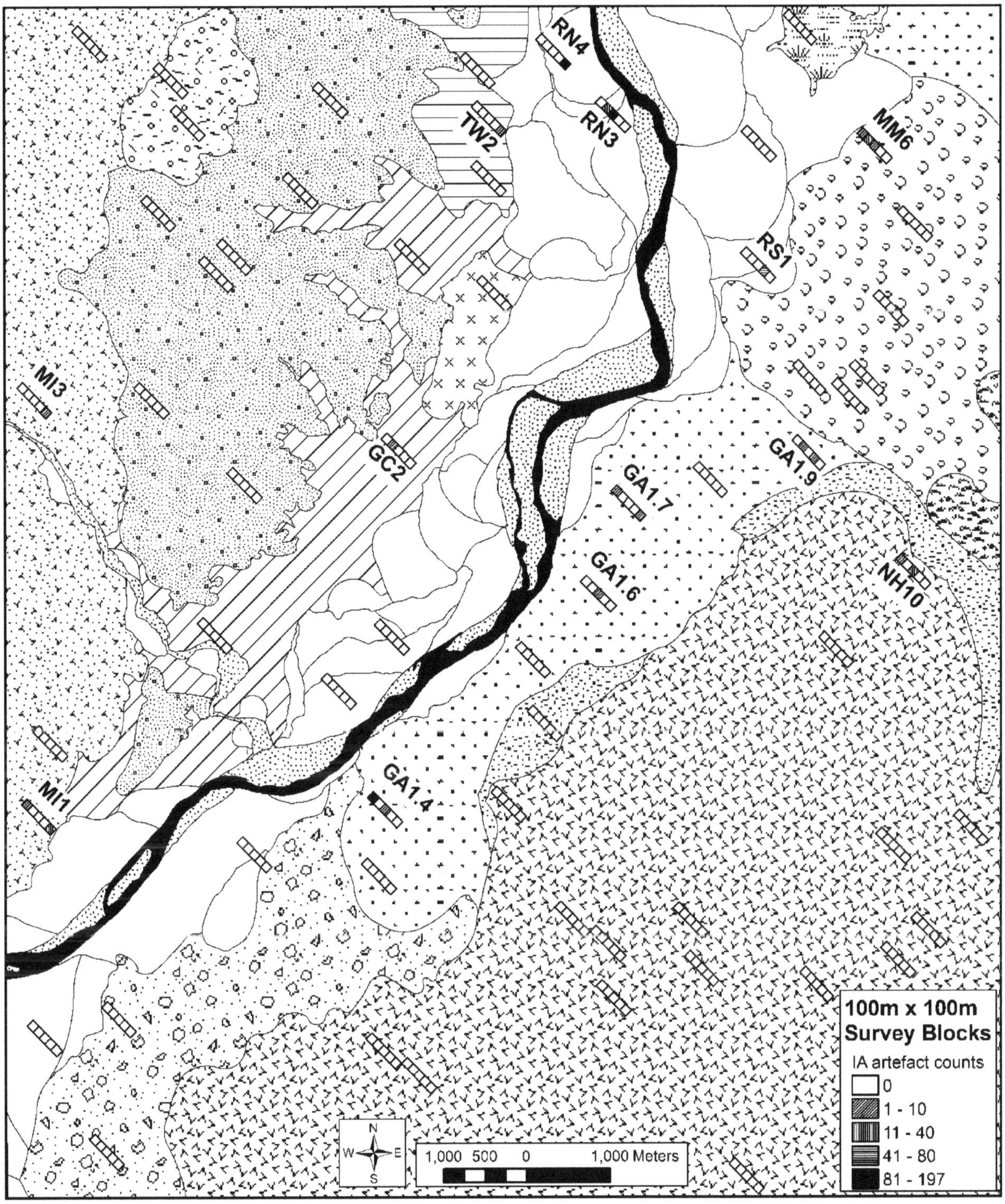

Figure 7.24: The breakdown of the Iron Age data into the 100x100m survey blocks illustrating the concentrated nature of the Iron Age material at some localities. The transects containing artefacts have been labelled for ease of identification in the discussion. The raw data for the artefact counts on each survey block are presented in appendix 7.10.

Transect number	Iron Age Artefact types				
	Pot	top stone/pestle	Wall fragment (daga)	Bottom stone/mortar	China (bone china ?)
GA1.4	104	2	.	1	3
GA1.6	1	.	.	.	.
GA1.7	15	.	.	.	.
GA1.9	10	.	.	.	.
GC2	1	.	.	.	.
MI1	28	.	.	.	.
MI2	1	.	.	.	.
MI3	1	.	.	.	.
MM6	2	.	.	.	.
NH10	31	.	.	.	.
RN3	180	1	.	.	.
RN4	214	.	3	.	.
RS1	86	.	.	.	.
TW2	1	.	.	.	.

Table 7.3: A breakdown of the artefact types present in the Iron Age cross-referenced by transect number

Overall, the locations of sedentary Iron Age communities visible in figure 7.24 might indicate that distance to water sources are the primary factor, as Iron Age material is concentrated in areas not far from the Luangwa River and its larger tributaries. These would have acted as extensive reliable water sources for sedentary groups, whereas the supplies from smaller streams in the area might not have sufficient to support larger groups (*pers obs*). A 1000m buffer was created around the Luangwa and its larger tributaries (wider than 3m), and the result was that 16 of the 21 (75%) survey blocks are located on or in the buffer, which only occupies 56% of the survey area. These statistics might be incidental and relate to the fact that soils more suited for agriculture are located near the Luangwa. It is known that the seasonally replenished sediments of the Luangwa Valley are suitable for cereal based crops; this is particularly true of the soils in tributary areas of the south-eastern bank of the Luangwa River that receive nutrients from both seasonal flooding and the weathering Archean bedrock (Trapnell 1996:9). Further away from the Luangwa and its larger tributaries the soils tend to be thinner and clast rich, often due to the combined effects of slope angle and elevation on hillslope processes in the Nchindeni Hills (appendix 6.1). In some areas soils are entirely absent, as is the case for the cobble zones (zones 7 & 8). Chichele Hill and Chipembele Ridge are also at higher elevations and have thin or negligible soils. Therefore elevation and slope angle are also factors that will affect the distribution of Iron Age activity. Considering that the best soils in the area are replenished by seasonal flooding and that reliable all year water sources would have been a necessity, proximity to the Luangwa and its tributaries is likely to be the primary factor affecting settlement location choices, and situations where there is one but not the other are not likely to be suitable.

7.7.3. Modelling Iron Age land-use

The distribution of Iron Age material indicates that settlement preferences are related to a combination of factors, the primary one being soils, which is in turn affected by hill slope and elevation, as well as distance to the replenishing floods of the Luangwa. A predictive model was constructed using the same technique used for the Later Stone Age model. The raster surfaces for geomorphological zones, distance to the Luangwa and its larger tributaries, elevation, and hill slope, were all reclassified to a common scale of one to five based on the prevalence of Iron Age material in each area. Table 7.4 records the scales used for each factor in the model.

7.7.4. Iron Age modelling results

The initial Iron Age model, similarly to the Later Stone Age model first produced, over emphasised the complexity of the predicted pattern. To simplify the model it was reclassified from fourteen grades of land-use intensity to five. This has the added advantage of providing a direct comparison to the land-use model for the Later Stone Age. The model created is illustrated in figure 7.25. The darker colours represent areas that are likely to have experienced more intensive occupation, and the lighter areas less or no occupation. The model accurately predicts the locations of the known concentrations of material with the highest counts of Iron Age material (41+ artefacts) all occurring in the highest scoring parts (scoring 5) of the landscape, and only three transects containing artefacts occurring in parts of the landscape scoring 4. Parts of the landscape scoring 3 or less contain no Iron Age material whatsoever.

Scale: 1 = Least suitable – 5 = Most suitable.		
Layer and justification for scale.	**Scale.**	**Category weights**
Geomorphological Zones. The weighting applied to each layer assumed that if there were no artefacts in the landscape then the probability of discovering more artefacts was low and a score of 1 was given. The middle ranking of 3 was given where a few artefacts are present as part of the background scatter. The highest score, 5, was given for zones where the highest density of artefacts was found.	1 2 3 4 5 6 7 8 9 10 11 12 13 14 15 16	1 1 5 3 5 5 1 1 1 1 3 1 5 5 1 5
Elevation. The weighting assumed that as artefacts were found at elevations below 640m that these areas were most suitable and were given a score of 5, the other three classes were then weighted accordingly, the least favourable class was not used as the data does not suggest a strong relation to elevation (see text)	498-575m 575-652m 653-728m 729-805m 806-881m*	5 1 1 1 1
Slope Angle. The steepest slopes have no artefacts present and were unsuitable for habitation so were given the lowest weighting, and the subsequent classes were graded equally, with the flattest areas receiving the highest score.	0 - 6° 7 - 12° 13 - 19° 20 - 25° 26° +*	5 4 1 1 1
Distance from streams. Statistical tests revealed that Later Stone Age material is more likely to be located near streams so areas closest to streams were given the highest weighting, and as distance increases the weighting was graded equally.	0 – 1041m 1042 – 2083m 2084 – 3125m 3126 – 4167m 4168 – 5208m*	5 4 1 1 1

Table 7.4: The weighting system used to create the predictive density model for the distribution of Iron Age artefacts.

* The classifications were set using the inbuilt functions of GIS to reclassify each layer into 5 equal intervals.

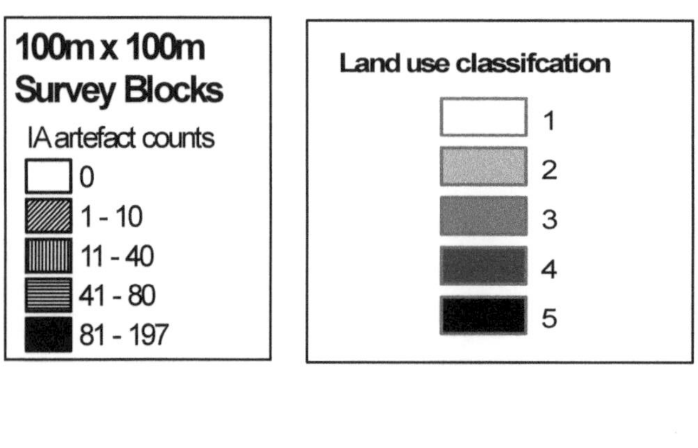

Figure 7.25: Iron Age land-use model using five grades of colour where 1 = lowest land use intensity, and 5 = the highest.

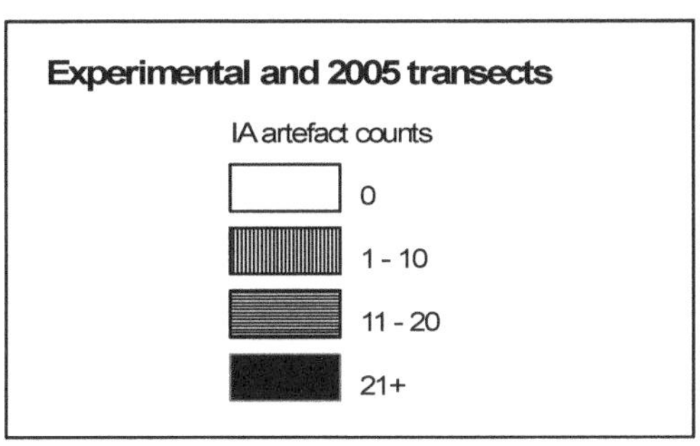

Figure 7.26: The Iron Age land-use model displayed with the experimental transects and the first season transects included. The extra transects are labelled. Figure 7.25 should be consulted for the key to determine artefact counts for the original transects.

7.7.4. Testing the Iron Age model

The experimental transects surveyed to test patterns seen in the field, and the first season's transects are displayed against the model in figure 7.26. Of the first season's transects 10 of the 103 survey blocks contain Iron Age artefacts, and all of these are contained within areas predicted at category 5 land-use intensity. Approximately half (49) of the 103 survey blocks occur in category 5 areas, and if the archaeological distribution was not well served by the model, or random, only half of the survey blocks containing artefacts would therefore be in category 5 areas. The experimental transects reveal the same pattern. Of the 30 survey blocks, 14 are in category 5 areas, with the rest in category four or lower areas. Only four survey blocks contained Iron Age artefacts, and all of these were in blocks in the category 5 area. The correlation between Iron Age archaeological occurrences and the model created agrees more reliably than the Later Stone Age model agrees with its data. This suggests that Iron Age settlement requirements presented more constrictions on where people could settle in the natural environment due to their requirements for good soils for crops, and are as a result easier to model with the environmental datasets available. Later Stone Age hunter-gatherers, on the other hand, relied heavily on more widely dispersed resources, so their use of the landscape combined with the lack of resolution on suitable cave sites, means their behaviour is less precisely predictable in the landscape.

7.7.5. Comparison of Later Stone Age and Iron Age land-use

One of the most valuable uses of the predictive land-use models produced here is the provision of a good visual comparison of the of Later Stone Age and Iron Age land-use, and this is useful as these two groups of people are known to have coexisted in Zambia until the 19th century (Musonda 1987:156). Comparing the two land-use maps (figures 7.21 and 7.25) reveals that the land-use intensity of the two groups differs, but overlaps considerably. Although the reduction in Later Stone Age activity close to the Luangwa is possibly an artefact of the recent erosion by the river the general pattern is still clear, and in places when the two maps are viewed together they almost appear as a positive and negative image of each other, whilst overlapping at other points. Both land-use models indicate that the steepest and highest areas of the Nchindeni Hills are not suitable for any activity by either group; this is not surprising as the steeply undulating landscape would not be suited to farming, setting camp, or even hunting and generally moving through the landscape. There is a definite overlap of activity in parts of the landscape approximately 1.5 – 2.5km either side of the Luangwa River on the flatter terrain, though further from the river hunter-gatherer activity becomes more common as Later Stone Age people used the moderate elevations of the Nchindeni Hills more intensively. The most noticeable difference is the use of the cobble zones, which are the most intensely used area in the Later Stone Age and one of the least intensely used areas of the Iron Age.

The overall impression is that both hunter-gatherers and sedentary farmers used similar areas of the landscape. These areas of overlap are of interest as it allows the data from this survey to be linked to the theories developed based on research elsewhere in Zambia regarding the interaction of the two groups. In brief there are three key theories (for detail see section 2.2.2.4): either there was no initial contact followed by the development of trade (Miller 1969b:87), or there was no contact at all between the two groups (Musonda 1987:156), and thirdly there is a developmental sequence form Later Stone Age to Iron Age (Robertson 2000:179). Taking into consideration the geomorphological record, it is likely that one of the first two theories is the more appropriate in the context of the survey. The third theory is unlikely as land surfaces such as those in zone 5 around SL7 have aggraded throughout the Iron Age and the land surface for at least parts of zone 5 is as recent as the 20th century (section 2.2.2.5). As Later Stone Age material is found on separate areas of this surface, the two groups of people must have coexisted until recently.

To determine which of the two remaining theories might be applicable it is necessary to consider the nature of the archaeological distributions. It is unlikely, considering the overlap of activity areas, and the fact that material is found on recent surfaces, that the two peoples were unaware of each other. However, without dating surface sites directly, it cannot be ruled out that the area was inhabited at different times. To use the survey data to illustrate the relationship between the two groups, it is necessary to examine the material culture of both groups when it occurs together on scatters that might be, or are, indicative of site-based rather than off-site activity. There are five survey blocks where Later Stone Age and Iron Age material was found at the same locality, and where artefacts from one or other period occurred in high quantities relative to the rest of the survey (> 10 artefacts): these are GC2D, MI1A, RS5E, and GA1.4C and E. The basic numerical data for the artefacts found at these localities are provided in table 7.5. It is clear that where Later Stone Age material is dominant, at GC2D, RS5E, and GA1.4C only one Iron Age artefact is present in each locality. The small amount of pottery suggests that Later Stone Age people inhabited sites that were abandoned by Iron Age people and were scavenging the pottery, or may even have been unaware of or uninterested in its presence. Alternatively there may have been limited trade with Iron Age people, with the result that there is relatively little pottery used by hunter-gatherers. Likewise the relatively low proportion of Later Stone Age artefacts present on Iron Age dominated localities suggests that Iron Age sites are situated on top of the general background scatter of material, rather than indicating that both Later Stone Age stone working and Iron Age activities were concurrent in one locality. Again, lack of precise chronological controls limits further development of this option.

Block code	Later Stone Age		Iron Age (pottery only).	
	Number of artefacts	Weight (g)	Number of artefacts	Weight (g)
	Sum	Mean	Sum	Mean
GC2D	41	6.8	1	44.0
MI1A	19	3.7	28	2.9
RS5E	56	6.8	1	4.0
GA1.4C	18	4.1	1	7.0
GA1.4E	5	12.3	102	17.5
NH20D			8	32.1

Table 7.5: A table of the artefact counts and average artefact weights for the 100x100m blocks that contained both Later Stone Age and Iron Age material

Possibly one of the most curious sites was discovered on the experimental transect NH20 (figure 7.27). This is a cave site situated approximately 170m from the Later Stone Age site of Poachers' Cave (SL3), where Later Stone Age artefacts can be seen on the surface of the site, and excavations revealed that the majority of material is Later Stone Age, though there were Iron Age pottery sherds present in the topmost level (Barham *pers comm.*). After a search around the NH20 cave site, no evidence of Later Stone Age activity could be found, and there was only evidence of Iron Age occupation in the form of pottery and a grinding stone (table 7.5 and figure 7.27). It is possible that the stability of the soils at the NH20 site are obscuring evidence for Later Stone Age occupation, which is more visible in the sediments eroding from Poachers' Cave. The proximity of this site to SL3 makes it unlikely that the occupants of each would have been unaware of the other's presence. The lack of Later Stone Age material on the surface of the sediments in the cave on NH20 hints at a lack of contact between the two groups, but may also be a factor of the caves not being inhabited at precisely the same time. This site, in combination with the data presented above, suggests that there might have been very limited contact between the two groups in the landscape, possibly with some limited trade or scavenging of Iron Age pottery by hunter-gatherer groups. This lack of contact has occurred despite the two groups using overlapping areas of the landscape. It is possible therefore, in similar scenarios reported elsewhere in Zambia, that relatively low population densities, the mobility of hunter-gatherer groups, and the different resource requirements of the groups meant that little interaction occurred until very late, possibly as late as the 19th century (e.g. Musonda 1987:155). Or alternatively, that the two groups were aware of each other and the similar requirements for land, either for farming or hunting, led to tension and minor conflicts (e.g. Miller 1969b:87). The limited amount of data from this survey data means that this is not entirely conclusive, and for the best resolution excavation would need to be undertaken at a number of localities to determine the stratigraphic relationship of the material.

Figure 7.27: The cave site on NH20D, no scale is shown but the height of the cave on the right is approximately two metres. Inset is a grinding stone found in the cave (not collected), with a pen for scale (approx 15cm).

7.8. Transects Outside the Survey Area

7.8.1. Introduction

During the second season there was time to survey eleven extra transects, six of which were designed to test and clarify some of the patterns seen in the survey area (see above), whilst the remaining five were designed to investigate the nature of the archaeological record outside the survey area, towards the centre of the park using the Kasweta Road for access. The archaeological data from these five transects are described and interpreted in reference to geomorphological descriptions for each area. Following this, the data from the transects surveyed along the O5 road in the first season are described and interpreted, and a geomorphological interpretation of the area is given based on the observations made during the survey. The rest of the data from the first seasons transects is then presented and discussed in section 7.9.

7.8.2. The Kasweta Road Transects

Figure 7.28 illustrates the locations of the Kasweta Road transects in relation to the main survey area and the Luangwa River and its large seasonal tributaries. The intervals used to place the transects along the Kasweta Road were chosen arbitrarily, by plotting positions on the maps of the area to give a roughly equally spaced coverage along the road. Two further transects were plotted beyond K4 that would have taken the survey to within 16km of the Muchinga Escarpment. However, it was decided after reaching K4 that it would be too great a risk to survey further due to the distance away from base camp, the condition of the road and, more importantly, the poor condition of the survey vehicle. K4 is situated 26km away from the Nchindeni Hills and 20km away from the Muchinga Escarpment, so the aim of sampling the central portion of the valley floor was still achieved.

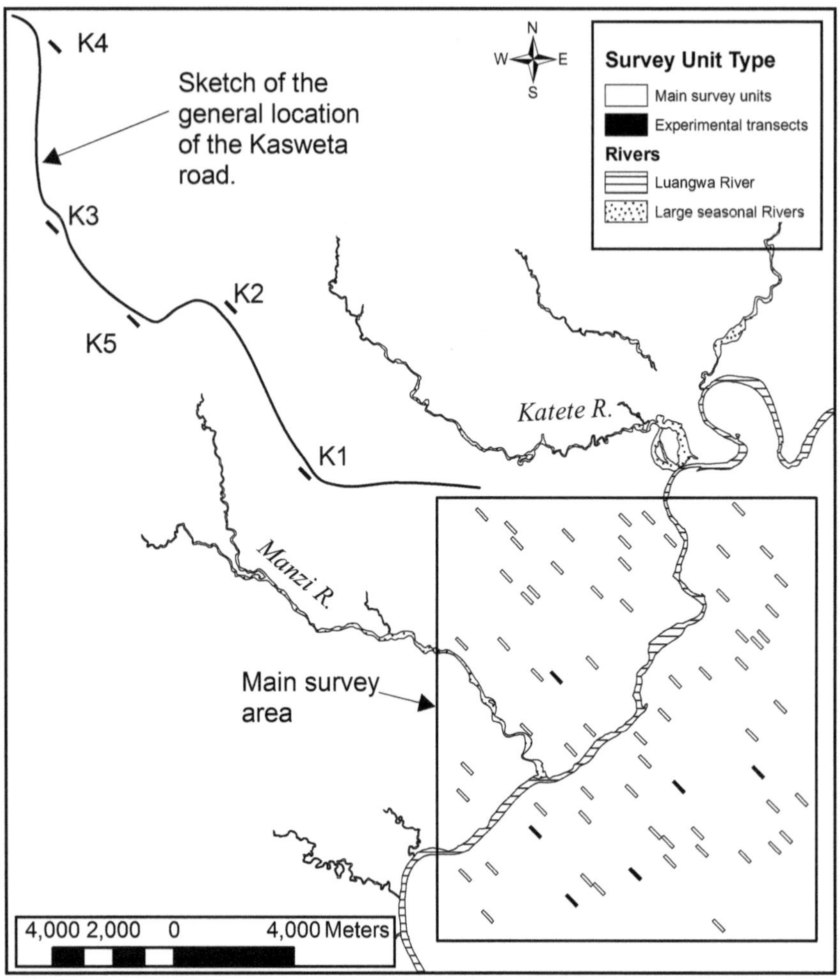

Figure 7.28: Location map illustrating the positions of the Kasweta Road transects in relation to the main survey area, the Luangwa River, and the main tributaries in the area.

7.8.3. Description of the Kasweta Road geomorphological terrain and archaeological data

The Kasweta Road, in the surveyed half of the valley, is located along the watershed boundary of the Manzi and Katete Rivers (figure 7.28). Much of the road is situated on gently sloping sandy substrate interspersed with seemingly random patches of gravels and cobbles. This gentle topography is interrupted by two large hills that form elongated ridges in the landscape rising to 720 and 760m high (figure 7.29), which are both covered in an armouring of gravels and cobbles. On the more gentle terrain the sand, in places, is up to five metres deep (e.g. figure 6.6), and is possibly even deeper in some areas. These sands date from recent geologic history as archaeological material is observed outcropping from the sands in some of the gullies surveyed (e.g. K5, see below). No bedrock was seen underpinning the cobbles and sands on the two ridges, but presumably, at an unknown depth, the Karoo sediments known to underlie the sediments of most of the valley are present here as well. The vegetation is exclusively of the miombo woodland type and is usually dense, though there are occasional patches of more open woodland. In some areas the sandy soils are stable, whilst in others the sand has been deeply gullied by recent erosion; vegetation in this area does not appear to be a factor in stabilising the substrate as in densely vegetated areas deep gullies are still found (figure 6.6). It is likely that proximity to the large tributaries is the controlling factor in the erosion of the area, as most gullied areas are closer to these tributaries, which are also usually at an elevation of at least 40m below the sediments found along the Kasweta Road (figure 7.29).

The five transects were numbered in the order in which they were surveyed, but are described and analysed according to their distance from the main survey area. Table 7.6 provides the data for the artefact numbers retrieved from Kasweta Road transects. No artefacts and no cobbles were discovered on K1, the substrate was recorded as fine to coarse sand with some granular grade material (2-4mm), and the ground was nearly flat at an elevation of 680m. The vegetation was dense, but ground visibility was high as only patchy dead grass was present at ground level. The vegetation on K2 was identical (figure 7.30), but the soil in places was primarily granular sand (figure 7.30 inset). K2 was situated at 720m and was dissected by two broad 2m deep gullies, one at each end of the transect, which appear as indentations on the 720m contour in figure 7.29. A total of 41 artefacts were recovered from K2, which are primarily of Early and Early/Middle Stone Age affiliation (70.7% in total, table 7.6). Definite Middle and Later Stone Age artefacts were rare compared with the Early Stone Age material. Although a proportion of Middle Stone Age artefacts could be in the indeterminate Early/Middle category the general impression is that the earlier material dominates. Transect K5 is dissected by numerous deep steep-sided gullies (see figure 6.6), which frequently contained cobbles eroding from the sands, some of which are exceptionally large when compared with the other surveyed areas, and can be over 25cm (boulder sized) (figure 7.31). K5 was at the highest elevation of the Kasweta Road transects, situated along the 740m contour, and it provided the largest sample from the Kasweta Road transects, totalling 112 of which 16, 5, 12, and 6 were classed as Early, Middle, and Later Stone Age, and Iron Age respectively (table 7.6). No artefacts were retrieved

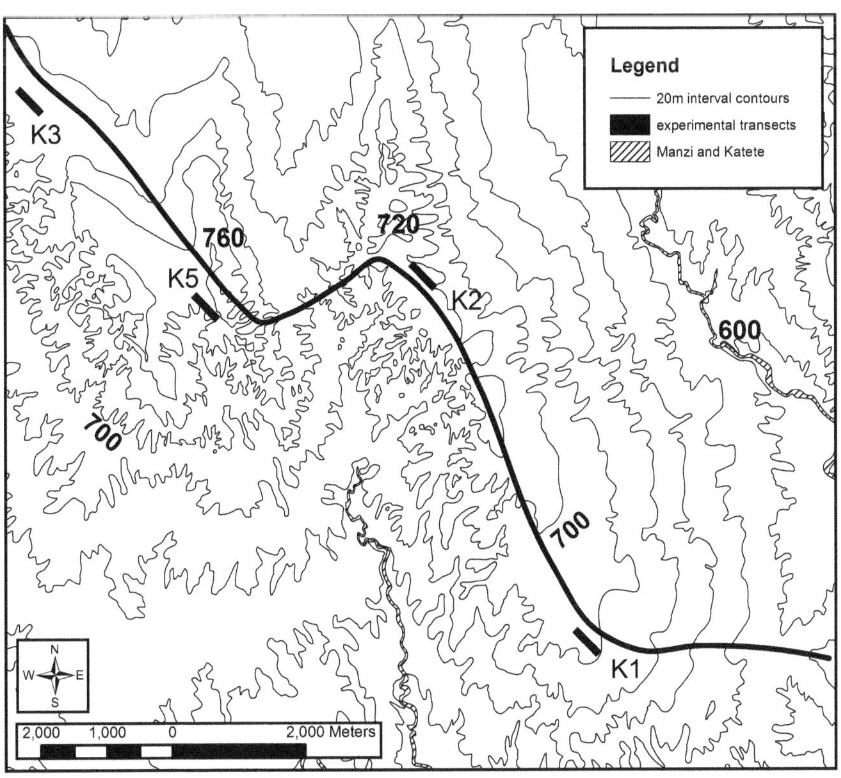

Figure 7.29: Detailed view of terrain along Kasweta Road for transects K1 – K3 and K5.

The contours and rivers for the figure were digitised from Zambian Survey map 1331B1 (copyright Zambia Government 1972).

from transect K3, which is situated on stable, flat uniformly sandy soil at 700m elevation. K4 is situated the furthest away, and is also at the lowest elevation along the Kasweta Road, 680m, on slightly undulating fine sands with occasional cobble scatters, supporting open miombo woodland. Of 17 artefacts found on this transect, two were Early Stone Age, one was Middle Stone Age, and one was Later Stone Age, whilst the rest were indeterminate (table 7.6). Unusually, the Later Stone Age appears to be the least represented of the Stone Ages in every sample along the Kasweta Road, whereas in the main survey area the later material nearly always provided the greatest number of artefacts (see appendix 7.3).

Tables 7.7 and 7.8 illustrate the percentages of artefacts by condition and raw materials cross-referenced by the Kasweta Road transect numbers and typological associations. In general, the distribution of the varying condition of the artefacts from each period appears similar to that of the main survey area, as the Early Stone Age contains the majority of the rolled artefacts (9 of the 14 rolled artefacts are Early Stone age) and the only heavily rolled artefacts. Transect K5 however, contains a low proportion of rolled Early Stone Age material (2 of its 16 artefacts) compared with K2, K4, and transects in the main survey area where usually over 50% of artefacts demonstrate abrasion of some extent (compare figure 7.5and table 7.7). In contrast, K5 contains the only Middle Stone Age abraded artefacts, whilst the rest of the material on K2 and K4 is fresh. Similar to the main survey area pattern all the Later Stone Age material is in an un-abraded condition. The raw material used in each period also shares characteristics with the main survey area. In the Early Stone Age a range of quartzites were used with no apparent preference for any particular quality. On K4 one of the Early Stone Age artefacts was manufactured using a tabular quartzite, which in knapping experiments proved difficult to work due to the tabular nature of the fracture planes in the raw material causing unpredictable results. Middle Stone Age artefacts are generally made on fine-grained quartzite and chert, though unusually there is one artefact made of coarse quartzite on K5. The Later Stone Age artefacts are manufactured exclusively using quartz or chert.

Tables 7.9a to 7.9c list the tool types recorded on each transect from each period, and in general it is clear that flakes and cores comprise the largest groups of artefacts from each period. In the Middle and Later Stone Ages, the more complex tool forms are rare or absent. For example, no prepared core flakes and only one prepared core are present in the Middle Stone Age; no specialised tools such as segments, and only one slit cobble are present in the Later Stone Age sample. This is presumably a reflection of the relatively small sample size for both periods rather than an expression of behavioural variability.

Typology	Transect number					
	K2		K5		K4	
	Total	Column Total %	Total	Column Total %	Total	Column Total %
Early Stone Age	14	34.1%	16	14.3%	2	11.8%
Early/Middle Stone Age	15	36.6%	40	35.7%	7	41.2%
Middle Stone Age	2	4.9%	5	4.5%	1	5.9%
Middle/Later Stone Age	5	12.2%	15	13.4%	5	29.4%
Later Stone Age	1	2.4%	12	10.7%	1	5.9%
Early Iron Age	.	.0%	.	.0%	.	.0%
Iron Age	.	.0%	6	5.4%	.	.0%
Later Iron Age	.	.0%	.	.0%	.	.0%
Unknown	4	9.8%	18	16.1%	1	5.9%
Total	41	100.0%	112	100.0%	17	100.0%

Table 7.6: Artefact counts for the Kasweta Road transects. K1 and K3 have been excluded as no artefacts were found on those transects.

Figure 7.30: The vegetation on transect K2 illustrating one of the shallow gullies present on the transect, and inset the coarse sand and granular nature of much of the soil surface.

Figure 7.31: An example of one of the quartzite boulders seen on transect K5 at the base of one of the gullies. It was un-worked. Scale is in cm (total 21cm).

Condition (degree of rolling)	Transect number									
	K2									
	Early Stone Age		Early/Middle Stone Age		Middle Stone Age		Middle/Later Stone Age		Later Stone Age	
Fresh	6	42.9%	14	93.3%	2	100.0%	5	100.0%	1	100.0%
Slightly Rolled	3	21.4%	1	6.7%	.	.0%	.	.0%	.	.0%
Moderately Rolled	3	21.4%	.	.0%	.	.0%	.	.0%	.	.0%
Very Rolled	2	14.3%	.	.0%	.	.0%	.	.0%	.	.0%

Condition (degree of rolling)	K4									
	Early Stone Age		Early/Middle Stone Age		Middle Stone Age		Middle/Later Stone Age		Later Stone Age	
Fresh	1	50.0%	6	85.7%	1	100.0%	5	100.0%	1	100.0%
Slightly Rolled	.	.0%	1	14.3%	.	.0%	.	.0%	.	.0%
Moderately Rolled	.	.0%	.	.0%	.	.0%	.	.0%	.	.0%
Very Rolled	1	50.0%	.	.0%	.	.0%	.	.0%	.	.0%

Condition (degree of rolling)	K5									
	Early Stone Age		Early/Middle Stone Age		Middle Stone Age		Middle/Later Stone Age		Later Stone Age	
Fresh	14	87.5%	39	97.5%	3	60.0%	15	100.0%	12	100.0%
Slightly Rolled	1	6.3%	1	2.5%	1	20.0%	.	.0%	.	.0%
Moderately Rolled	1	6.3%	.	.0%	1	20.0%	.	.0%	.	.0%
Very Rolled	.	.0%	.	.0%	.	.0%	.	.0%	.	.0%

Table 7.7: Artefact counts from each of the Kasweta Road transects tabulated by abrasion and cross-referenced by typological association. The percentage is recorded for each typological group on each transect.

Transect number and typological association		Raw material count and percentage for each typological													
		Fine Quartzite		Medium Quartzite		Coarse Quartzite		Chert		Quartz		Tabular Quartzite		Fossil Wood	
		No. of artefacts		No. of artefacts		No. of artefacts		No. of artefacts		No.of artefacts		No. of artefacts		No. of artefacts	
K2	Early Stone Age	.	.0%	8	57.1%	6	42.9%	.	.0%	.	.0%	.	.0%	.	.0%
	Early/Middle Stone Age	3	20.0%	8	53.3%	3	20.0%	.	.0%	.	.0%	1	6.7%	.	.0%
	Middle Stone Age	.	.0%	2	100.0%	.	.0%	.	.0%	.	.0%	.	.0%	.	.0%
	Middle/Later Stone Age	.	.0%	.	.0%	.	.0%	.	.0%	5	100.0%	.	.0%	.	.0%
	Later Stone Age	.	.0%	.	.0%	.	.0%	.	.0%	1	100.0%	.	.0%	.	.0%
	Unknown	.	.0%	1	25.0%	3	75.0%	.	.0%	.	.0%	.	.0%	.	.0%
K4	Early Stone Age	.	.0%	.	.0%	1	50.0%	.	.0%	.	.0%	1	50.0%	.	.0%
	Early/Middle Stone Age	4	57.1%	1	14.3%	2	28.6%	.	.0%	.	.0%	.	.0%	.	.0%
	Middle Stone Age	1	100.0%	.	.0%	.	.0%	.	.0%	.	.0%	.	.0%	.	.0%
	Middle/Later Stone Age	3	60.0%	1	20.0%	.	.0%	.	.0%	1	20.0%	.	.0%	.	.0%
	Later Stone Age	.	.0%	.	.0%	.	.0%	.	.0%	1	100.0%	.	.0%	.	.0%
	Unknown	.	.0%	.	.0%	.	.0%	.	.0%	.	.0%	1	100.0%	.	.0%
K5	Early Stone Age	2	12.5%	7	43.8%	7	43.8%	.	.0%	.	.0%	.	.0%	.	.0%
	Early/Middle Stone Age	11	27.5%	18	45.0%	5	12.5%	.	.0%	4	10.0%	2	5.0%	.	.0%
	Middle Stone Age	1	20.0%	1	20.0%	1	20.0%	2	40.0%	.	.0%	.	.0%	.	.0%
	Middle/Later Stone Age	1	6.7%	4	26.7%	.	.0%	5	33.3%	5	33.3%	.	.0%	.	.0%
	Later Stone Age	.	.0%	.	.0%	.	.0%	2	16.7%	10	83.3%	.	.0%	.	.0%
	Unknown	2	11.1%	5	27.8%	3	16.7%	3	16.7%	4	22.2%	.	.0%	1	5.6%

Table 7.8: Artefact counts from each transect tabulated by raw material and cross-referenced by typological association. Percentage figures are recorded by typological group on each transect.

Transect number	Early Stone Age							
	Flake		Broken Flake		Core		Total	
	Number of artefacts		Number of artefacts		Number of artefacts		Number of artefacts	
	Sum	Row Sum %	Sum	Row Sum %	Sum	Row Sum %	Sum	Row Sum %
K2	5	35.7%	1	7.1%	8	57.1%	14	100.0%
K5	10	62.5%	.	.0%	6	37.5%	16	100.0%
K4	.	.0%	.	.0%	2	100.0%	2	100.0%

Table 7.9a: Counts and percentages for Early Stone Age tool types found on each of the Kasweta Road transects.

Transect number	Middle Stone Age											
	Flake		Core		Blade Core		Blade		Prepared Core		Total	
	No. of artefacts		No. of artefacts		No.of artefacts		No. of artefacts		No.of artefacts		No. of artefacts	
	Sum	Row Sum %	Sum	Row Sum %	Sum	Row Sum %	Sum	Row Sum %	Sum	Row Sum %	Sum	Row Sum %
K2	.	.0%	.	.0%	.	.0%	1	50.0%	1	50.0%	2	100.0%
K5	2	40.0%	2	40.0%	1	20.0%	.	.0%	.	.0%	5	100.0%
K4	.	.0%	.	.0%	.	.0%	1	100.0%	.	.0%	1	100.0%

Table 7.9b: Counts and percentages for Middle Stone Age tool types found on each of the Kasweta Road transects.

Transect number	Later Stone Age									
	Flake		Core		Prepared Core Flake		Split Cobble		Total	
	No. of artefacts		No. of artefacts		No. of artefacts		No. of artefacts		No. of artefacts	
	Sum	Row Sum %	Sum	Row Sum %	Sum	Row Sum %	Sum	Row Sum %	Sum	Row Sum %
K2	1	100.0%	.	.0%	.	.0%	.	.0%	1	100.0%
K5	6	50.0%	4	33.3%	1	8.3%	1	8.3%	12	100.0%
K4	1	100.0%	.	.0%	.	.0%	.	.0%	1	100.0%

Table 7.9c: Counts and percentages for Later Stone Age tool types found on each of the Kasweta Road transects.

7.8.4. Interpretation of the Kasweta Road archaeological data and geomorphology

The geomorphological terrain of the area differs from the main survey area in that it is relatively uniform over the distance surveyed along the Kasweta Road in comparison with the variability closer to the Luangwa. The sediments in this area are essentially an extension of zone 13 (Chipembele Ridge Sands). In chapter 6 it was suggested that these sands might be distal alluvial fan deposits originating from the Muchinga Escarpment that have been periodically reworked resulting in the deposition of further fan deposits in the survey area (Cobble zone 7). During the survey work on the Kasweta Road it was clear that the land-surface was stable over large areas, up to $10km^2$, and this is visible in the smoother wider spaced contours of parts of figure 7.29, particularly in the areas of transects K1 and K3. On these two transects the ground surface is flat and no artefacts were retrieved; it is likely, based on the fact that artefacts were recovered from the other transects where the ground surface is less stable, that artefacts are present in the sub-surface of K1 and K3. The less stable areas of K2, K4, and particularly K5, are being eroded in the rainy season by gullies and streams that ultimately drain into the Luangwa, in this area it is clear that the down-cutting is slowed, but not halted completely, once the sand has been removed and the harder Karoo sediments are reached. The Karoo sediments are then eroded into and form deep channels cut through the landscape. Examples of these channels were observed approximately 1.7km south of the Kasweta Road along a line parallel to it between K1 and K2 (figure 6.7a & b). The down cutting of the unconsolidated sands above the Karoo sediments is the primary agent controlling the exposure of archaeological material in the area, and the distribution of artefacts therefore does not represent any behavioural choices in the landscape in this area. The inclusion of archaeological material confirms that the sediments are of Pleistocene origin. The lack of abrasion on the Middle and Later Stone Age artefacts suggests that the material discovered is not located far from its initial point of discard, though as they are exposed by erosion processes, their distribution does not reflect behavioural decisions in the landscape. A larger proportion of Early Stone Age material on transects K2 and K4 has been abraded, though in contrast the material on K5 is primarily unrolled.

The artefacts on K5 are the most significant in the entire survey with regard to the Early Stone Age, not only because the material is predominately fresh and numerous compared with the rest of the survey (13 artefacts), but also because six of the flakes and flake fragments, though not directly refitting, appear to derive from the same piece of quartzite (figure 7.32). This raises the question that this may be an Early Stone Age knapping site that is being exposed by recent erosion. To examine the distribution of the material on transect K5 more closely it is split into 100x100m survey blocks. Figure 7.33 shows the individual survey blocks on K5 and the density of the artefacts recovered. The highest density of finds is located on the top surface of an eroding sand promontory projecting into the end of one of the gullies (figure 7.34), the general position of the gully is clear from the position of the contour lines, and it is also clear from the density of streams that the area is being intensively eroded. Figure 7.35 illustrates a breakdown of the density of the archaeological material on K5 by typological association. It is clear that archaeological material from the Later Stone Age is concentrated on the same survey block as the Early Stone Age artefacts (K5D). One of the four Middle Stone Age artefacts was also found on the same block, whilst the rest of the Middle Stone Age artefacts are found on the adjacent block. Overall the majority of the material on K5D appears to derive from the earlier period, and 51 of the 103 artefacts are Early or indeterminate Early/Middle Stone Age (appendix 7.18). It is unlikely therefore that the larger material represents Mode 1 knapping undertaken in the Middle or Later Stone Age, where for some reason large cobbles were worked, and it is particularly unlikely as the definitively identifiable Later Stone Age presence is minimal (13 artefacts) compared with other transects in the survey. The Later Stone Age material was retrieved is in direct association and interspersed amongst the Early Stone Age artefacts, and there are two possible causes for this. The later material may be derived from higher elevation sands and is now in a secondary context after the supporting sands were washed away. Alternatively they may be genuinely *in-situ*, and the concentration of earlier artefacts acted as a raw material source that was an attraction to the locality. The Iron Age pottery discovered on transect K5D is also of interest, but is presumably displaced from its initial place of discard or associated with very recent activity, as it was discovered at the base of the promontory 2-3 metres below the earlier material discussed here. The material is located over 2.5km from the nearest permanent water source and on poor soils, and as hunter-gatherers range farther away from water sources than Iron Age people (see section 7.7) it is possible this small amount of pottery is associated with Later Stone Age hunter-gatherers.

The overall interpretation of the material on K5 has to remain speculative due to the geomorphic processes that are clearly disrupting the sequence. It is however clear that there is a concentration of archaeological material that has not been moved far since its discard. A large proportion of this material is probably of Early Stone Age origin, which makes it likely that this part of the landscape, at this level, has not been disturbed by geomorphological processes until the present day. Material deriving from the Early Stone Age in this well-preserved condition is rare in the valley, and further investigation should be a priority to try and confirm whether the material genuinely represents Early Stone Age material and whether it is, as is suspected, an *in-situ* knapping site.

Figure 7.32: Photographs of five flake fragments from K5 found within 2m of each other that, though they do not refit, appear to derive from the same piece of quartzite. All are presented from the ventral side and the scale in each photo is in cm. The flake on the right of the central photo, and possibly the flake on the left of the same, is the only whole flake. The rest have end breaks, siret fractures, or other breaks.

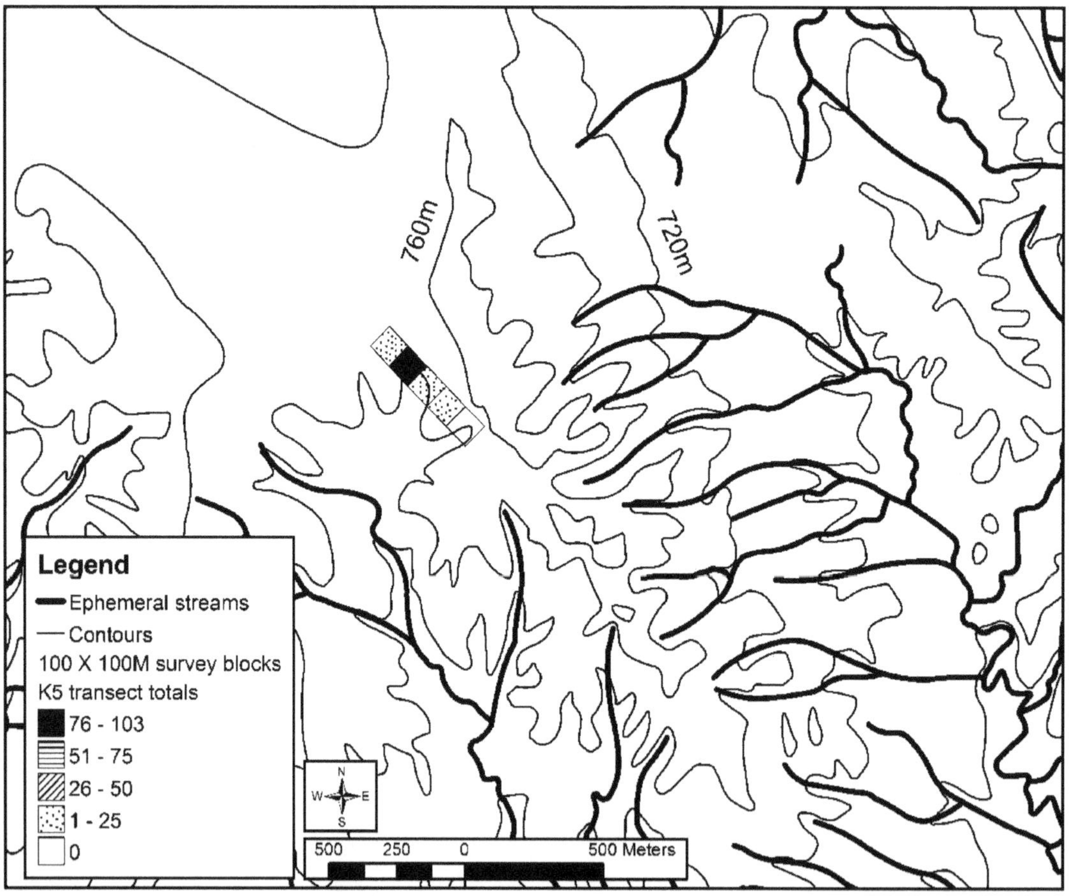

Figure 7.33: Detailed map of transect K5 showing the density of all artefacts collected in relation to the terrain and local drainage systems (raw data displayed in appendix 7.18. The contours are in 20m intervals and the highest point is to the right of the transect which is 760m. The contours and streams were digitised from Zambian Survey map 1331B1 (copyright Zambia Government 1972).

Figure 7.34: A photo of the promontory on which the majority of material was recovered (scale in cm). Note the sand pedestals that the artefacts are supported on, where the artefacts have protected the underlying sand from rainfall. Some of the flakes visible in the photo are those in figure 7.32.

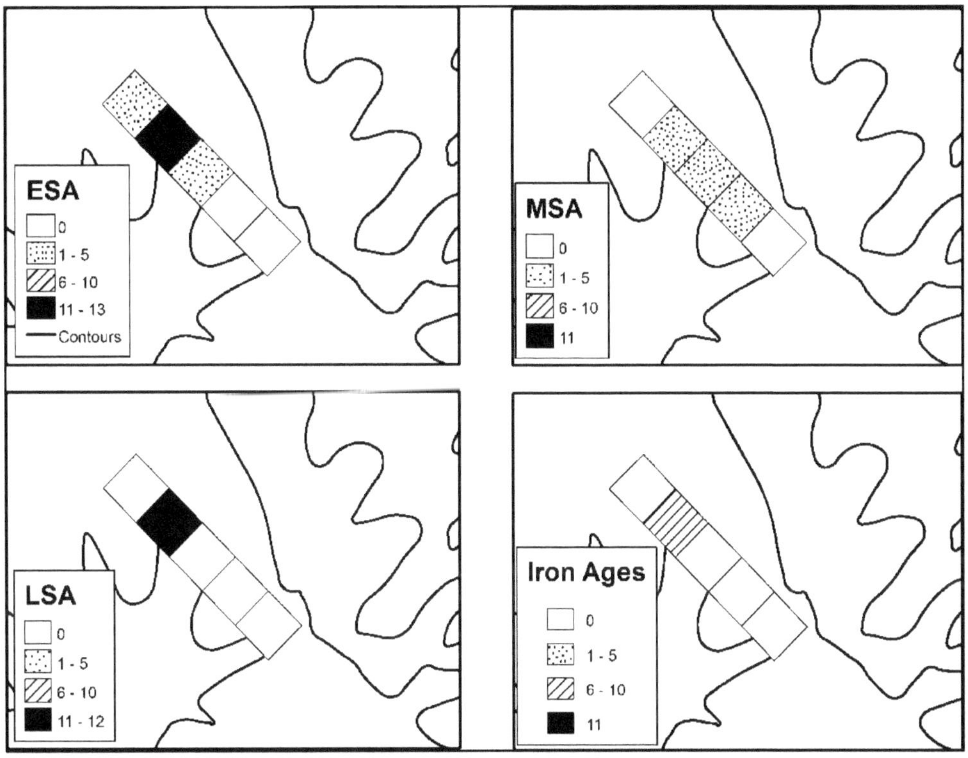

Figure 7.35: Detailed breakdown of the material on transect K5 by each period, the raw data is presented in appendix 7.18. The contours were digitised from Zambian Survey map 1331B1 (copyright Zambia Government 1972).

7.8.5. Introduction to the 'Fossil Wood' or O5 road first season transects

The second group of transects surveyed outside of the main survey area were surveyed near the O5 road during the first season. As noted above (section 5.5.4 and 7.6.5) the collection strategy used in the first season was not well defined, so the archaeological patterns cannot be used as a direct comparison to the main survey. Selective collection of artefacts was not practised at any point in the surveys, and artefacts were collected randomly. The data therefore provides a reliable indicator of the distribution of artefacts in the landscape, but only an indication of artefact density, as artefacts were more intensively collected at some localities usually as a result of time constraints. The 'fossil wood' in the title of this section relates to the location along the O5 road, which is in an area where a large amount of fossilised wood outcrops or lies on the surface. The fossilised wood is ~190 million years old and originally derives from the Late Karoo strata (Utting 1988:4). The transects were surveyed whilst a nearby site, designated SL9, was excavated as part of the PPE project; this is a typologically Early Stone Age (Mode 1) site predominantly devoted to knapping the fossil wood found at the locality (Barham *pers comm.*). The O5 road transects are approximately 70km directly north of the main survey area, their general location is illustrated in figure 5.1, and figure 7.36 illustrates the location of the transects in relation to the elevation and the local streams.

7.8.6. Description of the geomorphological observations and archaeological data from the O5 road transects

The vegetation in the area is predominantly miombo woodland with some patches of *Colophospermum mopane* woodland. The soils are primarily fine to coarse sands and, similarly to the Kasweta Road area, there are patches of cobbles randomly distributed in the area. The key difference is the presence of the fossilised wood that ranges in size from cobbles to large logs and even entire tree trunks. This material varies greatly in quality ranging from friable, easily broken material to hard homogenous glassy material with a crystal structure much like chert, making it very suitable for knapping. Of the two rivers running through the transects the Concrete River was still flowing in May when the transects were surveyed, this suggests that in climates similar to today's smaller streams in the area could have provided a continual supply of water throughout much of the year. Karoo sediments underpin the area and were seen outcropping as a sandstone ridge 2 – 5m high that

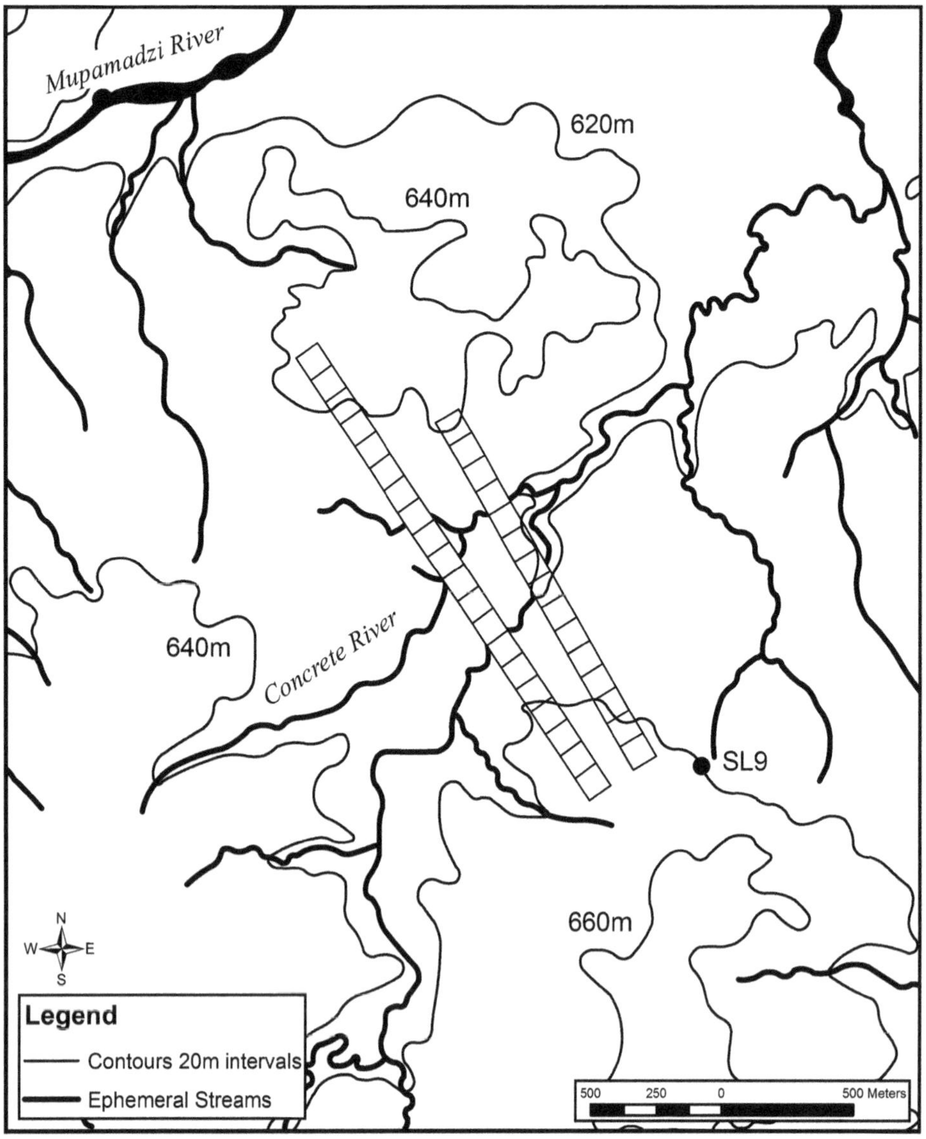

Figure 7.36: Detailed location of the two transects surveyed adjacent to the O5 road in relation to the contour lines and the seasonal streams. The Mupamadzi River (top left of the map) contains water throughout the year. The contours and rivers were digitised from Zambian Survey map 1231B3 (copyright Zambia Government 1972).

bisected the transects perpendicularly 250m from their southern end, approximately in line with the 640m contour. The geomorphological observations were undertaken at the same time as the survey, and were summarised at the end of the time spent along the O5 road in an interpretative cross-section. This cross-section is redrawn and presented here (figure 7.37).

A total of 97 artefacts were collected on the two transects (table 7.10), and on each transect a roughly equal number (10 – 12) of artefacts was recorded from each of the three Stone Ages, with slightly fewer in the Middle Stone Age on each (5-7 artefacts) (these figures include artefacts not collected, but recorded *in situ* on the transects). The distribution of artefacts is illustrated in figure 7.38, and it is clear that there are two clusters of material on each transect that are present in roughly the same positions on each. These clusters are present in each period, though there is some difference in the location of the densest material. There is a greater density of Early Stone Age

Transect number	Typology						
	Early Stone Age	Early/Middle Stone Age	Middle Stone Age	Middle/Later Stone Age	Later Stone Age	Unknown	Total
FS1	12	5	7	2	10	.	36
FW2	11	11	5	12	14	8	61

Table 7.10: Table recording the number of artefacts collected and recorded *in situ* whilst conducting the O5 road survey.

material in the northern-most survey blocks, whilst the Middle Stone Age, though producing lower numbers of artefacts, is densest in the two artefact concentrations to the south. The Later Stone Age material is present in high numbers on three of the four artefact concentrations. No Iron Age material was seen on or around the transects, which suggests a genuine lack of Iron Age activity in the area.

The condition of the artefacts is similar to that of the main survey area (table 7.11). The Early Stone Age material exhibits the most abrasion, and the Middle and Later Stone Ages proportionally less. There is one abraded Later Stone Age artefact, but this is not unusual as it is not unknown in the main survey area. A total of 13 Early Stone Age artefacts were collected from the two transects, three of which were of an unknown raw material, whilst the rest were knapped using fine to coarse quartzites (table 7.11). The definite Middle Stone Age artefacts are either produced in fine quartzite or fossil wood. The fossil wood is the most common raw material in the Middle/Later Stone Age category, indicating that the material is used relatively frequently in later archaeological time. In the Later Stone Age, quartz is overall the most commonly used raw material, though the other glassy textured raw materials such as chert and fossil wood do make up a smaller percentage of raw materials. There are also seven artefacts made from unknown raw materials (table 7.11). Appendix 7.20 lists the tool types identified on each transect by the typological association. The sample size is small, but matches the general trend of the artefact types in the main survey area. Prepared cores and flakes comprise the majority of the Middle Stone Age (Mode 3) artefacts, whilst Later Stone Age (Mode 5) is primarily represented by split cobbles.

Transect number		Typology									
		Early Stone Age		Early/Middle Stone Age		Middle Stone Age		Middle/Late Stone Age		Later Stone Age	
		Number of artefacts		Number of artefacts		Number of artefacts		Number of artefacts		Number of artefacts	
		Sum	Column Sum %	Sum	Column Sum %	Sum	Column Sum %	Sum	Column Sum %	Sum	Column Sum %
FS1	Fresh	4	33.3%	2	40.0%	4	57.1%	2	100.0%	9	90.0%
	Slightly Rolled	5	41.7%	1	20.0%	3	42.9%	.	.0%	1	10.0%
	Moderately Rolled	1	8.3%	2	40.0%	.	.0%	.	.0%	.	.0%
	Very Rolled	2	16.7%	.	.0%	.	.0%	.	.0%	.	.0%
FW2	Fresh	1	11.1%	3	27.3%	3	100.0%	9	75.0%	12	100.0%
	Slightly Rolled	1	11.1%	5	45.5%	.	.0%	2	16.7%	.	.0%
	Moderately Rolled	4	44.4%	3	27.3%	.	.0%	1	8.3%	.	.0%
	Very Rolled	3	33.3%	.	.0%	.	.0%	.	.0%	.	.0%

Transect number		Typology									
		Early Stone Age		Early/Middle Stone Age		Middle Stone Age		Middle/Later Stone Age		Later Stone Age	
		Number of artefacts		Number of artefacts		Number of artefacts		Number of artefacts		Number of artefacts	
		Sum	Column Sum %	Sum	Column Sum %	Sum	Column Sum %	Sum	Column Sum %	Sum	Column Sum %
FS1	Fine Quartzite	2	100.0%	2	66.7%	2	66.7%	.	.0%	.	.0%
	Medium Quartzite	.	.0%	1	33.3%	.	.0%	.	.0%	.	.0%
	Quartz	.	.0%	.	.0%	.	.0%	.	.0%	6	85.7%
	Fossil Wood	.	.0%	.	.0%	1	33.3%	2	100.0%	1	14.3%
FW2	Fine Quartzite	.	.0%	7	63.6%	1	20.0%	1	8.3%	.	.0%
	Medium Quartzite	6	54.5%	3	27.3%	.	.0%	.	.0%	.	.0%
	Coarse Quartzite	2	18.2%	.	.0%	.	.0%	1	8.3%	.	.0%
	Chert	.	.0%	.	.0%	.	.0%	1	8.3%	1	7.1%
	Quartz	.	.0%	.	.0%	.	.0%	1	8.3%	6	42.9%
	Fossil Wood	.	.0%	1	9.1%	1	20.0%	8	66.7%	.	.0%
	Unknown	3	27.3%	.	.0%	3	60.0%	.	.0%	7	50.0%

Table 7.11: The numbers and percentages of artefacts collected on each transect along the O5 road tabulated by typology, condition, and raw material.

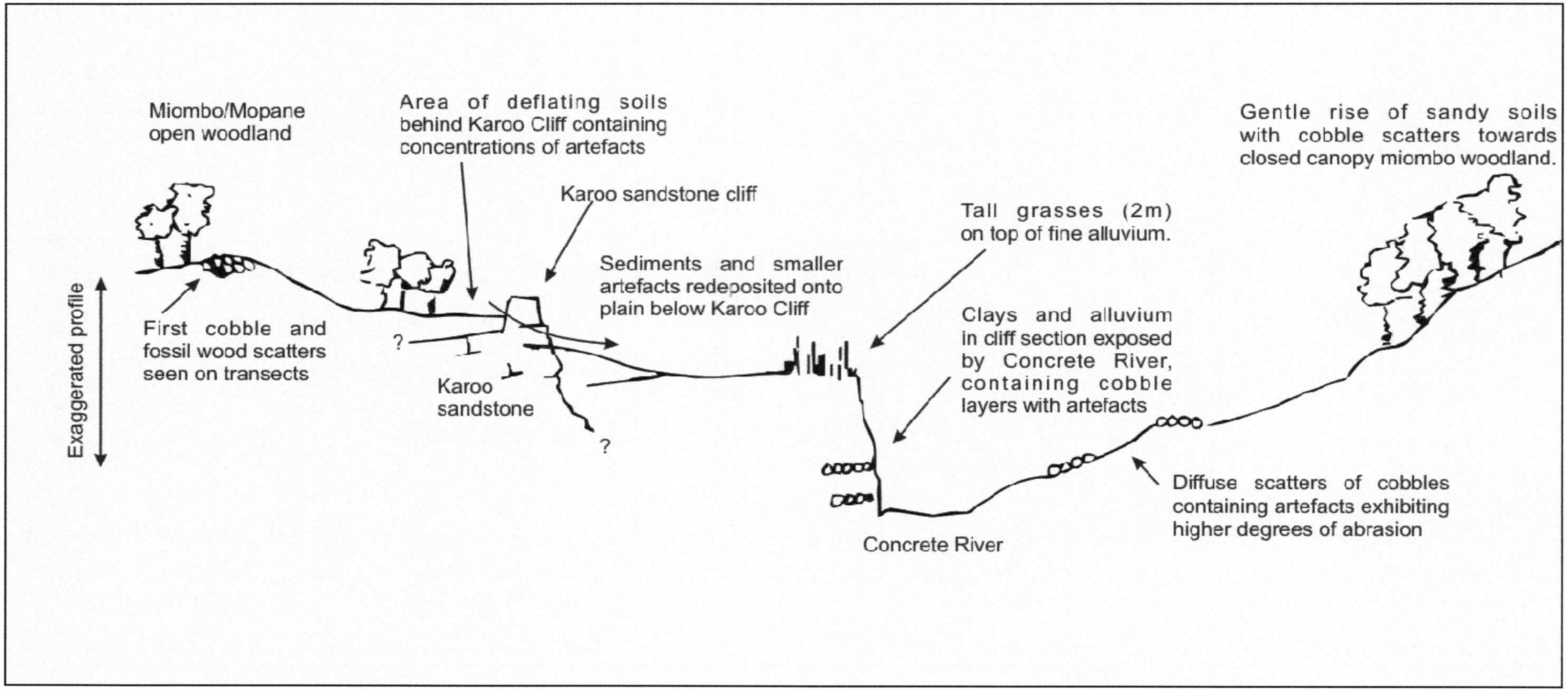

Figure 7.37: Field sketch summary of the geomorphology on the O5 road transects, the horizontal distance is approximately 2km, and vertical elevation is 40m (features not illustrated to scale). The question marks refer to uncertainty regarding the position of the Karoo bedrock, as it does not outcrop anywhere else on the transects.

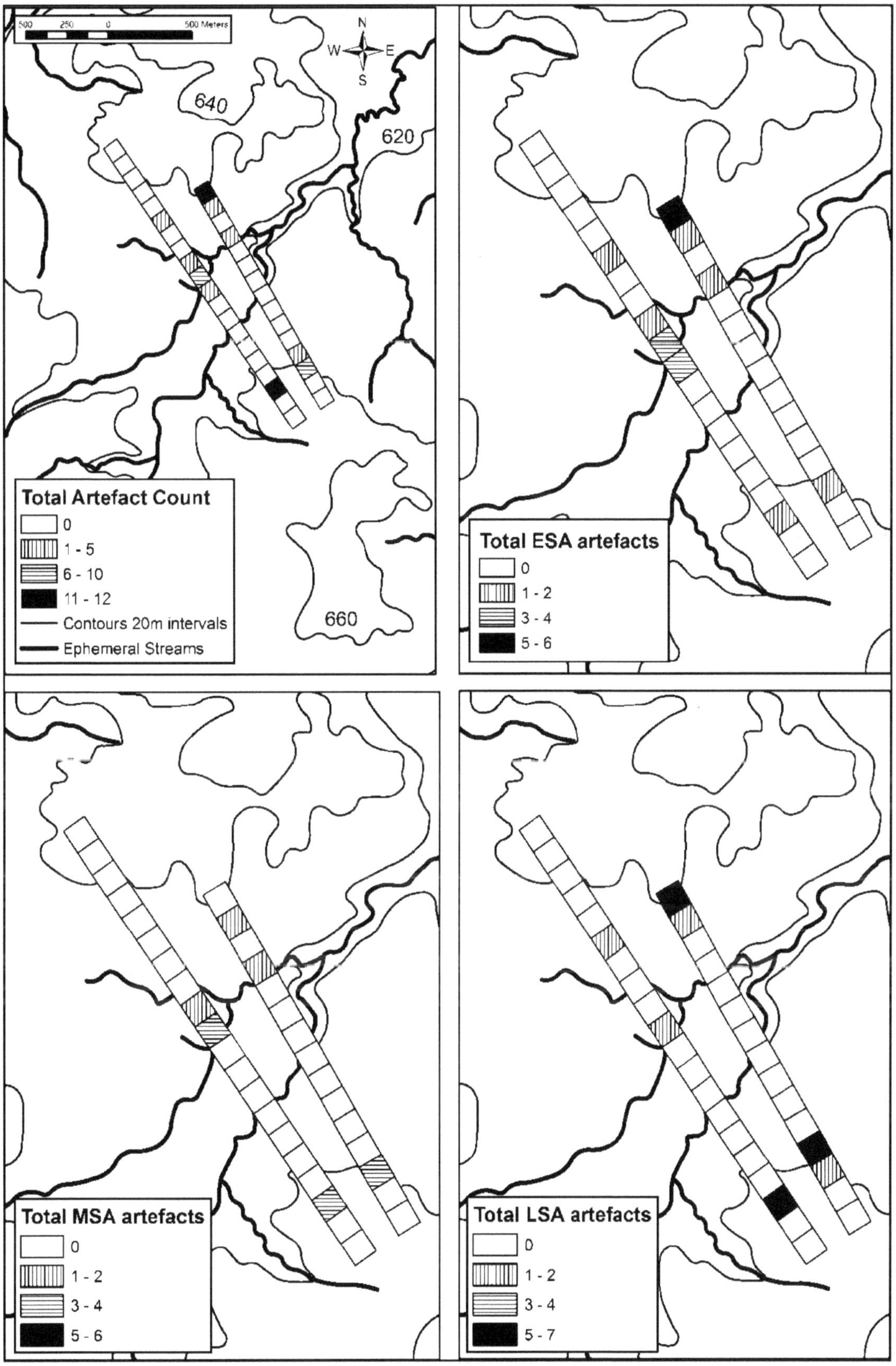

Figure 7.38: Maps illustrating the artefact density on the two O5 road transects. The total artefact counts are illustrated top left, and the streams and contours are provided for reference. The transect on the left was designated FS1 and the one on the right FW2, each survey block was assigned a letter, starting on each transect with A at the southern end. The next three maps sequentially illustrate the artefact density through the Stone Ages, the artefact density for these three maps is presented to the same proportions. The raw data is displayed in appendix 7.19. The contours and rivers were digitised from Zambian Survey map 1231B3 (copyright Zambia Government 1972).

7.8.7. Interpretation of the geomorphological and archaeological data from the O5 road transects

The cross-section of the geomorphology of the area (figure 7.37) provides a summary of the processes in the local area. The location of the Karoo cliff corresponds to the highest artefact concentrations on the southern portion of the two transects, and it is clear from figure 7.38 that this concentration was primarily made up of Middle and Later Stone Age material. The area directly below the ridge contains a rapidly decreasing number of artefacts that derive from the area above the ridge, as it was clear from the fan shaped deposits at the base of the channels dissecting the ridge that the soils here are being reworked from above the Karoo ridge. The two clusters of artefacts in the northern survey blocks are present close to the streams that bisect the transects and, as it seems likely that the streams would have affected the archaeological record, the condition of the artefacts is listed by the block code that they are found on (appendix 7.21). This data demonstrates that the rolled artefacts mainly occur nearer to the ephemeral streams, whilst the fresh artefacts derive from survey blocks at slightly higher altitudes. It is also clear that all the Middle and Later Stone Age material above the Karoo ridge is un-abraded (the single Early Stone Age artefact is slightly abraded). The area behind the ridge is therefore interpreted as surviving in a relatively undisturbed condition since the Middle Stone Age material was discarded here. The Mode 1 site SL9 is also situated behind a small ridge, suggesting a similar scenario for its preservation (Barham *pers comm.*). The higher density of artefacts near the top of the ridge compared with those of the survey blocks in the southernmost portion of the transects is likely to be a result of the deflating sands near the ridge exposing artefacts. It is therefore a preserved part of the landscape, but it is not possible to conclude that the ridge was a particularly heavily used part of the landscape in the Middle and Later Stone Age. The Concrete River that runs through the area cuts through the sediments that are washing down from the Karoo ridge, resulting in exposure of cobble layers in the sediment sections in the area. This cobble material included some artefacts, which presumably represent a combination of material that originated at the top of the ridge and lag deposits from the previous courses of the Concrete River. The majority of the rolled artefacts derived from the north side of the Concrete River, and it is also here that the majority of Early Stone Age artefacts are found; this surface clearly represents an area of greater disturbance relating to the river system.

The raw materials used in the area appear in a similar distribution to the main survey area. The key difference is with the appearance of fossil wood as a raw material, which replaces chert for some Middle and Later Stone Age knapping. There were no clast counts undertaken in this area so it is difficult to compare the archaeological remains with the background clast count. However, observations on the transects and at SL9 revealed that fossil wood is very common, and often occurs as large fossilised logs. The fact that it does not appear to be transported to parts of the landscape where it does not locally outcrop, for example the Kasweta Road and the main survey area, indicates that its use represents a localised response to the presence of a material with useful knapping qualities. Despite the prevalence of the fossil wood in the area, artefacts that can be definitively assigned to the Later Stone Age are predominately made of quartz suggesting that, for the more distinctive split cores and microlithic Mode 5 technology, quartz remained the preferred raw material.

The absence of Iron Age activity is surprising as it seems probable, based on the observations and interpretations of section 7.7, that the proximity of the area to a permanent water source and potentially fertile soils would have rendered it suitable for Iron Age settlement. However, the soil quality in the area might have been regarded as quite poor. Trapnell, in a summary of the traditional agricultural techniques of eastern Zambia, notes that one of the indicators used by traditional agriculturalists to identify suitable soils is freedom from iron concretions within two feet of the surface (Trapnell 1996:12). In parts of the survey area, mainly above the Karoo ridge, a large amount of haematite is seen outcropping in the deflating soils, often as narrow layers of small nodules (1 – 3cm). If Iron Age farmers were in the area and used this as a guide to soil quality, and were aware of the haematite nodules, it is likely that it would have would have been a factor influencing the decision not to locate in this area.

7.9. Data from the First Season (2005)

7.9.1. Introduction

The data from the first season's transects within the main survey area are fully described and analysed here. Some of the data from the first season was included in sections 7.6.5 and 7.7.4 as it provided a test for the Later Stone Age and Iron Age models of land-use. The O5 road transects from the first season were fully described and analysed above, as they provided an additional sample from outside the main survey area. The first season's data provides a comparison and an addition to the sample in the main survey area, and is used to determine whether there are any major differences in the data that conflict with the interpretations made above. As noted above, selective collection based on artefact type, raw material, or period was not undertaken. However, not all areas surveyed were equally sampled. This data has to be treated with caution, but it should still provide a useful comparison to the main survey data. Detailed geomorphological observations are not necessary here as the area are described and interpreted in appendix 6.1 and chapter 6.

7.9.2. Description of distribution of the archaeological data, artefact condition and raw material used

The location of the transects with reference to the geomorphological zones is illustrated in figure 7.39; it is clear from this that the LU1 transects cross three geomorphological zones (Recent Cut-off (3), Cobble zone (7), and Sand Bank zone (11)), and the CHO transect straddles two, the Sand and Silt zone on the south-east bank (5), and the Black Rock zone (9). The variability in the geomorphological terrains seen on each transect was one of the key reasons for redesigning the survey for the second season, so that overlaps between the archaeological data and the different zones would be prevented allowing for more straightforward analyses (see section 5.5.4). In this case the 100x100m blocks provide a good breakdown of the data between the different zones, and these were used to illustrate the distribution of artefacts.

Figure 7.40 illustrates the distribution of the Stone Age and Iron Age artefacts discovered on the LU1 transect. The Early Stone Age material is entirely located within the Cobble Zone (7), although one block shares a boundary with zone 11, the recording form reveals that the material originated from the cobbles in zone 7. The highest Early Stone Age count numbered eight artefacts, which is a figure within the range of the artefact counts in the main survey. The spacing of the blocks containing artefacts in the Cobble zone appears regular, however this is a result of the rushed collection strategy in the first season and not an indication of any behavioural pattern. Samples were taken at regular intervals in the cobble ridges (later properly mapped as Cobble zone 7) to gain samples from near the edge of the cobble ridges, and comparative samples from further away from the floodplain. The Middle Stone Age artefacts occur in the same area, and were retrieved from the Cobble zone (7). The Later Stone Age material appears in all of the samples collected from the Cobble zone (7), and also appears on the Recent Cut-off zone (3) occasionally in great quantity, up to 120 artefacts on one survey block, which to an extent reflects an intensive collection by the survey team. The collection on the floodplain (zone 3) was more systematic than the collection in the Cobble zone and the distribution, if not density, of artefacts is therefore more accurately represented. The Iron Age material was only present on zone 3, and no artefacts were seen or collected from the Sand Bank zone (11) or the Cobble zone (7). In general 1 – 10 Iron Age artefacts were recovered from survey blocks containing artefacts, although in one instance 35 were collected.

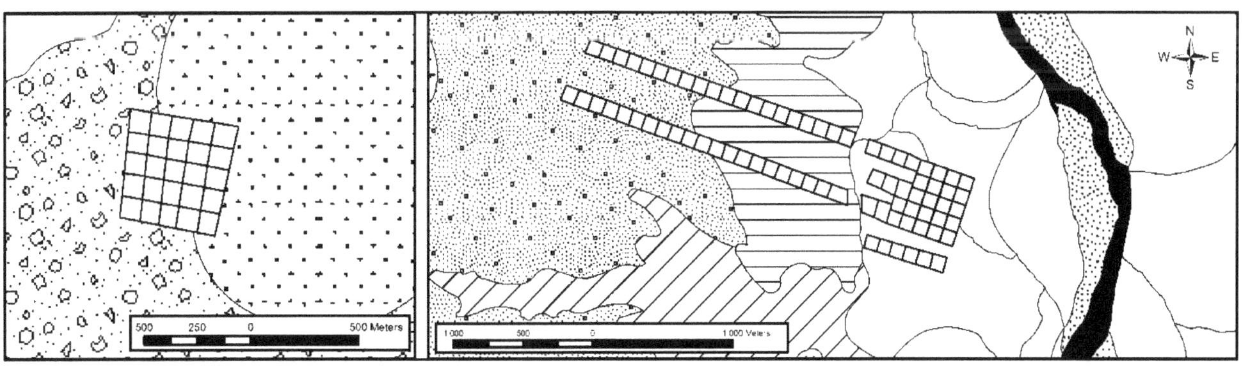

Figure 7.39: Maps illustrating the relationship of the first season's transects to the geomorphological zones (LU1 above and CHO right). An overall view of the transects location in relation to each other can be gained from figure 7.26.

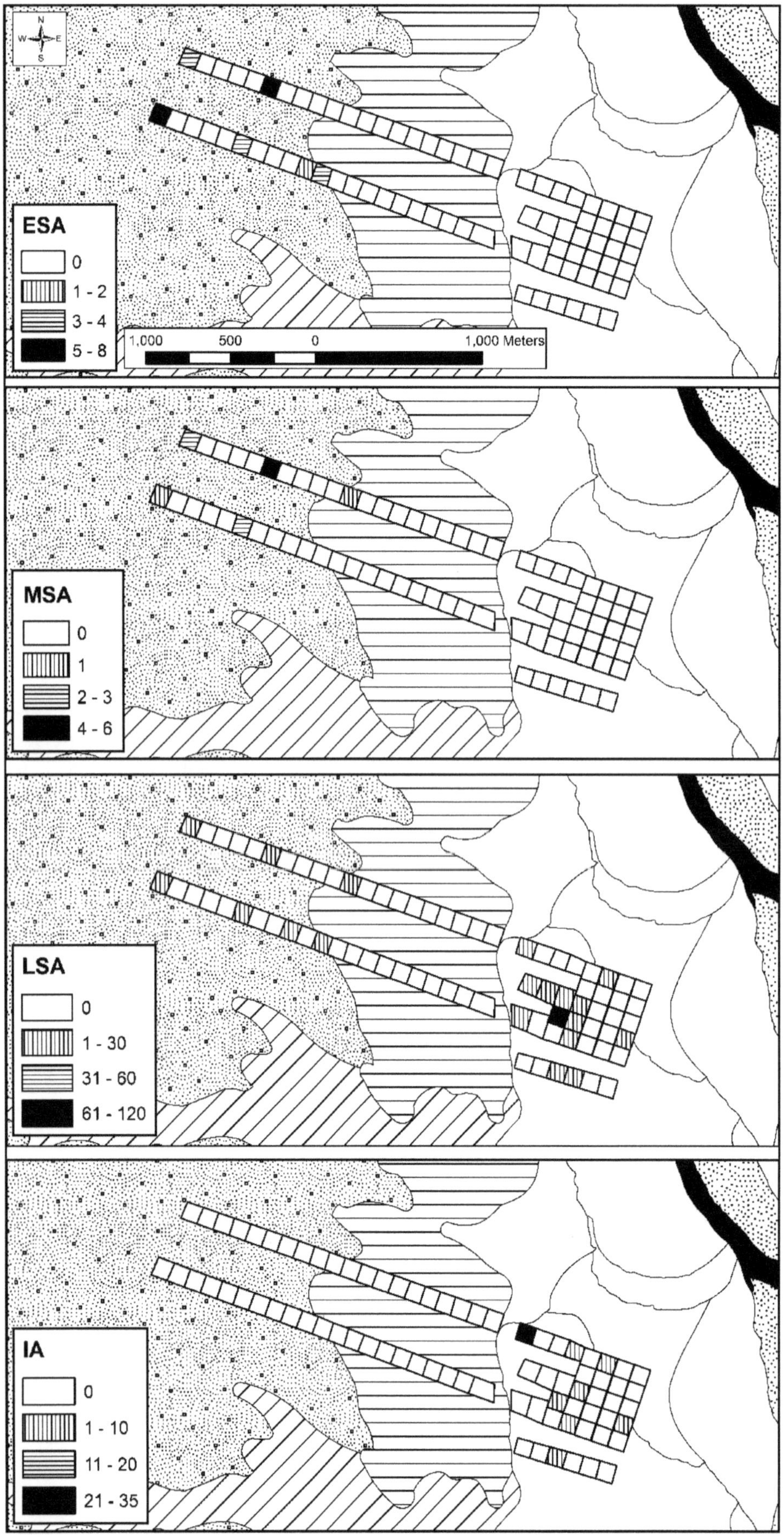

Figure 7.40: Maps illustrating the density of artefacts on transect LU1 for the Early Stone Age to Iron Age (the data for the Iron Ages has been combined). The raw data is presented in Appendix 7.22.

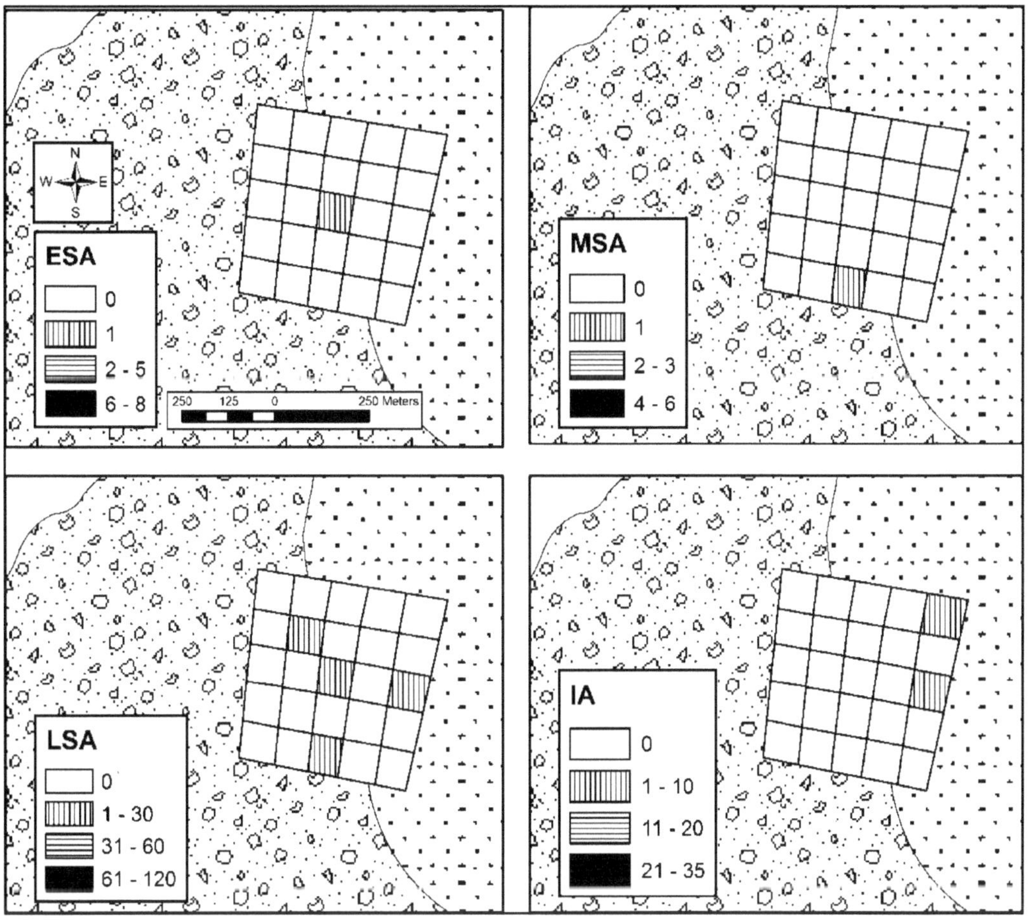

Figure 7.41: Artefact counts for the CHO transects. The raw data is presented in appendix 7.22.

The results for the transect designated CHO, surveyed on the south-east bank, are displayed in figure 7.41. There was only one Early Stone Age artefact retrieved from this area, and this was on the boundary between the Black Rock (9) and Sand and Silt zone (5). Similarly, only one Middle Stone Age artefact was recovered from this transect, and this was in the Black Rock zone. The Later Stone Age artefacts were recovered from both zones, but in low quantities comparative to the LU1 transect. Iron Age material was also recovered in low quantities, but only from the Sand and Silt zone.

The Early to Middle Stone Age material from LU1 is usually abraded to some extent (table 7.12), and a relatively low number of artefacts are fresh (16 of the total of 94 artefacts in the two periods (see appendix 7.22)). Only one slightly rolled artefact was recorded from the Later Stone Age for both LU1 and CHO, though the Middle/Later Stone Age data contains a large proportion of rolled artefacts on LU1, which includes one heavily rolled artefact.

The raw material of the Early Stone Age artefacts found on transect LU1 is primarily coarse quartzite, with fine and medium quartzite accounting for only 24% of the sample (table 7.13). The only Early Stone Age artefact recovered from CHO was knapped in silcrete, a rare raw material in this part of the valley that was not present in any of the clast counts. The Middle Stone Age artefacts were primarily knapped using fine quartzite or chert, with lower proportions of medium and coarse quartzite. The Later Stone Age artefacts are almost exclusively quartz, though there is also a small proportion of chert and other raw materials on LU1. The indeterminate categories for the artefacts appear to contain a mixture of raw materials, which reflect the raw materials used for the artefacts from the known Stone Ages either side.

7.9.3. Interpretation of distribution of the archaeological data, its condition and raw material used

The collection regime in the first season was not as well designed and controlled as it was in the second season. This resulted in the density of material being less proportionally representative as it is in the second season though, with the exception of the LU1 in the Cobble zone (7), the overall distribution of artefacts is representative. The artefacts from the LU1 and CHO transects yield similar patterns in the data to those seen in the second season. Material from all three Stone Ages is present in the Cobble zone (7), and the fact that the Early and Middle Stone Age material is often rolled is indicative of the alluvial fan formational processes outlined in chapter 6. Compared with the Middle Stone Age data from the main survey area a greater proportion of artefacts are rolled to some extent, suggesting that the degree of stabilisation of the fan suggested in section 7.4.2 above was not present over the whole of the fan. Differential amounts of activity on alluvial fans is typical, and as the sample in the Cobble zone (7) was spatially restricted in the first season, the data does not contradict the conclusion above that overall fan activity was reduced in the Middle Stone Age. The fresh condition of the Later Stone Age artefacts in the Cobble zone, and over the rest of the land surface, further indicates that the land surface over much of the survey area had stabilised by this time. The presence of a wide distribution of Later Stone Age material across the survey blocks on the Recent Cut-off zone (3) seems inconsistent with the observations of the second season, where virtually no Later Stone Age artefacts were recovered in zone 3. However, the artefacts were all present on one meander cut-off suggesting that this is a localised phenomenon. This meander must therefore have been isolated from the Luangwa before the last century, preserving the Later Stone Age activity in the area, suggesting that Later Stone Age hunter-gatherers, as suggested in section 7.5.2, did not specifically regard the landscape in zone 3, as unsuitable, and the rapid movement of the Luangwa River has destroyed a large amount of occupational evidence. Only a few Early and Middle Stone Age artefacts were recorded on the second season transects in the Black Rock zone (9), and similarly a small number appear on the edge of zone 9 on the CHO transect, indicating that the debris flow deposits here were present and possibly accumulating throughout the Pleistocene. The Iron Age material only appears on the alluvial soils of the Recent Cut-off zone (3) and the Sand and Silt zone on the south-east bank (5), which is the same pattern described and modelled in the second season's data. The raw material used in each of the stone ages corresponds to the general pattern described and interpreted in the main survey. There appears to be a preference for quartzite in the Early Stone Age, whilst the Middle Stone Age artefacts are primarily produced using finer raw materials such as chert and fine quartzite. The Later Stone Age artefacts are almost exclusively produced using quartz.

Table 7.12: Condition of the artefacts found on transects LU1 and CHO. Artefact numbers and percentages for each typological grouping are listed.

Condition (degree of rolling)	Transect number									
	LU1									
	Early Stone Age		Early/Middle Stone Age		Middle Stone Age		Middle/Later Stone Age		Later Stone Age	
	Number of artefacts		Number of artefacts		Number of artefacts		Number of artefacts		Number of artefacts	
	Sum	Column Sum %	Sum	Column Sum %	Sum	Column Sum %	Sum	Column Sum %	Sum	Column Sum %
Fresh	2	8.7%	10	17.5%	4	28.6%	12	57.1%	259	99.6%
Slightly Rolled	9	39.1%	22	38.6%	8	57.1%	8	38.1%	1	.4%
Moderately Rolled	9	39.1%	21	36.8%	2	14.3%	.	.0%	.	.0%
Very Rolled	3	13.0%	4	7.0%	.	.0%	1	4.8%	.	.0%

Condition (degree of rolling)	CHO									
	Early Stone Age		Early/Middle Stone Age		Middle Stone Age		Middle/Later Stone Age		Later Stone Age	
	Number of artefacts		Number of artefacts		Number of artefacts		Number of artefacts		Number of artefacts	
	Sum	Column Sum %	Sum	Column Sum %	Sum	Column Sum %	Sum	Column Sum %	Sum	Column Sum %
Fresh	.	.0%	1	25.0%	.	.0%	1	100.0%	16	100.0%
Slightly Rolled	.	.0%	3	75.0%	1	100.0%	.	.0%	.	.0%
Moderately Rolled	1	100.0%	.	.0%	.	.0%	.	.0%	.	.0%
Very Rolled	.	.0%	.	.0%	.	.0%	.	.0%	.	.0%

Raw material	Transect Number									
	LU1									
	Early Stone Age		Early/Middle Stone Age		Middle Stone Age		Middle/Later Stone Age		Later Stone Age	
	Number of artefacts		Number of artefacts		Number of artefacts		Number of artefacts		Number of artefacts	
	Sum	Column Sum %	Sum	Column Sum %	Sum	Column Sum %	Sum	Column Sum %	Sum	Column Sum %
Fine Quartzite	3	13.0%	16	27.1%	5	35.7%	4	19.0%	3	1.2%
Medium Quartzite	2	8.7%	6	10.2%	2	14.3%	2	9.5%	3	1.2%
Coarse Quartzite	16	69.6%	24	40.7%	2	14.3%	3	14.3%	.	.0%
Vein Quartz	.	.0%	1	1.7%	.	.0%	.	.0%	.	.0%
Chert	.	.0%	10	16.9%	5	35.7%	5	23.8%	15	5.8%
Quartz	2	8.7%	1	1.7%	.	.0%	7	33.3%	238	91.5%
Tabular Quartzite	.	.0%	1	1.7%	.	.0%	.	.0%	.	.0%
Black Rock	.	.0%	.	.0%	.	.0%	.	.0%	.	.0%
Heat Treated Chert	.	.0%	.	.0%	.	.0%	.	.0%	1	.4%

Raw material	Transect number									
	CHO									
	Early Stone Age		Early/Middle Stone Age		Middle Stone Age		Middle/Later Stone Age		Later Stone Age	
	Number of artefacts		Number of artefacts		Number of artefacts		Number of artefacts		Number of artefacts	
	Sum	Column Sum %	Sum	Column Sum %	Sum	Column Sum %	Sum	Column Sum %	Sum	Column Sum %
Fine Quartzite	.	.0%	1	25.0%	1	100.0%	1	100.0%	.	.0%
Medium Quartzite	.	.0%	.	.0%	.	.0%	.	.0%	.	.0%
Coarse Quartzite	.	.0%	1	25.0%	.	.0%	.	.0%	.	.0%
Vein Quartz	.	.0%	2	50.0%	.	.0%	.	.0%	.	.0%
Chert	.	.0%	.	.0%	.	.0%	.	.0%	.	.0%
Quartz	.	.0%	.	.0%	.	.0%	.	.0%	16	100.0%
Tabular Quartzite	.	.0%	.	.0%	.	.0%	.	.0%	.	.0%
Black Rock	1	100.0%	.	.0%	.	.0%	.	.0%	.	.0%
Heat Treated Chert	.	.0%	.	.0%	.	.0%	.	.0%	.	.0%

Table 7.13: Raw material of the artefacts from transects LU1 and CHO. Artefact numbers and percentages of artefacts from each typological grouping are listed.

7.10. Summary

The interpretation of the geomorphological processes in the main survey area provided the necessary context for interpreting the archaeological record, which at times and in places has been significantly altered or destroyed as a result of these processes. The Early and Middle Stone Age artefacts occur in a similar distribution to each other, and are primarily located in both primary and secondary contexts in zones characterised by alluvial fan deposits (zones 7, 8, and 14) that accumulated throughout the Pleistocene, and within Holocene terraces of the Luangwa River (zone 5). The un-abraded Early Stone Age artefacts indicate that some material is in a primary context. Overall, however, the behavioural record has been virtually destroyed. As a result, it is unlikely that any sites remain intact and any that do will be difficult to locate, and, settlement patterns and tool use behaviour cannot therefore be reconstructed in any detail. Outside the main survey area, toward the centre of the Luangwa Valley, there is evidence on transect K5 suggesting that Early Stone Age sites may survive undisturbed and are only now being exposed by geomorphological agents. The stability of the sediments over large areas and the limited number of transects in this landscape makes it impossible to know the extent of these scatters, or to model where more of these scatters might outcrop. Considering the uniqueness of the scatter, or site, seen and sampled on K5, future investigations into the Early Stone Age of the Luangwa Valley should regard this location as a priority for excavation.

In light of the geomorphological history of the area outlined in chapter 6, the data for the Middle Stone Age are surprising as it was expected that the artefacts would demonstrate similar abrasion patterns to the Early Stone Age sample. However, the Middle Stone Age material from the second season's survey exhibits less abrasion, indicating that the alluvial fans that deposited the cobbles were less active. The younger age of the Middle Stone Age material was considered to be a factor in the lower degrees of abrasion seen. However, the data from the LU1 transect in the first season revealed that material from this age could be rolled to the same extent as Early Stone Age material, and that localised parts of the fans were still as active during the Middle Stone Age. The disruption to the archaeological record limits behavioural interpretations of the data, but, there is evidence of raw material selection in the landscape. Though it is not possible to say for certain that raw material at the two localities discussed was the primary factor influencing site location, it is likely that chert, at the minimum, was used for the majority of the more complex Mode 3 technologies. The presence of a small amount of fresh Middle Stone Age material on the Sand and Silt zone (5) suggests that there may be relatively undisturbed material from this period contained within, or possibly near, the base of these sands that are being eroded. As a result of the small sample size in the Middle Stone Age no further factors affecting land-use can be identified without more extensive survey work.

The planned survey plan was redesigned based on the experience of the first season, and the methodology and data effectively provided a pre-test for the main survey. The data did not contribute to the initial interpretation of the archaeological record as the first season transects did not use the same randomised sampling strategy and their inclusion would have skewed the results. The first season's data were analysed as they still provide a useful comparison with the second season's results. Overall, the first season data supported the interpretations derived from the main survey, and in some cases provided enhancements, with the increased component of rolled artefacts from the Middle Stone Age in the Cobble zone (7), and the Later Stone Age artefacts found on parts of the Recent Cut-off zone (3).

The Later Stone Age and Iron Age contain the most complete datasets in the landscape. In the sand and silt zones (4 & 5) the landscape has in places aggraded until recently (e.g. SL7), suggesting that the scatters present on the surface represent the very final stages of hunter-gatherer and sedentary land-use before the 20th century. Later Stone Age land-use was distributed across much of the landscape, and it is also apparent that some sites are present in addition to the off-site material. The predictive model created in GIS aimed to model land-use intensity, rather than the location of sites, which is the more usual quest. Judging by the qualitative assessments of its reliability presented in this chapter, it has been successful. Although the model does not always reliably predict site location, its key achievement has been the provision of a comparison between Later Stone Age and Iron Age land-use intensity. The Iron Age survey results reveal that the requirement for good soils and water supplies severely limited location choices, with the result that the rocky or bare terrains of the Nchindeni Hills and the cobble zones were not suitable for habitation. Distance from water or the presence of unsuitable soils also appear to be limiting factors for Iron Age activity along the Kasweta and O5 Roads. The land-use model of the Iron Age data correlated well with the surveyed distribution of both off-site scatters and sites. The comparison with the Later Stone Age model allows the survey work and the archaeology of the valley to be linked with research conducted elsewhere in Zambia regarding the relationships of the two groups of people, and the data and models presented here indicate that a developmental sequence between the two is unlikely, and that Later Stone Age people would have either used pottery from abandoned Iron Age sites or engaged with limited trade with the Iron Age farmers.

8. Summary and Discussion

The primary aim of the research was to develop an understanding of the geomorphological processes that have affected the archaeological record in the South Luangwa National Park in Zambia, and determine the extent to which genuine behavioural residues survive. In the final design of the survey a sample of the Luangwa Valley was selected for investigation, this included part of the valley floor, the Luangwa River, and the Nchindeni Hills that form the south-eastern margin of the valley. The archaeological distribution was surveyed using a stratified off-site strategy, which provided spatial data of the density, condition, and typology of archaeological material in the landscape. The previously un-described Pleistocene geomorphology of the Luangwa Valley has now been described and interpreted for the survey area. Holocene processes have also been examined, and those described and modelled by Gilvear *et al.* (2000) for the Luangwa River have been further researched. Descriptions of the geological sections and sediments in the survey area, mapping from aerial photography, and a clast count analysis provided vital data to develop an understanding of the deposits, which could then be related to the distribution of archaeological material.

The experience gained from the first season's research confirmed the findings made in the case studies analysed in chapter 2, which indicated that particular regard needs to be paid when working in Africa to the working environment, and to be flexible with the sampling methodology, though not to the point that the survey is compromised. Fortunately, surveying in two seasons provided a test of the survey methodology, which could then be suitably revised. The methodologies for both the archaeological and geomorphological research were successful in meeting the projects primary aims. This project demonstrates that a simple well-designed survey combining geomorphological and archaeological research, using the current generation of GPS units and portable computers to manage the survey can yield high quality data in unknown and logistically difficult environments. This strategy is of value for any research project where resources such as finance and manpower are limited, and could in practice be used in any landscape environment. The key limitation to the technologies used is access to reliable sources of power to recharge the equipment, although if solar powered battery rechargers become more affordable, they might provide a viable solution. This proviso may make it unsuitable to a number of African environments, but, despite this, local heritage organisations such as the National Heritage Conservation Commission in Zambia would benefit from these techniques where an assessment of the archaeological record is required for management, or planning concerns.

The geomorphological investigation included the mapping of sixteen geomorphological zones within the survey area. Based on the map three key areas of investigation were defined: to determine the past and present action of the Luangwa River (encompassing zones 1 – 5 & 15); to develop an understanding of the erosion processes on the Nchindeni Hills (zones 6 & 9); and to interpret the environment of deposition of the cobble deposits that occur in the region (zones 7, 8, & 14). The geomorphological evidence gained in this project indicates that the Luangwa River has been constrained by the geology of the valley, with the result that within the main survey area the river has actively avulsed and reworked its current floodplain throughout the Pleistocene. The sediments in the Manzi sequence and terrace deposits of zones 4 and 5, which are above the current level of the Luangwa, indicate that this activity has included both aggradation and degradation of the floodplain. Past climates can be reconstructed from records in and around the southern African region, and these indicate that the Luangwa River would have been much reduced during glacial episodes, and presumably absent in the harshest hyper-arid conditions recorded in the Lake Malawi cores in the cooler sub-stages of MIS5. Older drier climate cycles related to the precessional cycle might have been characterised by hyper-aridity for which, at present, there are no records. The cobble deposits on either side of the Luangwa's floodplain derive from alluvial fans that were deposited in semi-arid or arid glacial episodes. The archaeological evidence suggests that they were reactivated during a number of glacial cycles forming an Early to Middle Stone Age sequence. These fan deposits are dissected by Holocene streams that have removed a large proportion of the deposits leaving cobbles and the associated archaeological record isolated on the Karoo or Archean bedrock in zones 7, 8, and 12. Zone 14 is similar, but partially buried by overbank deposits of sand and silt from the Luangwa. Later Stone Age occupation is evident on this dissected surface. The Nchindeni Hills have been subject to gradual erosion and weathering, but have essentially remained an unchanged feature throughout the Pleistocene.

The Early Stone Age record of the main survey area does not retain any spatial behavioural data, and is primarily contained within secondary contexts. Where large proportions of un-abraded artefacts occur, transect GR2 for example, much of the associated material has been disturbed or removed. It is therefore difficult to suggest locations of high integrity sites. The only behavioural evidence that could be discerned was a preference by Early Stone Age hominins for using quartzite for Mode 1 knapping, and that possible raw material constraints inhibited the use of Mode 2 practices. Geomorphological processes have affected the Middle Stone Age record in a similar manner, but there is evidence for preferential raw material selection as Mode 3 technologies are primarily knapped using fine-grained raw materials. There is also limited spatial behavioural data present in the distribution of Mode 3 technologies, as they occur most frequently in restricted areas of the landscape that might relate to outcrops of chert.

In the wider context of the PPE project and central African archaeology the Early and Middle Stone Age occupation of the valley is important to understand, as the Valley may have served as a refuge for fauna and flora during glacial episodes when central and western Zambia was uninhabitable (Barham 2001:71-76), and potentially acted as a dispersal corridor linking central and eastern Africa (Barham 2001:71). During glacial episodes, the Luangwa Valley and its catchment areas would have experienced cooler and drier conditions overall, but the precessional forcing of the monsoon continued to influence the climate and would have caused alternations between comparatively wetter and drier conditions. The presence of un-abraded stone tools within the Pleistocene alluvial fan deposits suggests that the valley may have at the least been inhabited during parts of glacials. However, it is also possible that these tools were manufactured and deposited during an interglacial, when the deposits were exposed and the fans inactive, so this hypothesis remains to be tested, and will require a robust chronological framework. The potential Early Stone Age site located along the Kasweta Road requires further investigation because if it is confirmed as Early Stone Age, it represents a rare class of site in the Luangwa Valley which could potentially provide a dateable stratigraphic context and valuable environmental data regarding the earliest occupation of the valley. Further survey work along the Kasweta Road would also provide data that could be used to map potential outcrops of such archaeological material in the area. The results of this survey support the theory that the Luangwa Valley could have acted as an ecological corridor and possibly a refuge, linking western and central Zambia with eastern Africa during harsher climatic regimes that precipitated the abandonment recorded at sites such as Mumbwa Caves and Twin Rivers (Barham 2000:49 & 236).

The Later Stone Age and Iron Age archaeological records retain intact spatial behavioural data in the landscape, evident by the large amount of off-site data and the small number of sites discovered. Later Stone Age hunter-gathers primarily used small river worn cobbles for the majority of their split cobble microlithic technology, and transported this raw material at least short distances to occupation sites. Later Stone Age and Iron Age land-use can be modelled using GIS and, although the models cannot be statistically developed or tested in the manner usually proscribed by GIS practitioners, the models fit the known distribution of archaeological material from the main survey and the test transects surveyed, with the exception of some Later Stone Age sites. The GIS models and detailed analysis of the data provide a test for some of the theories developed elsewhere in Zambia to explain the Later Stone Age to Iron Age transition. It is clear that there is no developmental sequence between the Later Stone Age and the Iron Age, as the two groups of people coexisted in the landscape. Despite this coexistence and overlapping land-use, the small dataset presented in this research indicates that a large amount of contact did not occur, as Iron Age sites contain very few lithics, and Later Stone Age sites contain relatively few fragments of pottery. It is likely though, that the main survey area is a fruitful area for further research into the interactions of these two groups.

This project has provided the first data for the Pleistocene geomorphology and archaeological landscape of the area, and this research opens a number of new avenues of investigation:

1. The Later Stone Age and Iron Age distributions in the valley provide the best potential for further research into the land-use strategies of two coexisting and contrasting social groups in Zambia. Further research could include more detailed mapping of features in the landscape, such as the individual granite inselbergs which would have acted as shelters, and thus draws to specific parts of the landscape for hunter-gatherers. In the case of the Iron Age populations detailed mapping of soil types could provide further resolution to possible land-use strategies.
2. An intensive survey of the survey area specifically targeting the Later Stone Age and Iron Age populations, combined with sub-surface sampling to determine the depth of some archaeological deposits, could further resolve the interrelations of the two groups.
3. The survey area covered a relatively small but ecologically and geomorphologically diverse area of the South Luangwa Valley, but many geomorphological terrains remain unexplored. Assuming that it is logistically possible, the environments that include the Muchinga Escarpment and the alluvial fan deposits at the base require investigation to characterise the geomorphological and archaeological deposits. Potentially sites of any period may survive in good condition.
4. An intensive survey of the Middle Stone Age archaeological record in the main survey area could further elucidate the spatial data regarding the distribution of Mode 3 technology, which is primarily produced on fine-grained or crystalline raw material. This could further improve the understanding of the behavioural pattern seen in some localities in the Cobble zone (7) and Cathedral Mopane zone (14).

This research comprises the first landscape study in Zambia for at least 20 years, and has successfully integrated the off-site approach developed in the late 1970s with 21st century navigational and computational technology. This provides a fast and effective way of surveying an area where it is only feasible to use a small team, and is a technique that can be easily adopted across Africa. The Luangwa Valley clearly has a rich variety of geomorphological environments and archaeological landscapes that with future research could further illuminate the long human occupation of the area.

9. Appendices

9.1. Appendices for Chapter 5

Appendix 5.1: Calculations for first season's fieldwork times

Assuming an area of 5 square kilometres per transect on each strata, using a rectangular transect measuring approximately 1km by 5km.

For transects spaced at 30m apart this would equate to 33 transects to cover a 1000 meter wide transect. At 5 km length for each this would then equate to a total distance to be walked of 165km. Assuming an average walking speed of 4km/h, and this is probably an optimistic estimate, this would take 41.25 hours to cover. The working conditions in the field mean that there is only a 5.5-hour effective working day. This results in the total time for one person surveying being 7.5 days at the minimum to cover each transect.

Assuming that there would be a large amount of time needed for recording and collecting of artefacts a realistic estimate was made of 14 days for each 1km x 5km transect (this would include one or two days off).

The total area covered by 3 sampling squares in the landscape is 15 km^2 and would take 42 days or 6 weeks.

Appendix 5.2: Calculations of site survey adequacy

Calculations based upon those outlined in Sundstrom, L. (1993).

The probability of discovering a target (archaeological site) is a function of its size and shape, and the spacing of the parallel transects.

P is the probability that sites of size *s* (the sites perimeter edge) are intersected, and the transects are *d* (distance apart):

$$P = s \div \pi d$$

The perimeter *s* is calculated:

$$S = 2\pi \sqrt{} a^2 + b^2 \div 2$$

Where *a* and *b* are the semiaxes of the ellipse the formula for *P* can be reduced to:

$$P = (2\sqrt{}(a^2 + b^2) \div 2) \div d.$$

Or where the target is circular, and *r* is the radius of the target:

$$P = 2r \div d$$

So following these calculations; assuming that an elliptical target is 30m long by 15m and transects are spaced at 10 m then the probability of intersecting targets of this size in the field is 4.7, or 470%.

Assuming a circular target of 15 m diameter (7.5m radius), then the probability of intersection is 3, or 300%.

Appendix 5.3: Standard measurements of sediment and clast sizes

Sediment and clast sizes used to describe sediments (After Tucker 1991:11 and Stow 2006:109). For quick determination of sediment size in the field a grain-size comparator card was used (in this case from Stow 2006: inside cover).

Size in mm	Class terms	
256+	Boulders	Boulders
128 –256	Large cobbles	Cobbles
64 – 128	Small cobbles	
32 – 64	Very large pebbles	Pebbles
16 – 32	Large pebbles	
8 – 16	Medium pebbles	
4 – 8	Small pebbles	
2 – 4	Granules	Granules
1 – 2	Very coarse sand	Sand
0.5 - 1	Coarse sand	
0.25 – 0.5	Medium sand	
0.125 – 0.25	Fine sand	
0.062 – 0.125	Very fine sand	
0.031 – 0.062	Coarse silt	Silt
0.016 – 0.031	Medium silt	
0.008 – 0.016	Fine silt	
0.004 – 0.008	Very fine silt	
-0.004 mm	Clay	Clay

Appendix 5.4: Recording form for season 2

The diagram on the left hand portion of the form provided an easy visual method for recording features, locations of collections, and navigation. Each block represents $100m^2$, and the circles with crosses on them allowed easy reference to be made to the location on the transect at any one time, without the need for zooming out on the screen of the PDA. The lowest block has been deleted here so that the form can be displayed.

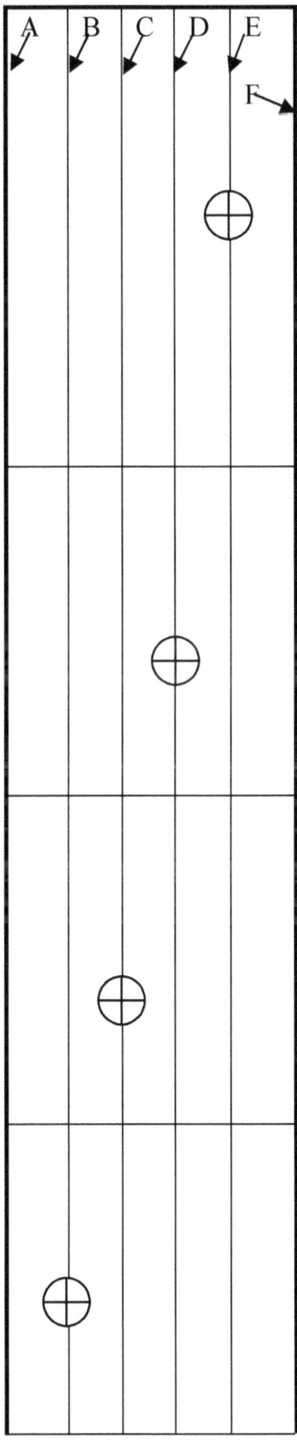

Transect No:	Page.	Date:
Start:		Finish:
Tree Canopy ☐ 0-25% ☐ 25-50% ☐ 50-75% 75-100% ☐ Ground Cover ☐ 0-25% ☐ 25-50% ☐ 50-75% 75-100% ☐		
Soil type: ☐ Vegetation Type: ☐		
Notes and observations GPS postions of individual finds + sketches of features. Bag finds and number according to transect and find no.		

9.2. Appendices for Chapter 6

Appendix 6.1: Description of the geomorphological zones mapped

Figure A6.1 illustrates the geomorphological map that was created during the first field season and improved in the second using a combination of the methods outlined in section 5.6. The zones numbered in the map legend are described in this appendix along with the key geological sections and observations made in each zone. The positions of the key archaeological and geological sites are displayed in figure A6.1. Interpretations of the geomorphological processes occurring today in some of the zones are given in this appendix where necessary, as this provides the context for the interpretations of the past processes in the area. Two cross-sections are given at the end of this section to provide a summary of the relationships of the key geomorphological zones to each other (figure A6.1, and A6.3 and A6.4).

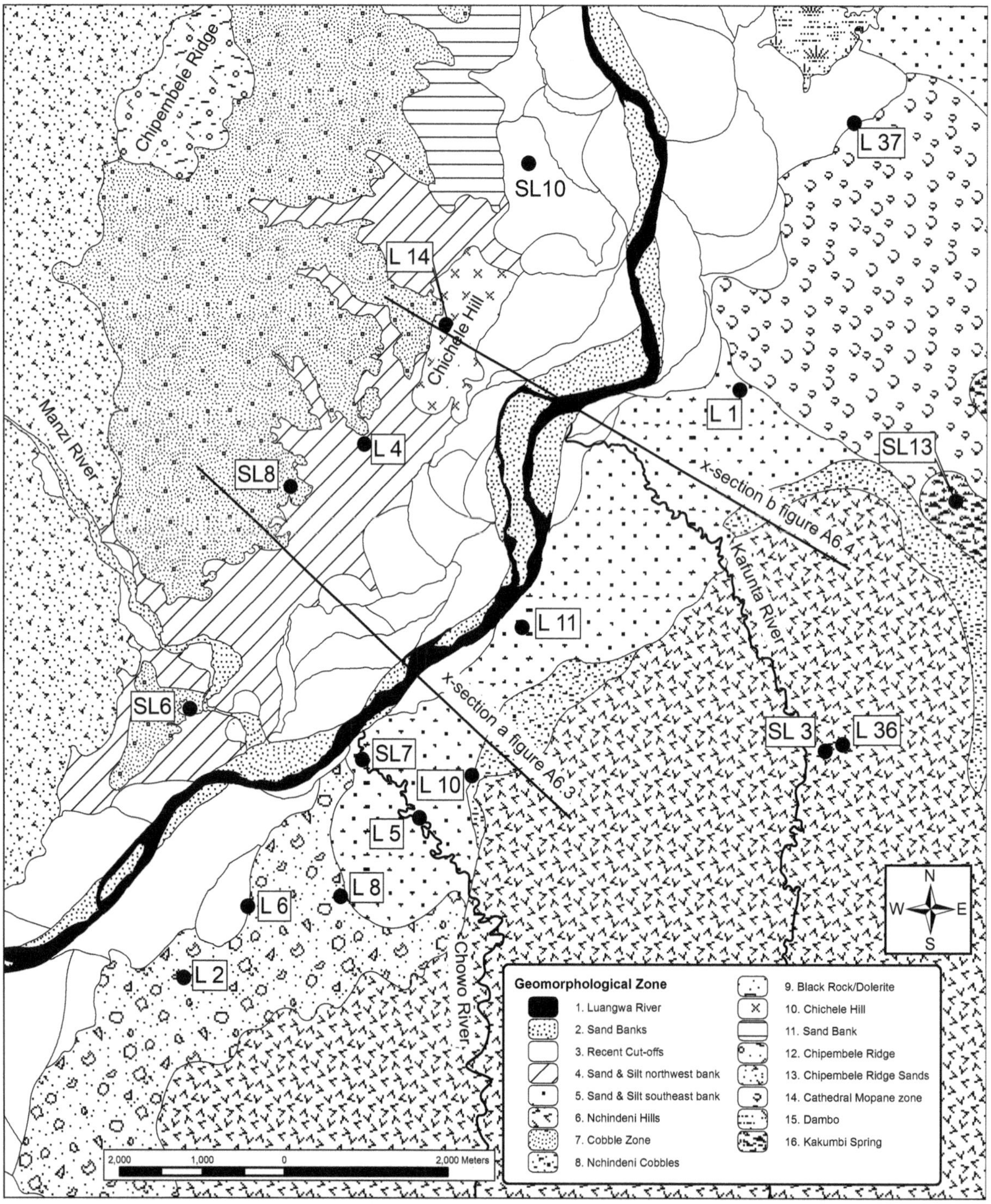

Figure A6.1: The geomorphological zones mapped in the survey area, displayed with the locations of sites (SL) and geological localities (L) discussed in text. The positions of two cross-sections that provide an overview of the vertical relationship of the zones with each other are also displayed (figure A6.3 and A6.4).

Zone 1. The Course of the Luangwa River (1999)
This zone represents the course of the Luangwa that was mapped from the aerial photography taken in June 1999 (the middle of the dry season); it represents the active channel that contains water throughout most of the year. Although the river never dries up completely, it can become little more than pools connected by a large stream towards the end of the dry season (late October). The channel sections on the banks of the Luangwa all appear to comprise recent sand and silts characteristic of zone 3 (only one or two could be visited for safety reasons). There is no evidence indicating the depth of the channel sediments, however, based on observations in zones 4 and 5, the underlying bedrock is presumably Karoo age sediment.

Zone 2. Sandbanks of the 1999 channel
These are recent sand banks that were also mapped from the 1999 aerial photography. They are for the most part devoid of vegetation except for occasional dry season grasses. During the wet season they are submerged by the main flow of the Luangwa and can be completely restructured by fluvial action.

Zone 3. Recent Cut-off zone, land surfaces from abandoned channels
Observations.
The abandoned channels of the Luangwa's previous courses are clearly visible from aerial photographs and each meander could be accurately mapped. Some of these extinct meanders still retain remnants of abandoned channels, and in the wet season these can be reactivated with high flow rates and erosive power. The soils on the surface range from sandy clays to grey sand and silt. The section walls of the Luangwa River and the stream channel walls in zone 3 are comprised of recent fluvial sand and silt, in which silted up channel sections can occasionally be seen. Underlying sediments did not outcrop at the base of any of the sections visited in zone 3. The vegetation can be sparse or absent in areas up to $3km^2$; however, it can also occur in dense woodland and thicket. The vegetation typically consists of trees and shrubs that include *Combretum fragrans*, *C. imberbe*, *Faidherbia albida*, *Kigelia afticana*, *Lonchocarpus capassa*, and *Sclerocarya birre*. There are also grasslands that usually contain various species of the genera *Andropogon*, *Diheteropogon*, *Hyparrhenia*, and *Hyperthelia* (Astle 1995).[7]

Interpretations.
The variability in vegetation cover appears to arise from a combination of proximity to water, the amount of past human clearance of the landscape, and overgrazing by elephant and hippo in certain areas. Some large areas (c. $1km^2$) have been bioturbated by elephant and hippo, with the result that in the dry season there are extensive areas with dried 30cm deep footprints, which are further cut with deep desiccation cracks. Most of this surface is very recent, as the Luangwa has a high erosion rate and frequently avulses, cutting new channels or reactivating old ones (see section 6.2). A radiocarbon date retrieved from a small test excavation near Later Stone Age material (SL 10) provides a date of 400BP for the surface of one of the cut-offs (at 10cm depth) (Barham *pers comm.*). This meander cut-off has been cut into a number of times indicating that it is relatively old, and the radiocarbon date provides a minimum age for the abandonment. The underlying sediments of zones 4 and 5 are Karoo, therefore the underlying sediments of zone 3 must presumably be Karoo or older sediments as zone 3 is at a lower elevation than 4 and 5.

Zone 4. Fluvial Sand and Silt on the north-west bank.
Observations.
The nature and origin of the sediment in zone 4 was not known until it was examined in the field; observations revealed that the soil is similar to the sediment of zone 3 consisting of fluvial sands and silts that are occasionally bioturbated. It was not possible to map the boundaries of this zone from aerial photography alone, necessitating some of the boundaries to be recorded in the field, which included the boundaries with zone 11, and parts of zones 7 and 10. Zone 4 is at a slightly higher elevation (c. 2 - 5m) than the adjacent zone 3, and in places has been dissected by small seasonal streams that flow from the Cobble zone (7) towards the Luangwa. Occasionally, near the boundary with the Cobble zone there are localised scatters of quartzite cobbles present on the surface of the sand and silt. Most of the stream cuts were relatively shallow and the eroded sections contained only the recent sand and silt of zone 4. Only one stream section visited revealed the depth of the sand and silt deposits, and at the base of this section Karoo sediments are found at a depth of three metres (locality 4). The elevation of the Karoo is relatively high compared to the height of the Luangwa River and presumably the base of the sediments in zone 3. Vegetation tends to be sparse or absent for the same reasons outlined for zone 3, but where present it varies from large patches (approximately $1km^2$) of single species woodland of *Colophospermum mopane* trees, grading to that of riverine and riparian woodlands consisting of *Accacia nigrescens Berchamia discolor*, *Combretum fragrans*, *Cordyla africana*, *Garcinia livingstonei*, and *Ziziphus abyssinica*.

Interpretations.
The higher elevation of zone 4 suggests that the Luangwa River, which deposited zone 3, has cut down into Karoo or older sediments to a lower elevation (see summary cross sections in figures A6.3 & A6.4). This, combined with the fact that the sands and silts in the zone 4 are similar to zone 3, suggests that it consists of meander cut-offs that are older than zone 3. These cut-offs have been abandoned at a slightly higher elevation as the Luangwa has eroded a deeper channel, and have lost the defining boundaries visible in the aerial photographs of zone 3 through a combination of erosion of the surface in some areas, and aggradation in others due to deposition of

[7] All scientific names listed in this section are from Astle (1995). The lists here are shortened to the most common present in the applicable zones; Astle's (1995) vegetation map and Astle *et al.*'s (1997a & b) dictionary of scientific names should be consulted for more comprehensive species lists of the vegetation in the area.

sediment from overbank flooding of the Luangwa. The discrete scatters of quartzite cobbles have been deposited onto the sands and silts recently by the seasonal streams emanating from the cobble zone.

Zone 5. Fluvial Sand and Silt on the south-east bank
Observations.
This zone is virtually identical in character to zone 4, with a few minor differences. Fieldwork revealed that like zone 4 the area comprises fluvial sand and silt, presumably from overbank deposits aggrading and filling in previous channels of the Luangwa. The vegetation of the area is also similar, though large areas of open canopy, bare soils, and bioturbation are rare. In contrast to zone 4, the stream sections present often reveal the depths of the sand and silt deposits; these vary in thickness and there is sometimes as little sediment as 50cm (locality 1), and as much as four metres plus (localities 5, 11 and SL7). In some stream sections there are a layer of cobbles at the base of the sand and silt that frequently contain Early and Middle Stone Age artefacts; this layer lies directly above Karoo sediments. One of the deeper sections in the Chowo River contained an archaeological site, SL7, which yielded radiocarbon dates at the base of the sequence indicating that two metres of sediments accumulated over the past 1,600 years (Barham & Lie Jarman 2005:117, and section 2.2.2.5).

Interpretations.
The Karoo sediments at the base of the sand and silt in the sections are at an elevation above the modern day level of the Luangwa River (see summary cross section in figures A6.3 and A6.4). It is therefore likely that the sand and silt material in both zones 4 and 5 represent abandoned terraces, and contain older sediments than in zones 1 to 3. These sediments have aggraded due to seasonal floods depositing sand and silt. The cobble deposits at the base of some of the sections appear to be channel lag deposits, as at a number of localities the clasts show imbrication that suggests a fluvial origin. The cobbles were probably deposited in these locations by previous courses of the Luangwa's tributaries or the Luangwa itself, and were then buried by sediments deposited by overbank events during floods. These lag deposits frequently contain Early and Middle Stone Age artefacts (Modes 1 and 3) indicating that the deposits have accumulated during archaeological time. A possible source of the deposits are the nearby cobble deposits at the base of the Nchindeni Hills (zone 8). It is clear therefore that the main channel of the Luangwa has not been present in this area and at this elevation since this time. It is difficult to determine more precisely how recently this abandonment occurred, however, it is known that the sediments at SL7 began accumulating 1,600 years ago (Barham & Lie Jarman 2005:117), which provides a minimum age for initial abandonment. The Middle Stone Age artefacts that are contained in the lag sediments elsewhere in zone 5, provide a maximum age of approximately 250,000 years for some of the lag deposits. The Luangwa has now cut down into the Karoo sediments in zones 1 to 3, as the Karoo outcrops at and above the modern average high water mark of the Luangwa in both zones 4 and 5, the base of the river must therefore be at least two to four metres below the base of the sand and silt in zones 4 and 5 (see summaries in figures A6.3 & A6.4).

Zone 6. Nchindeni Hills
Observations.
The hills that extend into the southern part of the survey area are an extension of the Nchindeni Hills that form the southern margin of the Luangwa Valley. At this point in the valley they rise 400m from the floodplain at 520m to a maximum 900m altitude above sea level. The hills have been geologically mapped as an uplifted block of an Archean complex of granites and metamorphic rocks (Thieme & Johnson 1981) (figures A6.3 & A6.4). Numerous northward flowing streams and rivers deeply incise the hills, with the result that the Nchindeni undulate to a large degree despite retaining a sharp break of slope with the valley floor. The rivers are prone to flash floods during the rainy season, and are dry throughout the dry season with occasional puddles, though water can be found below the surface of the sand on deeper meanders throughout the year (*pers obs* and K.Kakumbi *pers com*). The boundary of zone 6 with the flatter valley floor zones (5, 8, & 14) is only partially identifiable on the aerial photographs, so it was mapped with additional field observations of the changes in slope and soil type.

Field observations revealed that the Archean bedrock is usually overlain by brown soils of varying depths which contain frequent angular clasts of the weathered bedrock; in many places these clasts are the key constituent of the soils and vary in size from pebbles to boulders. The soils can be as shallow as 5cm, but can attain a depth of more than 2m on the lower slopes and valley floors. Quartz veins are present in the hills, and though the vein material is rarely visible *in-situ*, the weathered surface material can exceed 50cm in length and 15cm width. Groups of large outcrops of granite (often 6 or so large 20m x 20m weathered boulders forming small inselbergs) are common in the eastern part of the Nchindeni; they occur in a one to three kilometre tract and frequently form natural shelters (locality 36). The vegetation of the zone varies from open grassland, to dense thicket, and areas of mixed woodland. Astle's vegetation map (Astle 1995) and the other vegetation studies (Astle *et al.* 1997a & 1997b), do not extend into the Nchindeni Hills so a precise species list is not available. There is a mixture of denser riparian and riverine woodlands similar to zone 5, located near many of the seasonal streams emanating from the hills.

Interpretations.
Understanding the sediment feeds from the Nchindeni is the second key aim of the geomorphological work, and the majority of the interpretation is given in section 6.3 and is based on the clast count results presentation in appendix 6.2. The rivers and streams that flow from the Nchindeni have possibly eroded along fault lines or boundaries between different rock types; this is likely as a number of the valleys in the Nchindeni are incised deeply

and along straight parallel valleys. The rainy season flash floods, judging by the material present in the riverbeds, can move large boulders and considerable amounts of material. The varying thickness of the soils are a result of hillslope processes, which is where on slopes and hilltops material is gradually transported downhill over time by rainsplash, sheetwash, dry ravel, and bioturbation (Huggett 2007:171-173). The result is that the soils are usually thin on the Nchindeni hilltops and slopes and are deeper at the base of the hills.

Zone 7. Cobble zone

Observations.

This zone consists of gently undulating hills, occasional longitudinal ridges, and globular mounds that vary in size from low relief <5m high 20m long and 10m wide, to relatively large and prominent 20-30m high hills that are sometimes in excess of 200m long by 50m wide. The top surfaces of the hills are covered by an armouring of quartzite cobbles, this layer varies in thickness from a single layer of cobbles in some areas, to two metres in thickness. Field observations suggest that the deposits are unlikely to be deeper than five metres in the survey area, with the individual cobbles typically ranging in size from 2cm to 20cm in length, with some larger clasts in rare cases (figure A6.2). At a minimum the zone is elevated by 10m at the boundary with zone four (average 20m); beyond this the landscape gently undulates and rises, eventually to 40m above zone 4 at the boundary with zones 12 and 13.

An excavation (at SL8), directed by Lawrence Barham, revealed that this cobble armouring lies directly on the Karoo bedrock; additional observations around the bases of the hills and across the zone revealed that this was the case for the entire area. The section at SL8 (figure A6.2) revealed that there is a matrix of coarse sands and gravels present, but this matrix does not separate the clasts and the deposits appear to be hyper-concentrated (see section 6.4.2). The Karoo rocks at the base of the SL8 sequence are relatively soft mudstones, but observations around the base of the hill reveal that it is partially situated on hard sandstones, suggesting the Karoo sediments can vary in the area. Between the mounds there can be gaps up to 500 metres wide filled with fluvial sands (similar to zone 4) that grade, after the first few undulations away from the Luangwa, to cobble scatters and bare Karoo surfaces that can cover a kilometre square (these areas were not mapped separately due to time constraints). There is no evidence of sand and silt overlying the cobbles in zone 7, with the exception of the boundary with zone 11.

The boundary between zones 7, 4 and 11 could be mapped relatively easily using the aerial photography and, as one of the key roads in the area frequently crossed the boundary, it was improved in the field. Tributary streams of the Luangwa dissect the cobble zone at certain points and have clearly removed some of the cobble material creating a sinuous boundary between zone 7 and 4. Elsewhere the action of heavy seasonal rain is evidenced by the presence of small runoff streams that are often choked with cobble material. The boundary with zone 13 was mapped from aerial photography and is based on a visible vegetation change, and when checked in the field this did relate to a change from the cobble capped topography to the flatter topography of zone 13, although the nature of the boundary could not be determined as there were no sections cut into the terrain in this area. The vegetation varies little, there are areas of bare ground with sparse grasses, but the majority of the landscape is covered by stunted mopane trees (*Colophospermum mopane*) with the occasional baobab tree (*Adansonia digitata*).

Interpretations.

The interpretation of the formational processes for zone 7 is given in section 6.4 and relates to the third key objective outlined in section 6.1. The sedimentology of all of the cobble zones (including 7, 8, and 14) relates to geomorphological processes that are no longer active, although modern processes are currently eroding zone 7. The lack of sand and silt on top of the cobbles suggests that the Luangwa has not been present above this elevation, and any lighter sediments have been weathered away. Judging by the nature of the contact between zones 7 and 4 it appears that the Karoo bedrock rises from below the deposits of zone 4, with the cobbles capping the undulating Karoo bedrock. Though no geological sections are exposed along this boundary, the impression is gained that zone 4 has been cut into, and thus has an erosive boundary with, the Karoo sediments (see figure A6.3

Figure A6.2: Excavated section at SL8 illustrating the nature of the cobble deposits found on the cobble capped hills. Note the lack of sorting or any structure in the material; there is also little obvious imbrication. The majority of the clasts are sub-rounded or sub-angular (see section 6.3 clast count 28). There is also a textbook well-rounded clast in the top centre of the section (note: photo taken 2 years after excavation, and the profile has weathered since excavation) (The compass case is 8cm in length).

Zone 8. Nchindeni Cobble zone

Observations.

The Nchindeni Cobble zone is similar in form to zone 7 on the opposite north-west bank and consists of gently undulating cobble capped mounds on the south-east bank of the Luangwa, between the Sand and Silt zone (5) and the base of the Nchindeni Hills (zone 6). In zone 8 the mounds are not as large and the area is less extensive. There are few streams originating in the zone and it appears less dissected than zone 7. Field observations and the excavation of a test pit at locality 10 revealed that the cobble layer is thin, approximately 30 – 50cm thick and lies directly on top of the granites and metamorphic rocks found in the Nchindeni zone (6) (see figures A6.3 & A6.4). Interestingly, no archaeological material was contained within the test pit. This was unexpected as the density of the Later Stone Age material found in zone 7 was high (in particular at SL8), if zone 8 was identical it is likely that a test pit would have recovered material if it were present (see chapter 7). The vegetation is identical to zone 7.

Interpretation.

The zone appears stable, as the cobble material is not currently being reworked. The experimental trench revealed that the deposits in general appear to be similar to zone 7, and the depositional environment for these zones is interpreted in section 6.4.

Zone 9. Black Rock / Dolerite zone.

Observations.

This zone is situated at the base of the Nchindeni Hills (zone 6) towards the southern portion of the survey area. It extends from the floodplain at 520m, into the Nchindeni Hills up to approximately 650m, after which it grades into zone 6. This zone consists of a surface of broken angular clasts, ranging in size from pebble to boulder size. A proportion of the rocks is similar to those found in the Nchindeni Hills (see appendix 6.2), but, the dominant rock type is a black rock that is assumed at present to be dolerite from dolerite dykes that are found in the area (there has been no mineralogical work to confirm this). There are also a number of metamorphic rocks present, as well as small quantities of quartz, chert, and a small percentage of other unidentified minerals that all occur in large angular chunks (appendix 6.2). Bedrock exposures are rare, but at higher altitudes it appears to be similar to the granites and metamorphic rocks of zone 6. At lower elevations observations from two localities revealed that Karoo sediments underlie the Black Rock zone where it meets zone 3 (localities 2 & 6 figure A6.1), and the clast surface often abruptly gives way to large areas of plain brown soil, with occasional scatters of angular clasts. The vegetation is similar to that of the rest of the Nchindeni Hills (zone 6). The boundaries of the zones were not well defined initially, as though the boundary with zone 3 was obvious on the aerial photography, the boundary with zones 5 and 6 could not be determined. This was resolved in the field during the second field season, but the diffuse nature of the boundaries means that their position may not be entirely correct, as they had to be inferred in locations away from the roads and routes walked to archaeological transects.

Interpretations.

The observations at localities two and six that the Karoo sediments are visibly in contact with the recent sediments of zone 3 suggests that the Luangwa, which deposited zone 3, has laterally cut into zone 9. No sand and silt is found on top of the sediments in zone 9, suggesting that the Luangwa has not recently affected the deposits in this zone. The surface material appears stable in this zone, so besides any hillslope processes there is little mass movement occurring today, although a stream section at locality 11 revealed that ancient debris flow deposits underpin parts of the Black Rock zone (see section 6.3 and figure 6.3).

Zone 10. Chichele Hill.

Observations.

Chichele Hill consists of a large outcrop of Archean basement rocks characteristic of the Nchindeni Hills, and some hard post-Karoo sandstone on the southwest slope of the hill (figure 6 in Utting 1988, and figure A6.4 here), there are also outcrops of chert present (Barham *pers comm.*). The hill rises to 574m above sea level above the floodplain sediments of zone 3 (520m a.s.l), and is 1.9km long and between 600m and 800m wide. On the north-east hillside facing away from the Luangwa there is a small 500m x 300m blanket of cobbles. The cobbles are similar in nature to those found on the cobble zones (7&8) however; they contain a far greater proportion of metamorphic and granitic clasts more akin to the material on Chichele Hill (see appendix 6.2). A section cut into this area of cobbles for a road cutting reveals that there are channels originating from Chichele Hill that have carried material from the hill towards the cobbles (locality 14). The blanket of cobbles in this area was deemed too small to be delineated separately and, as the clasts appear to have originated locally from Chichele Hill, it was included in this zone. The vegetation present is comparatively rare over the rest of the valley floor and includes *Kirkia acuminata*, *Sterulia africana*, and *Gyrocarfus americanus*.

Interpretations.

Overall the zone appears stable today, although there are some small stream channels running off the hill where roads cut into the soil, and hillslope processes are presumably removing sediment from the top and sides of the hill. The boundary between zones 10 and 3 is clear on the aerial photography, and the hard rocks of the hill appear to have provided greater resistance to erosion by the Luangwa than the Karoo sediments that underpin zones 7 and 9 (see section 6.2).

Zone 11. Sand Bank

Observations.

This zone is formed by a steep sand bank rising 10 to 20m from the floodplain (zone 4) in the north central part of the survey area, and the zones northwest boundary is adjacent to the Cobble zone (7). The bank shallows to the southern part of its extent, but uniformly consists of fluvial sand with little or no finer silt and soil component. There are no sections cut into the zone, so it is not possible to determine the nature of the boundary between either zone 4 or 7, and it is possible that the sands at the margin of the zone seven have partially buried the cobble mounds, or have been deposited in a channel cut into the edge of zone 7. The vegetation is mixed woodland and thicket, which in places is so dense it is impassable. Larger trees include *Accacia nigrescens*, *Schinziophyton rautanenii*, *Xeroderris stuhlmannii*, with smaller shrubs and trees of *Combretum obofvatum*, *Diosphyros quiloensis*, and *D. senensis*.

Interpretation.
In general the zone appears stable, there were no streams present, and the dense vegetation appears to stabilise the sand so that little material is being eroded. The presence of fluvial sands suggests that this zone my be a terrace of the Luangwa, but as there are no stream sections to examine this is difficult to determine for certain.

Zone 12. Chipembele Ridge.
Observations.
Chipembele Ridge rises 40m above the cobble ridge zone; it extends for 2km south-west from the northern edge of the survey area and is approximately 800m wide. The ridge is armoured with a layer of cobbles of unknown thickness, apparently similar in nature to the cobble zone (7). On the southeast side of the ridge the slopes have been deeply incised by streams draining the hillside, in which Karoo sediments outcrop below the cobbles and soil horizon. On the northwest side of the ridge facing a way from the Luangwa there is a gentler 20m slope down to a sandy soil zone (13). This is a distinct feature in the landscape and has been described and surveyed as a separate entity. Its prominence above the Cobble zone means that it might have experienced different geomorphic processes or, conceivably differential exposure throughout archaeological time. There may as result different archaeological periods represented in the zone. The vegetation primarily consists of mixed deciduous shrubland including *Brachystegia stipulata*, *Jubernardia globiflora*, and *Colophospermum mopane*. Grasses present include *Aristida scabrivalvis*, *Brachiaria serrata*, *Pogonarthria squarrosa*.

Interpretation.
The incised streams are an active erosive feature today and, when active in the wet season, remove material from the top and side of Chipembele Ridge. This material is not deposited on to the cobble ridges in zone 7 below, but is transported by tributary streams towards the Luangwa River.

Zone 13. Chipembele Ridge Sands.
Observations.
The zone behind Chipembele Ridge extends for a significant distance beyond the survey area (approx. 10km), the elevation increasing from 560m to 660m and possibly higher towards the Muchinga Escarpment. The sediment is unconsolidated fluvial sand that appears to be relatively deep; observations of this zone beyond the survey area reveal that the sediment can be in excess of six metres deep (see figure), often with discrete cobble layers within it. The zone extends down the west side of the survey are on the west bank of the Manzi River to one kilometres distance away from the Luangwa. Nearer the river the vegetation is similar to zone 5, but grades to miombo woodland away from the river, which consists *Brachystegia*, *Monotes*, and *Ochna* of various species, as well as that of *Burkea africana*, and *Isoberlinia angolensis*.

Interpretations.
In the area directly behind Chipembele Ridge the sands appear to be well stabilised by vegetation. Nearer the Luangwa River where the vegetation grades into zone 5 vegetation the soils are visibly deflating today, and in the rainy season it appears that overland water flow may erode and redistribute the surface of the soil.

Zone 14. Cathedral Mopane zone.
Observations.
This zone comprises grey to yellow fluvial sands and silts supporting mature single species *Colophospermum mopane* woodland. Occasionally low relief cobble mounds protrude through the sand and silt, suggesting the fluvial deposits overlie a cobble terrain similar to zones 7 and 8. The cobbles in zone 14 are at a lower elevation than zones 7 and 8, so presumably the sand and silt covering the cobbles in the area has derived from overbank deposits from seasonal floods from the Luangwa and its tributaries.

Interpretation.
The development of the mature woodland suggests that the ground has been stable for a relatively long period of time – compared to zone 3 and 4 for example. The cobbles in this zone are presumed, due to their similarity to the cobbles in zone 7, to have been deposited in a similar depositional environment (see section 6.4).

Zone 15. Dambo.
Observations.
This zone was not initially identified as a separate zone from the recent meander cut-offs (zone 3), but it was discovered when an archaeological transect in the area was surveyed. The decision was made to map it separately to zone 3 due to the different nature of the soils, grey bioturbated and desiccated clays, as well as differing vegetation. The vegetation consists of short grasses of various species of *Echinochloa*, as well as *Caperonia fistulosa*, *Duosperma quadrangularae*, and *Rhamphicarpa fistulosa*.

Interpretation.
Examination of the aerial photographs reveals that the area was originally a recent cut-off of zone 3. However, as it is lower lying the area becomes seasonally saturated and is as a result unsuitable for the vegetation in the rest of zone 3.

Zone 16. Kakumbi Spring.
Kakumbi Spring forms a small zone between zones 8 and 14. It is a large spring mound approximately 200 x 200m, which supports single species grass (locality and plant species not listed by Astle (1995) or Astle *et al.* (1997a & b)). On a prominence extending 20m into the spring mound is a dry soil mound that contains a deep sequence of Iron Age material, with evidence for periods of Later Stone Age occupation, suggesting the site has been a focus of activity for at least the duration of Later Stone Age and Iron Age and possibly to the Middle Stone Age (Barham *pers comm.*). Nchindeni granites and metamorphic rocks can be seen in outcrop at some points on the edge of the spring, and elsewhere cobbles from zone 14 form the banks encircling the mound. There are also outcrops of chert around the spring, which would have been a useful raw material resource (Barham *pers comm.*).

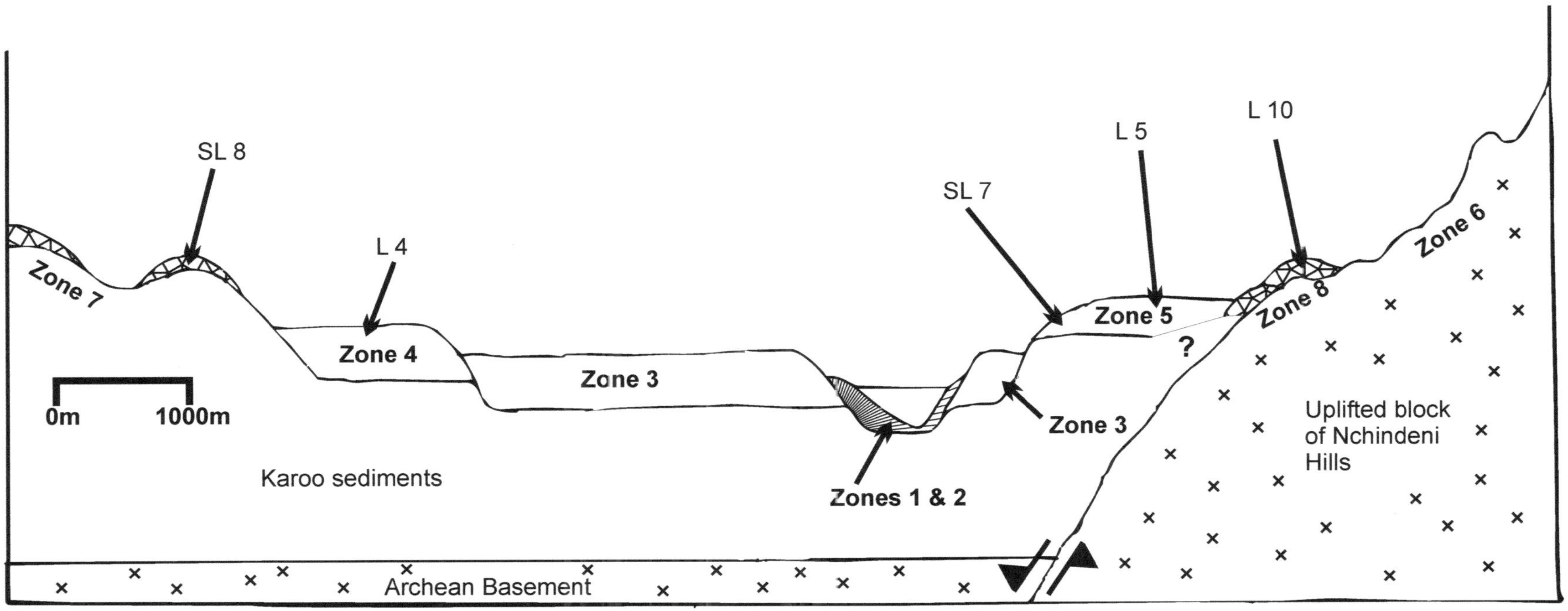

Figure A6.3: Cross-section of the survey area between zone 6 and 7, the position for which is given in figure 6.1. The vertical axis is not to scale and is exaggerated. The L numbers refer to geological localities, and the SL numbers refer to archaeological sites discussed in text. The cross section provides a summary of the relationship between the geomorphological zones, the Luangwa and the underlying Karoo sediments and Archean bedrock. The position of the Archean bedrock is based on Utting (1988: figure 6). Note the interpretation of the vertical relationship of the different zones on the floodplain, the Luangwa is at the lowest elevation surrounded by the recent sediments of zone 3. Zone 4 and 5 are higher elevation terraces, there is no evidence for the presence of the Luangwa at higher elevations. The question mark represents the uncertainty regarding the nature of the boundary between the Karoo sediments and zones 5 and 8.

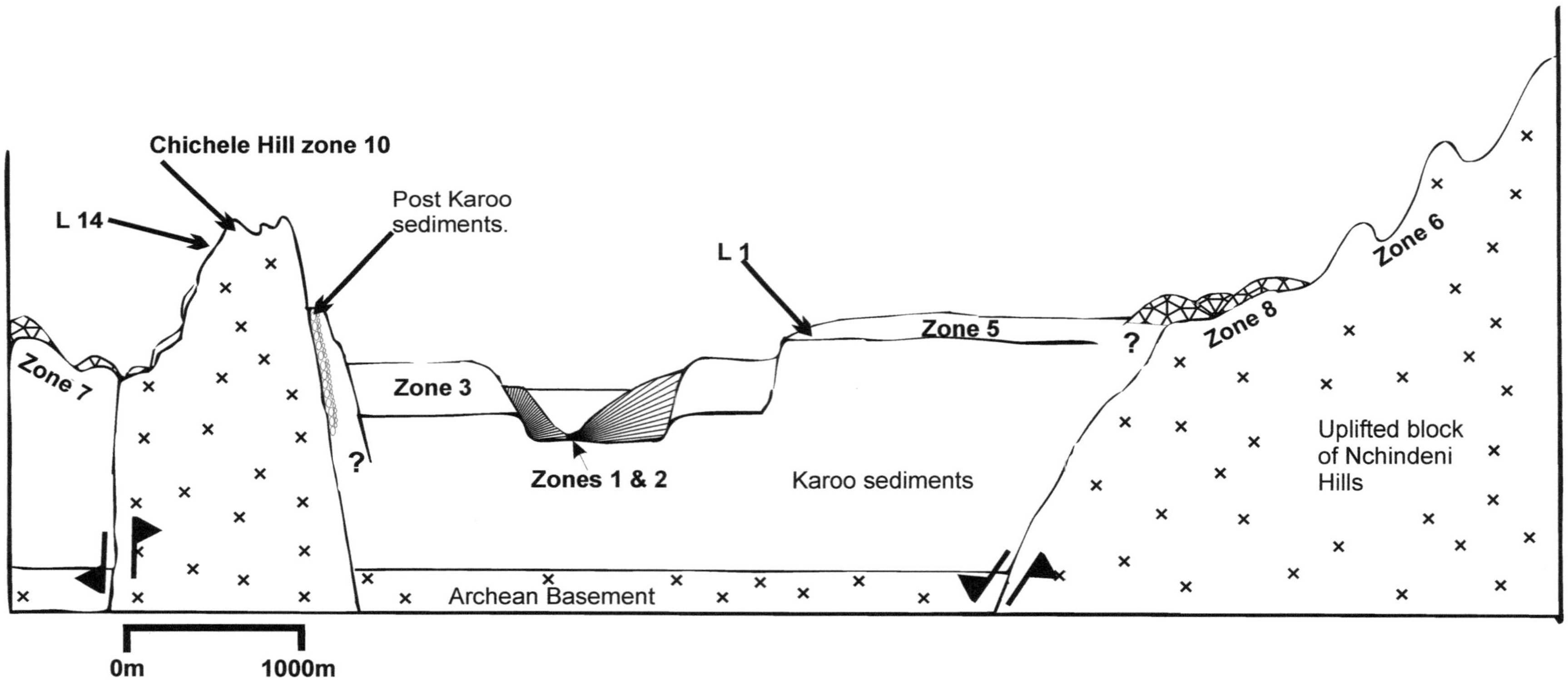

Figure A6.4: Cross section of the survey area that includes Chichele Hill. The vertical axis is not to scale and is exaggerated. The positions of the Archean bedrock and post Karoo sediments are based on Utting (1988: figure 6). The question marks refer to areas where the relationships of the sediments are unknown. In addition to the lack of evidence for fluvial evidence beyond zones 1, 2, 3, and 5 in this figure the uplifted block of Archean basement rock that forms Chichele Hill (zone 10) has formed an impenetrable barrier to the movement of the Luangwa at this point in the valley.

Appendix 6.2: The Clast Count Results

Table A6.1 below provides a brief description of the localities from which clast counts were taken, and tables A6.2, A6.3, and A6.4 provide the raw data for lithology, clast size (b-axis), and angularity. The locality numbers for the clast counts refer to locations that were visited in the field, and they are listed in the order in which they were recorded. Where there are numbers missing a locality has been recorded in the field notebook for other purposes, often the locality of a photograph or field-sketch of a sediment section. In the following section the patterns in lithology, clast size, and angularity of the clasts in the landscape are described. These data provide information to aid interpretations of the geomorphology in reference to aims two and three of the geomorphological research (section 6.1). With reference to aim one, the sand and silt sedimentology of the Luangwa River and its flood plain precluded the use of clast counting in the area as there were no clasts to analyse.

Lithological results description

Figure A6.5 illustrates the results from the lithologic section of the clast count. In general there are three broad groups of material: quartzites are predominant on the cobble ridge zones either side of the river and in the Cathedral Mopane zone; the material on the Nchindeni Hills comprises metamorphic rocks and granites; a third group of highly variable lithologies is associated with the Black Rock zone adjacent to the Nchindeni.

The cobble zones

The cobble zone on the north-western bank of the Luangwa (zone7), where the cobbles are most extensive, consists primarily of quartzite cobbles (counts 21, 30, 28, 22, 23, and 12, all at roughly 530m elevation). However there are some small differences across the zone. As noted in appendix 6.1, there are large areas of zone 7 where the Karoo bedrock forms the land surface. Count 32, at 550m elevation, was taken in such an area, and as a result 30% of the clasts were Karoo rocks, derived from the Karoo bedrock eroding and weathering locally. This occurrence is only expressed where the Karoo is sandstone rather than the mudstone which weathers to clay minerals directly, producing few large clasts on the process, except where concretions of hardened mudstone are present, Additionally, nearer to the Luangwa, in samples 21, 30, and 28, there are occasional metamorphic, chert, and Karoo clasts which comprise up to 5% of the total.

The smaller cobble zone on the south-eastern side of the Luangwa (zone 8) is lithologically more variable than zone 7, but with quartzites still making up the majority of samples. Clast counts 10 at 535m elevation, and 16 at 550m contain 73% and 85% quartzite respectively. The remainder comprise mainly of metamorphics, granites, and quartz, with 2% chert at count 16. The presence of these clasts may be due to localised inclusion of weathered fragments of the basement rocks that underpin zone 8 (see appendix 6.1 above), or due to inclusion of the rocks eroding slightly further away that have been transported from the Nchindeni Hills.

The Nchindeni Hills

The clast counts undertaken in the Nchindeni Hills (zone 6) revealed that the counts made at higher elevations, such as 34 and 35 (640 and 570m respectively), have no trace of the quartzite found on the cobble zone. Furthermore, one of the smaller streams emanating from the Nchindeni, near the Luangwa floodplain, had no trace of quartzite in its bedload or channel wall clast counts (counts 8.1 and 8.2 at 520m). These counts primarily contain metamorphic and granite clasts (approximately 60 – 90% in total, see table A6.2 and figure A6.5). There is some quartz (14%) in the form of vein quartz at count 35, which is a constituent part of the granitic and metamorphic rocks of the Nchindeni. At lower elevations in the Nchindeni Hills, up to 40 metres above the level of the Luangwa at counts 17, 18, and 19 (540 – 560m above sea level), a high proportion of appears to be in the form of banded quartzites that are likely to be derived from the metamorphic bedrock. The lineation of the crystals, larger joints, and weaknesses in the banded quartzite cause it to erode to flat tabular clasts rather than the rounded clasts on the cobble zone. However, due to the homogeneous nature of many quartzites it is difficult to prove the difference between them without geochemical analysis, so no separate categories were used in the clasts counts.

Clast counts in the larger seasonal rivers that flow from deep within the Nchindeni, such as 9.1 and 9.2 in the Chowo River, and 33 in the Kafunta River (all at 530m) reveal that the major constituent of the material in the current bedload and the section walls of the river banks are either metamorphic or granitic clasts, similar to the rocks seen in the Nchindeni Hills. Quartzite in these cases makes up less than 50% of each sample, and in the Chowo these quartzites appeared to be predominantly similar to the material found in the cobble zones, whilst the sample from the Kafunta was dominated by tabular quartzite. Count 11, which was undertaken in a small stream draining into the Luangwa, that does not extend far into the Nchindeni Hills, also has a metamorphic component (16%). This is much lower than the larger streams, and it is likely therefore that most of its material has originated from the Nchindeni Cobble zone nearby, hence the high proportion of quartz and quartzite.

Locality	GPS	Data recorded	Zone	Description of locality
2	357223 8537572	Lithology.	9	Material lying on top of outcrop of Karoo clays adjacent to recent cut of Luangwa, and current floodplain at 525m.
7	358404 8537904	Lithology. B-axis. Angularity.	9	Surface in Black Rock / Dolerite zone near a small granite hill, elevation of location approximately 560m.
8.1	359388 8538180	Lithology.	9	Modern channel of stream flowing from Nchindeni Hills. 540m.
8.2	359388 8538180	Lithology.	9	Channel wall of stream in 8.1. 540m.
9.1	359897 8539458	Lithology.	5	Channel Wall in Chowo River, a large ephemeral stream draining from the Nchindeni Hills. 530m.
9.2	359897 8539458	Lithology.	5	Modern Channel of Chowo River. 530m.
10	360737 8539942	Lithology. B-axis. Angularity.	8	Low relief cobble capped hill in the Nchindeni Cobble zone. 535m.
11	361332 8541672	Lithology.	5	Clast count from small stream flowing towards the Luangwa from the Nchindeni Hills. Narrow band of cobbles present at base of fluvial sand c 2.5m deep from ground surface, and above Karoo bedrock. 520m.
12	360056 8545318	Lithology. B-axis. Angularity.	7	Gravels and cobbles in section in a small quarry in the cobble zone near Chichele Hill. 530m.
13	356481 8548465	Lithology. B-axis. Angularity.	13	Cobbles behind Chipembele Hill at 570m.
14	360417 8545212	Lithology. Angularity.	10	Channel, with hyper-concentrated gravel and cobbles, in section wall of road cutting below Chichele Hill at 538m.
16	366422 8542952	Lithology. B-axis. Angularity.	8	Surface cobbles on the boundary of the Nchindeni Cobbles (zone 8) and the Cathedral Mopane zone (14). 550m.
17	365395 8541974	Lithology. B-axis. Angularity.	6	Cobbles in Stream bed 2km west of Poachers' Cave (SL3)in the Nchindeni Hill zone. Approx 540m.
18	365149 8541076	Lithology. Angularity.	6	Cobbles in Stream bed 800m west of Poachers' Cave in the Nchindeni Hill zone. Approx 550m
19	364282 8540390	Lithology. B-axis. Angularity.	6	Kafunta River channel clast count. Approximately 400m east of Poachers' Cave in the Nchindeni Hills zone. Approx 560m.
21.	357341 8540660	Lithology. B-axis. Angularity.	7	On the cobble surface behind the Manzi River section, SL6. 525m.
22.	359499 8544710	Lithology. B-axis. Angularity.	7	Cobble zone on the northwest bank of the Luangwa. 530m.
23.	359713 8545486	Lithology. B-axis. Angularity.	7	Cobble zone on the northwest bank of the Luangwa. 530m.
24.	360288 8545160	Lithology. B-axis. Angularity.	10	Cobbles below Chichele Hill. 530m.
26.	347629 8554908	Lithology. B-axis. Angularity.	13	Clast count from the Kasweta Road. 720m.
27.	341626 8563782	Lithology. B-axis. Angularity.	13	Clast count from the Kasweta Road. 660m.
28.	358519 8543312	Lithology. B-axis. Angularity.	7	SL8, an excavated locality on the edge of the cobble zone on the north bank of the Luangwa. 530m.
30.	357288 8540710	Lithology. Angularity.	7/4	Modern channel sediments in the Manzi River, at the base of the SL6 cliff face. 520m.
31.	365176 8544812	Lithology. B-axis. Angularity.	14	Gravel mounds in the mature mopane woodland. 530m.
32.	357163 8546518	Lithology. B-axis. Angularity.	7	Count taken from within the cobble zone towards Chipembele Ridge. Approx 550m.
33.	362711 8543760	Lithology. B-axis. Angularity	5	Clast count in Kafunta River approximately 400m from confluence with Luangwa. 530m.
34.	359921 8536551	Lithology.	6	Clast count in the Nchindeni zone, 640m elevation.
35.	361928 8537609	Lithology.	6	Clast count in the Nchindeni zone, approx 570m elevation.

Table A6.1: A table listing the locations of the clast counts undertaken in fieldwork, with the zones and elevations where the counts were undertaken.

Locality	Lithology									
	Quartzite	Chert	Metamorphic	Quartz	Black Rock	Granite	Karoo	Sandstone	Unknown	Total
2.0	39	0	7	0	46	2	4	0	2	100
7.0	31	0	1	0	58	7	0	3	0	100
8.1	4	0	19	0	36	41	0	0	0	100
8.2	0	0	22	0	15	56	0	5	2	100
9.1	48	0	33	0	8	10	1	0	0	100
9.2	35	0	39	0	5	12	6	0	3	100
10.0	73	0	21	3	0	3	0	0	0	100
11.0	83	0	16	0	0	0	0	1	0	100
12.0	95	2	0	3	0	0	0	0	0	100
13.0	96	1	0	3	0	0	0	0	0	100
14.0	0	0	36	0	0	64	0	0	0	100
16.0	85	2	11	0	0	1	0	1	0	100
17.0	63	0	29	8	0	0	0	0	0	100
18.0	81	0	19	0	0	0	0	0	0	100
19.0	28	0	72	0	0	0	0	0	0	100
21.0	95	2	2	0	0	0	0	0	1	100
22.0	99	0	0	0	0	0	0	0	1	100
23.0	99	1	0	0	0	0	0	0	0	100
24.0	11	1	87	0	0	1	0	0	0	100
26.0	100	0	0	0	0	0	0	0	0	100
27.0	99	0	0	0	0	0	1	0	0	100
28.0	94	0	5	0	0	0	0	1	0	100
30.0	98	0	2	0	0	0	0	0	0	100
31.0	93	1	6	0	0	0	0	0	0	100
32.0	71	0	0	0	0	0	29	0	0	100
33.0	46	0	53	0	0	1	0	0	0	100
34.0	0	0	80	8	0	12	0	0	0	100
35.0	0	0	76	14	10	0	0	0	0	100

Table A6.2: Lithology clast count data.

Locality	B axis class								
	2.0 - 2.9	3.0 - 3.9	4.0 - 4.9	5.0 - 5.9	6.0 - 6.9	7.0 - 7.9	8.0 - 8.9	9.0 - 9.9	10 +
7.0	23	39	12	6	7	5	3	3	2
10.0	57	17	12	7	5	0	1	0	1
12.0	58	25	6	6	3	1	0	0	1
13.0	34	23	18	10	9	4	1	1	0
16.0	69	20	11	0	0	0	0	0	0
17.0	62	26	9	3	0	0	0	0	0
19.0	18	14	19	11	9	10	6	4	9
21.0	44	26	16	12	1	0	0	1	0
22.0	59	26	7	7	0	1	0	0	0
23.0	31	51	13	3	2	0	0	0	0
24.0	57	30	10	1	1	1	0	0	0
26.0	28	33	19	9	3	3	2	0	3
27.0	39	30	21	9	0	0	1	0	0
28.0	46	32	11	7	1	2	0	1	0
31.0	55	28	13	3	0	1	0	0	0
32.0	35	31	20	7	4	1	0	1	1
33.0	35	39	15	5	6	0	0	0	0

Table A6.3: Clast count b-axis data.

Locality	Angularity					
	Very angular	Angular	Sub angular	Sub rounded	Rounded	Well rounded
7.0	4	54	21	18	3	0
10.0	1	53	34	11	1	0
12.0	0	13	38	31	18	0
13.0	0	18	30	33	19	0
14.0	0	15	40	39	6	0
16.0	0	29	45	21	5	0
17.0	5	70	20	5	0	0
18.0	1	54	41	4	0	0
19.0	0	23	45	18	14	0
21.0	1	9	45	27	18	0
22.0	0	10	46	31	13	0
23.0	0	3	39	35	22	1
24.0	2	36	51	10	1	0
26.0	0	29	37	28	5	1
27.0	0	28	43	25	4	0
28.0	0	9	39	32	18	2
30.0	0	2	39	37	22	0
31.0	0	13	41	33	12	1
32.0	0	20	43	26	8	3
33.0	2	21	50	15	12	0

Table A6.4: Clast count angularity data

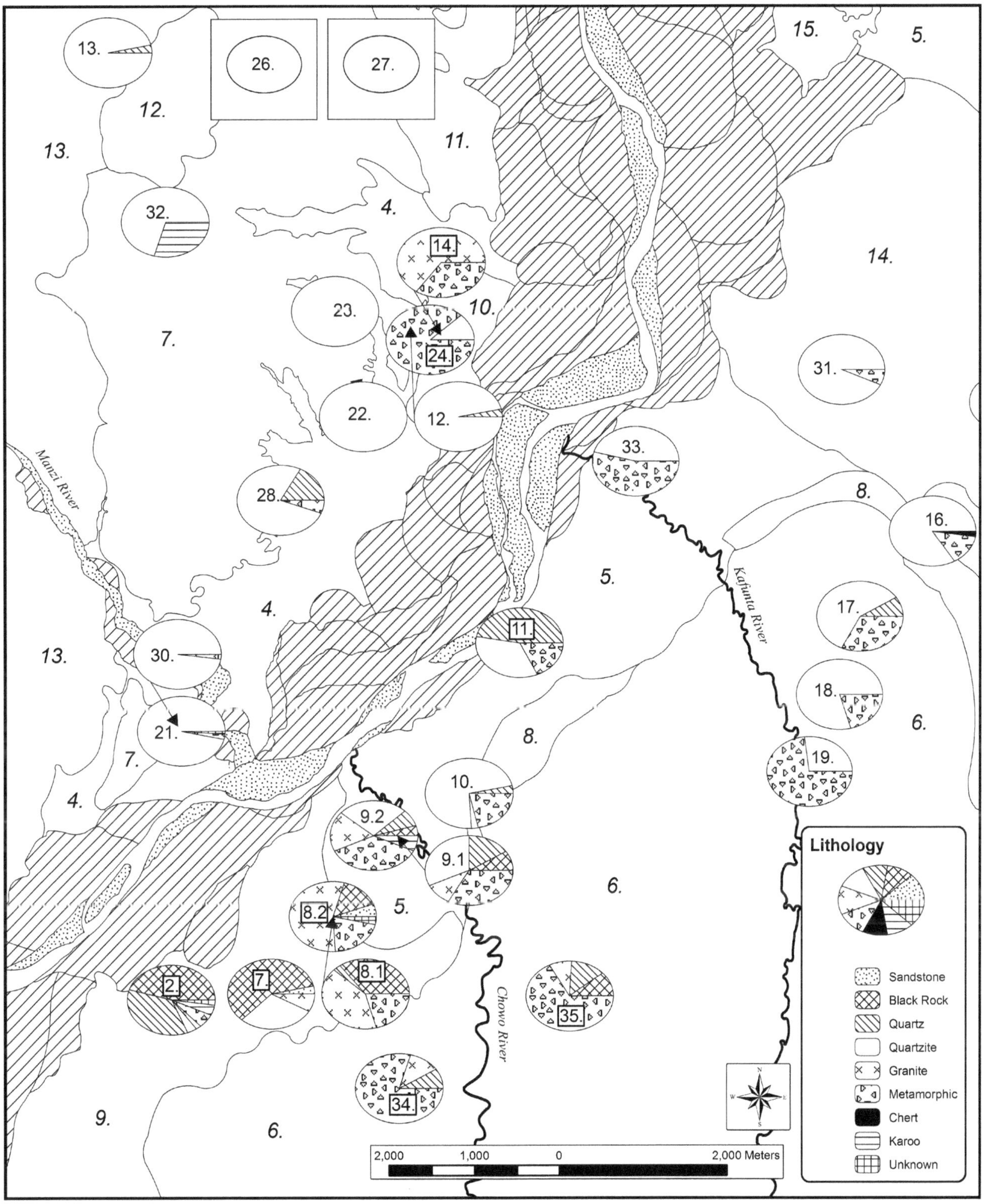

Figure A6.5: Pie charts illustrating percentages of different lithologies present at each locality where clast counts were undertaken. The number on each pie chart refers to the locality numbers listed in table A6.1. Zones 1, 2, and 3 are illustrated using the same symbols as figure 5.4, whilst for clarity, italicised numbers are used to refer to other geomorphological zones. The raw data for the clast counts is provided in table A6.2.

The Black Rock zone (9), which lies between the Nchindeni zone and the floodplain, contains the most variable clast counts of the survey. Counts 2 and 7 (530 and 560m) have a large number of rock types present (minimum of 5 and a maximum of 7), with the largest component of both counts being the black rock that defines the area (46% & 58% respectively). Except at its lowest levels, this zone appears not to have been affected by fluvial activity, and so it is likely that the mixture of lithologies have a local origin. Quartzite similar to that in the cobble zone is rare and, though there is some quartz present, it is most similar to that seen in the quartz veins eroding from the Nchindeni Hills. The material here is locally derived and does not appear to provide any material for the geomorphological zones at lower elevations. This is demonstrated by counts 8.1 and 8.2 at 530m that were conducted in the streambed and channel wall of a 2 to 4m wide stream draining the Black Rock and Nchindeni zone. These two counts contained a high proportion of granites (41% and 56% respectively) compared to the counts taken higher in the Black Rock zone. Analysis of the channel wall sediments revealed that the granites contained within the wall were highly weathered and would have not survived a large amount of transport (see figure 6.3).

Clast counts from other areas

There are two areas where clast counts were undertaken that were not directly related to the cobble zones and the Nchindeni Hills. In the Chichele Hill zone, two clast counts were taken, one from a palaeochannel dissected by a road cutting (14), and one on the surface of the cobble ridges at the base of the hill (24). The Chichele Hill deposits are unique among the clast counts conducted on the northwest bank, both counts 14 and 24 at approximately 540m contained a high proportion of granitic and metamorphic rocks (100% & 88% respectively), whereas elsewhere on the northwest bank they are rare, forming less than 5% of the total (see above). This is probably due to the fact that Chichele Hill itself is an outlier of the Nchindeni basement rock (Utting 1988: figure 6). It is therefore likely that the Chichele Hill clast counts indicate a localised source for the cobbles of that area. This supports the decision made early on in the project to classify the area as a separate zone (see appendix 6.1).

The second area is behind the cobble zone in the Chipembele Ridge Sands (zone 13). In this zone, count 13 at 570m elevation contains 96% quartzite, and 4% quartz and chert. The quartzite in this area is identical in nature to the quartzite in the cobble zones, suggesting a similar source for the material. Samples were also taken from further into this zone, beyond the survey area, to assess how the geomorphological processes toward the centre of the valley may differ. This was undertaken whilst surveying the experimental archaeological transects along the Kasweta road, the course of which is orientated to the north-west of the survey area. Time permitted only two clast counts and they were taken at equal intervals 11km apart along the road. Both clast counts contained 100% quartzite of a similar type to count 13 and the rest of the cobble zones. Count 26 was the nearest to the survey area 11km north-west of count 13 (the scale of the map does not allow counts 26 or 27 to be displayed in their correct position so they are inset in figure A6.5), and at an elevation of 720m (200m above the Luangwa), whilst count 27, a further 11km northwest is at a lower elevation of 660m.

Clast size (b-axis) results description

The clasts were measured to the nearest two millimetres in the field, however for the presentation of the maps the measurements were grouped into one-centimetre classes. In general the clast size graphs appeared to have relatively similar profiles (figure A6.6), with the highest proportion of clasts nearly always in the 2 – 2.9cm class. Samples with a large quantity of clasts greater than 6cm were rare. The counts throughout the cobble zone, behind Chipembele Ridge, and along the Kasweta road, fit the general pattern of relative similarity, however, in counts 13, 26, and 27 there are roughly double the number of clasts in the 4 – 4.9cm class compared to those nearer the Luangwa (12, 21, 22, 23, and 28). The relative sizes of clast material can give an indication of the direction of transport in certain depositional regimes (see section 6.4) so it is important to determine whether these differences were statistically significant. A statistical test was carried out to determine whether this represented a significant difference in the sizes of clasts in the two groups, or insignificant random variation in the data. The null-hypothesis was that there was no difference between the two groups. The test used was an independent samples t-test procedure for comparing means of two groups of cases (Table A6.5). The test was carried out using the raw measurements, rather than the 1 cm classes they were split into for the presentation of the bar charts and maps. In view of the fact that the aim was to specifically understand the nature of the quartzite cobble material in the area, count 32 was excluded to preclude the incorporation of the large amount of locally weathered Karoo material. The test results indicate that the difference in clast size is significant to a 95% confidence interval, and that the clasts behind Chipembele Ridge are statistically larger than those nearer the Luangwa. The significance of this observation is discussed in section 6.4.

On the south-east bank, count 31 in the Cathedral Mopane zone, and 10 and 16 in the Nchindeni Cobbles, appear to be very similar to the Cobble zone clast counts close to the Luangwa on the north-west bank (12, 21, 22, 23, and 28). A t-test to compare the means was carried out (table A6.6), this time with the null-hypothesis that there was a difference between the two sets of measurements. The result confirmed that there was no significant difference between the two groups to a 95%

confidence interval, and that the cobbles either side of the Luangwa are neither larger nor smaller than each other. The sizes of the cobbles in the Chichele Hill zone appear to be identical to the counts in the Cobble zone surrounding it, however no test was undertaken to compare the means, as the different lithologies indicate that the Chichele Hill deposits have a separate origin to the Cobble zone deposits.

Count 19, taken in the Kafunta River in the Nchindeni Hills zone appears to be very poorly sorted, having the greatest mix of clast sizes, with no weighting towards any particular class. In contrast count 17, which was taken in a smaller stream approximately two kilometres away, has a similar profile to most others in the survey area. This is probably due to it being a smaller stream at a lower elevation, which as a result has less competency to entrain larger clasts. Count 33 was taken in the Kafunta River approximately one kilometre away from the Luangwa, and four kilometres downstream from count 19. At this point most clasts were clustered in the two smallest classes, suggesting that over the distance and on reaching the flatter floodplain of the Luangwa, the competency of the river was reduced and unable to entrain the larger clasts. Count 7, in the Black Rock zone, has greater variation in clast size than most of the samples, with a large element of 4.0 cm plus clasts (38%). As noted above, this count that contains a local lithology, probably represents relatively localised transport of material, which explains the poor sorting.

The data for clast size and lithology is cross-tabulated in table A6.7. This data is useful when considering raw material selection (chapter 7), and it illustrates that, although the sample size for chert is small, there is a similar distribution in clast sizes for quartzite and chert (77% and 70% of clasts are below 3.9cm respectively). Overall the metamorphic, granite, and Karoo lithologies contain a larger proportion of larger clasts, presumably reflecting their origin from the local bedrocks and negligible transport. Quartzite clasts are the most numerous over the 4cm category, and are thus easy to locate where quartzite is common (zones 7, 8, 12, and 14). Quartz is overall the smallest of the lithologies present in the sample, and 76.5% of the clasts are smaller than 2.9 cm. Chert does occur as large clasts (20% are 5cm – 6.9cm, and 10% are over 10cm), but in general it is very rare in the landscape. There are localised outcrops where it is very common, for example at Kakumbi Spring (zone 16).

Angularity results description

The angularity measurements fall into two general groups, with most of the counts having a majority of sub-angular and sub-rounded clasts, whilst those on or near the Nchindeni Hills are dominated by sub-angular and angular clasts (figure A6.7). This suggests that in general the clasts have not been transported a great distance from their source. The clast counts from the cobble zone and the area behind Chipembele Ridge have a high proportion of sub-angular and sub-rounded clasts. However, there is a subtle difference in that the counts closer to the Luangwa; 12, 22, 23, and 28, have a relatively high proportion of rounded clasts (13 – 22 clasts), whilst the counts on the other side of Chipembele Ridge; 13, 26, and 27, have a greater angular component (18 – 29 clasts). Counts 10 and 16 in the Nchindeni Cobbles also have a high proportion of angular clasts. Count 31 in the Cathedral Mopane zone appears most similar to the cobble ridges on the northwest bank of the Luangwa, with few clasts above 5cm.

All the counts in the Nchindeni Hills have a high angular and sub-angular content, particularly in the smaller streams (counts 17 – 18). In contrast the larger Kafunta River has fewer angular, and more sub-angular clasts (counts 19 and 33), suggesting that though larger clasts had not been transported as far as count 33, the transported clasts had not been significantly abraded over the relatively short distance between the two points. Count 7, in the Black Rock zone, has a high angular and sub- angular content suggesting that the clasts have not travelled very far, further supporting the conclusion that the material is derived relatively locally.

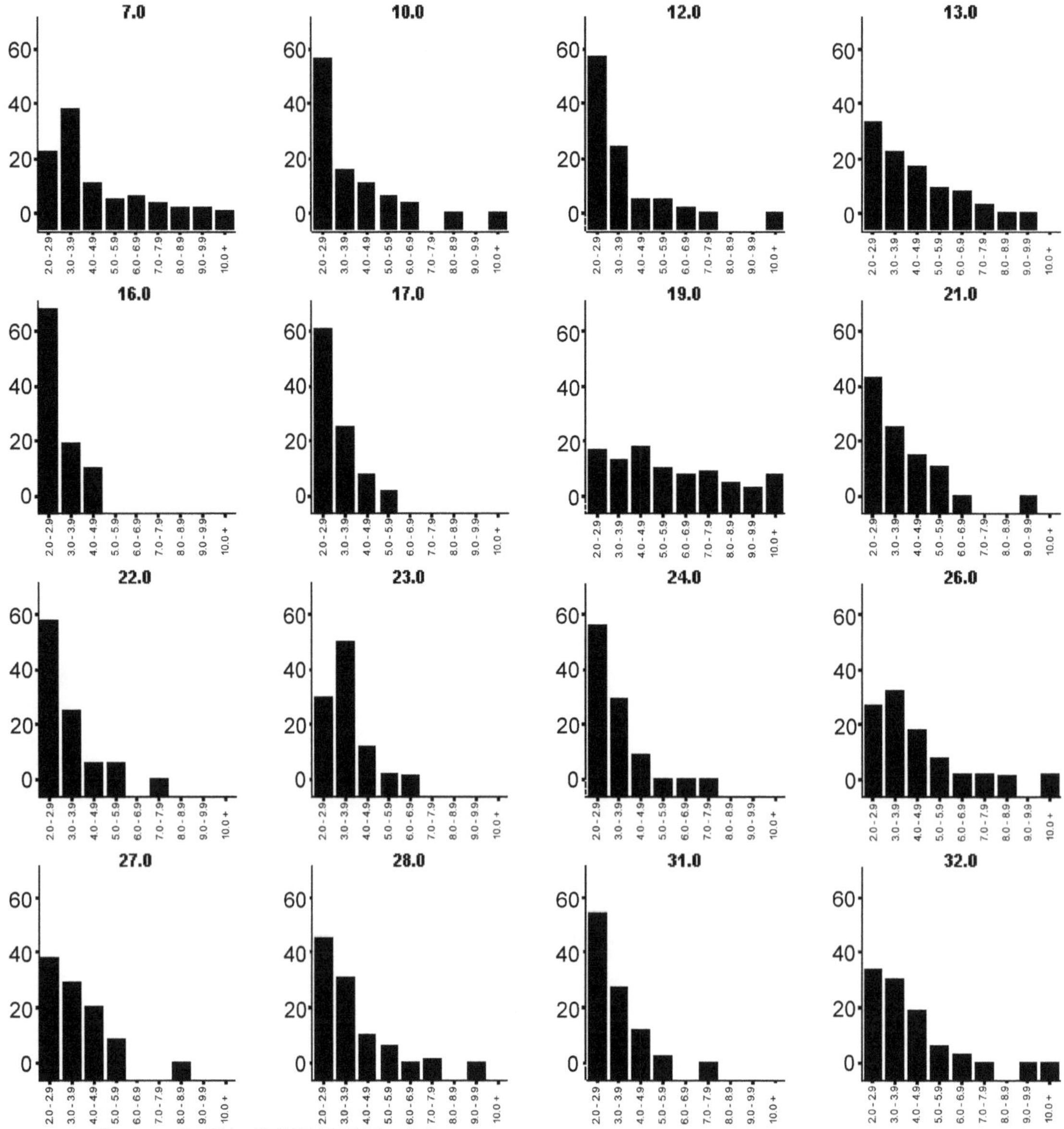

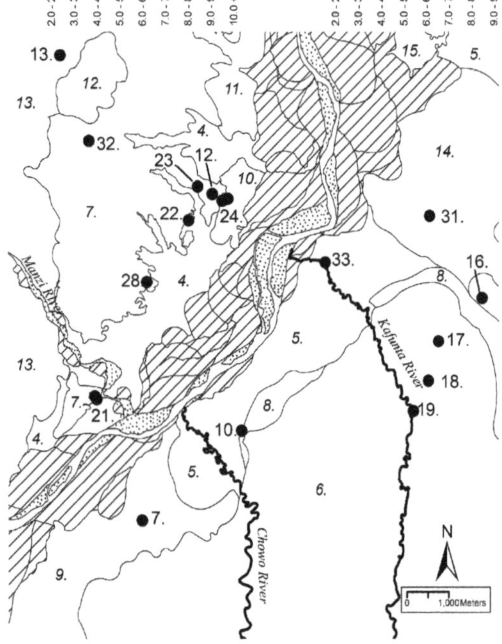

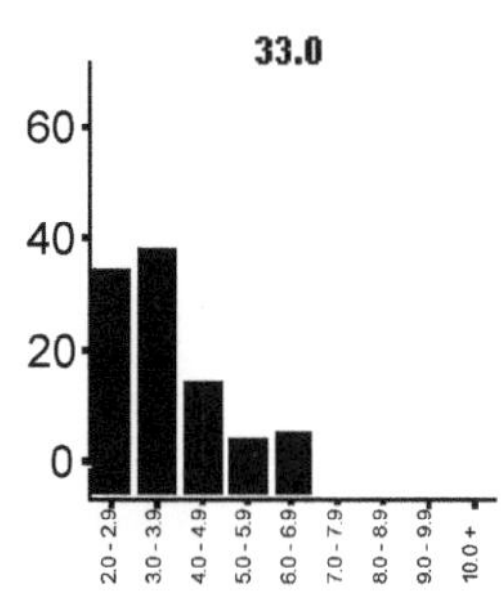

Figure A6.6: Bar charts illustrating clast sizes (b-axis measurements) recorded at each locality. Each sample is divided into 1cm classes (x-axis), and each class is represented as a percentage of each sample (y-axis). The inset map provides a key to the sample locations for this figure and figure A6.7 (numbered dots), and the zones (italic numbers). Counts 10 and 30 are not marked as they are behind counts 24 and 21 respectively. Counts 26 and 27 were undertaken along the Kasweta road, and are located between transects K1 and K5, and next to K3 respectively (figure 7.28).

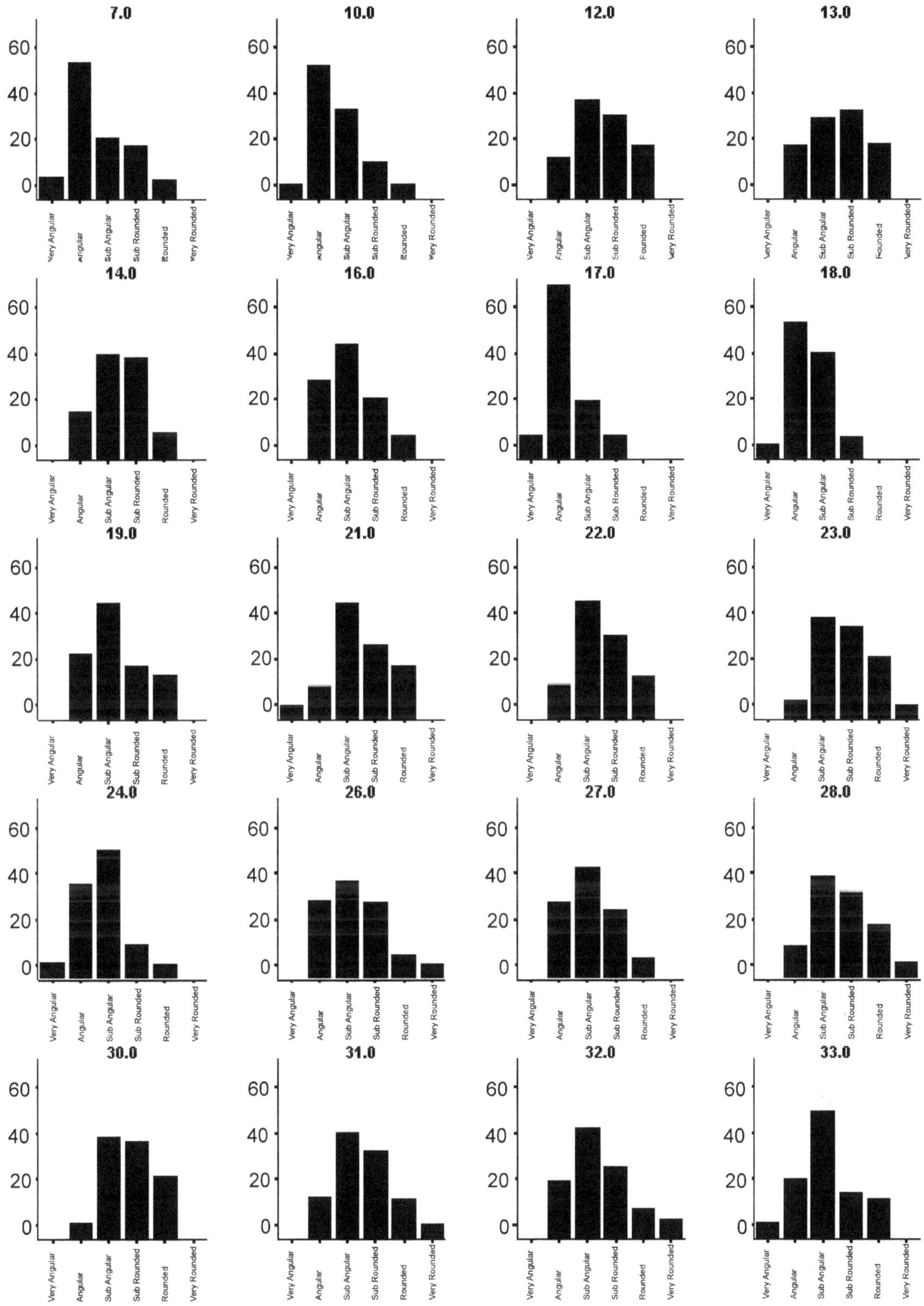

Figure A6.7: Graphs illustrating the angularity data from localities where angularity data was recorded (the inset map in figure A6.6 illustrates the locations of the counts). The bars represent percentage of total count at each locality.

Table A6.5 and A6.6: Significance tests undertaken to determine the significance of differences noticed in the b-axis measurements data. The tests were undertaken in SPSS 13, and are independent samples t-test procedures for comparing means of two groups of cases. The interpretation requires that the Levene test is first consulted; if the values in the Levene test is high (typically greater than 0.05), then the results that assume equal variances for both groups is used, and if it is low then the results that do not assume equal variances for both groups is used. If there is a low significance for the t-test (typically less than 0.05), then this indicates that there is a significant difference between the two groups. If the confidence interval for the mean does not contain zero, this also indicates that the result is significant, however if the significance is high and the confidence interval for the mean difference contains zero then it cannot be concluded that the difference is significant.

B-axis groups	N	Mean	Std. Deviation	Std. Error Mean
1. Cobbles behind Chipembele Ridge. Counts 26,27 & 13	300	3.817	1.6948	.0978
2. Cobbles on the cobbles near the Luangwa. Counts 21,22,28,23 & 12	500	3.277	1.1900	.0532

	Levene's Test for Equality of Variances		t-test for Equality of Means						
	F	Sig.	t	df	Sig. (2-tailed)	Mean Difference	Std. Error Difference	95% Confidence Interval of the Difference Lower	Upper
Equal variances assumed	23.600	.000	5.282	798	.000	.5403	.1023	.3395	.7411
Equal variances not assumed			4.851	477.060	.000	.5403	.1114	.3215	.7592

Table A6.5: Significance test of two groups of cobbles – both groups are on the north-western bank of the Luangwa River, one distant (behind Chipembele Ridge), and one proximal (near the Luangwa). The results of the test indicate that there is a significant difference in the sizes of the clasts further away from the Luangwa (they are on average larger).

B – axis groups.	N	Mean	Std. Deviation	Std. Error Mean
1. Cobbles on the Nchindeni side of the Luangwa. Counts 10 &31.	200	3.172	1.2554	.0888
2.Cobbles on the northwest bank of the Luangwa. Counts 28,21,22,12,& 23.	500	3.277	1.1900	.0532

	Levene's Test for Equality of Variances		t-test for Equality of Means						
	F	Sig.	t	df	Sig. (2-tailed)	Mean Difference	Std. Error Difference	95% Confidence Interval of the Difference	
								Lower	Upper
Equal variances assumed	.731	.393	-1.043	698	.297	-.1055	.1012	-.3041	.0931
Equal variances not assumed			-1.019	349.756	.309	-.1055	.1035	-.3091	.0981

Table A6.6: Significance test of two groups of cobbles – This test compares the cobbles on the south-east bank of the Luangwa River at the base of the Nchindeni Hills, with the cobbles on the north-west bank. The result of this test indicates that there is no significant difference in the size of clasts either side of the Luangwa.

B axis class	Lithology																	
	Quartzite		Chert		Metamorphic		Quartz		Black Rock		Granite		Karoo		Sandstone		Unknown	
	Number (N)	Column N %	N	Column N %	N	Column N %	N	Column N %	N	Column N %	N	Column N %	N	Column N %	N	Column N %	N	Column N %
2.0 - 2.9	612	47.9	3	30.0	99	34.5	13	76.5	9	15.5	4	30.8	8	26.7	1	20.0	1	50.0
3.0 - 3.9	368	28.8	4	40.0	83	28.9	0	.0	25	43.1	3	23.1	5	16.7	2	40.0	0	.0
4.0 - 4.9	178	13.9	0	.0	39	13.6	1	5.9	5	8.6	1	7.7	6	20.0	2	40.0	0	.0
5.0 - 5.9	72	5.6	1	10.0	21	7.3	1	5.9	4	6.9	2	15.4	5	16.7	0	.0	0	.0
6.0 - 6.9	22	1.7	1	10.0	19	6.6	1	5.9	4	6.9	1	7.7	3	10.0	0	.0	0	.0
7.0 - 7.9	12	.9	0	.0	10	3.5	1	5.9	4	6.9	1	7.7	1	3.3	0	.0	0	.0
8.0 - 8.9	7	.5	0	.0	4	1.4	0	.0	2	3.4	1	7.7	0	.0	0	.0	0	.0
9.0 - 9.9	3	.2	0	.0	3	1.0	0	.0	3	5.2	0	.0	1	3.3	0	.0	1	50.0
10 +	4	.3	1	10.0	9	3.1	0	.0	2	3.4	0	.0	1	3.3	0	.0	0	.0

Table A6.7: A table cross-referencing lithology and b-axis measurements for all clast counts where the relevant measurements were taken. The clast counts and the percentage of each lithology in each b-axis class are provided.

9.3. Appendices for Chapter 7

Appendix 7.1. Data collection procedures

Location data and measurements taken for artefacts

The archaeological data was stored in a SPSS database. This provided the capability to easily manipulate the raw data to produce complex tables of totals, averages, and other statistics, which is not possible with the simple dBase database system used by ArcView. The tabulated data could then be exported into a text file format allowing it to be copied into the dBase system of ArcView, which provides the capacity for the geo-spatial presentation and analysis of the data. In the SPSS database each artefact's basic geographic information was recorded (transect number and individual 100m x 100m block on each transect), as this is required to link the data produced to the individual transects in the ArcView database. In each instance, the transect and zone from which the artefact was retrieved, as well as an additional identifier that allows artefacts to be assigned to smaller 100x100m blocks on each transect, was recorded. This allows relationships such as artefact density in the landscape to factors such as hill slope and elevation to be analysed in detail where necessary.

The first step of the analysis was to determine the artefact type, the identification for which was based on the knowledge of Stone Age and Iron Age technologies used in this area of Africa (outlined in chapter 2; see figures 2.1 – 2.6). As it would not be possible to visit the artefact collection again during the course of this research, a policy of maximum data retrieval was pursued so that analyses not originally conceived could be undertaken if the material were suitable. As a result of this policy, some of the artefact parameters outlined here were not used in the final analyses as they were not required.

All stone artefacts were measured using a standardised method to the nearest millimetre. Cores were measured by taking the longest possible measurement for the artefact for length, and then the second longest measurement for breadth, and then the shortest possible measurement for thickness (see Andrefsky 1998:12 - 17). The total number of visible flake removals from the core, and the number of episodes that were used to remove them were also recorded (this is where one run of flake detachments ceased, and a new sequence of removals was begun from a new platform on the core). Whilst total number of flakes removed was a relatively simple observation, it was difficult to determine the number of episodes in which the removals were made and, as a result, it was not possible to record them for the majority of the cores. Flake sizes were measured using one of the standard methods of recording size outlined by Andrefsky (Andrefsky 1998:97-99). They were measured by taking the maximum length, which in this case was taken from the point of percussion on the butt, paralleling the flake's axis of percussion, to the most remote point on the distal end of the flake. For breadth, the maximum width of the flake was measured at the widest point of the flake perpendicular to its length, and the maximum thickness of the flake was taken from the dorsal to the ventral side. Broken flakes were also measured, though often not analysed as the original dimensions of the flake are unknown. These measurements may have been needed to provide data on natural processes, and it is a simple process in SPSS to exclude the measurements when not needed.

Flakes can contribute an understanding of core technology (Ashton & McNabb 1996:203), for example dihedral and multifaceted butts can be an indicator of prepared core technology. The technological features of flakes recorded included butt type of every flake with a complete butt, and the scar pattern and cortex index of each complete flake. The classifications for butt type, cortex index, and dorsal scar pattern were recorded using the appendices listed by Ashton and McNabb (1996:242 - 243). These features were easy to identify and are designed to favour a technological interpretation and allow a more accurate determination of knapping strategies if required, and thus avoid an arbitrary typological interpretation.

All the artefacts large enough to be measured were also weighed to the nearest gram. Where large proportions of undiagnostic chunks and very small debitage were present, the artefacts were not individually measured, but grouped together and a total weight was taken. Undiagnostic pottery sherds were counted and weighed in the same manner. The large collections of pottery made measuring every single piece impractical so pottery was only measured and weighed if the fragments were diagnostic rim sherds.

Raw material and artefact abrasion

The raw material was determined for every stone artefact in the collection by visual analysis (table A7.1), occasionally assisted by a hand lens. Unlike the geomorphological analysis of the rock types found in the area, the quartzites and quartzes were split into categories. In the geomorphological clast analysis, thin sections or geochemical analyses would have been necessary to differentiate the quartzite present in the area to provide a geological definition of each to assist in determining provenance (see section 5.6.4). These techniques were not available to this survey and their use was not necessary for the artefact analysis as determining their geological provenance was not an aim of the project. The definition of the quartzite classes was based on the texture of the raw material. This textural definition was created to determine whether any behavioural preference exists for quartzites with different knapping properties. Small exercises in experimental knapping revealed that finer grained homogenous quartzites were easier to work, as the material fractured more consistently and predictably than coarser quartzites.

The degree of wear caused by abrasion and rolling of the artefacts was recorded so that the relative effects of past geomorphological activity on archaeological material from different periods could be determined. An approach similar to Clark's (Clark 2001:103) at Kalambo Falls was used. Clark's abrasion classification was aimed at aiding the typological classification of artefacts in the mixed 'rubble' horizons of Kalambo Falls (*ibid.*) i.e., more heavily rolled artefacts were assumed to derive from an earlier period. Contrastingly, in this project artefact condition was not a deciding factor when attempting to determine age, as this may have unfairly skewed the results towards an abraded Early Stone Age sample that may not be accurately representative of the degree of abrasion experienced by Early Stone Age artefacts. Initially three degrees of rolling were defined: fresh, slightly rolled, and heavily rolled. Fresh condition was recognised by an absence of visible erosion or weathering of the artefact; slightly rolled, had minor abrasion to the arêtes and ridges between flake scars (on core surfaces and dorsal surfaces of flakes) with some damage to the artefact edges; heavily rolled artefacts had significant abrasion to the flake scar surfaces as well as the arêtes, ridges, and artefact edges. This system was slightly altered in the field as it was noticed that there was a rare, but nevertheless present class of artefacts that were abraded to a noticeably higher degree. To include these artefacts a fourth category was added, and the previous heavily rolled category was reassigned to a new category of moderate rolling. The heavily rolled category was redefined as an artefact that has very rounded features and edges, as well as an appearance of developing a natural cortex on surfaces that would not have had cortex after the knapping process was finished. This classificatory system was not based on any scientific measurement system and was achieved by visual study. To aid consistency, a standard reference group of artefacts from the collection was kept on display. In this case only one researcher was undertaking the analysis, which will have eliminated any inter-observer error that may have affected consistency, an issue that had to be considered in Clark's Kalambo Falls research (Clark 2001:103).

Raw material	Description
Fine Quartzite	Fine homogenous quartzite (fine sugary appearance), few unintended fractures created in the knapping process. Grain size less than 1mm.
Medium Quartzite	Intermediate quality, in that there may be some failures along faults in the material created when knapping. Grain size 1mm – 3mm.
Coarse Quartzite	Blocky often poorly cohesive and flake and/or core may fracture unpredictably. Grain size 3mm + or consists of angular quartzite.
Vein Quartz	Difficult to identify unless there is a remnant of the country rock of the quartz vein still adhering to the artefacts natural surfaces.
Chert	Cryptocrystalline silica, and occasionally chalcedony in this survey area. Usually of diagenetic origin in sedimentary rocks (Prothero & Schwab 1997:285), it may originate from veins running through the Karoo bedrock (*pers obs* of vein material present in the landscape but not *in situ.*). When knapped it is usually an easy to work raw material, and produces sharper edges than the quartzite.
Quartz	Glassy and homogenous, quartz, or single crystals of quartz, with no, or few, obvious crystal boundaries. Occasionally an overlap could occur between this and fine quartzite, however, if the material had a glassy texture it was assigned to quartz.
Tabular Quartzite	This classification refers to a structural difference in the rock, usually similar to fine – medium quartzite, there is a parallel metamorphic foliation in the rock (similar in form to bedding), that makes it liable to fracture parallel to the foliations rather than with the usual concoidal fracture needed when knapping.
Black rock	A black rock similar in appearance to dolerite, though no thin section analysis was undertaken to confirm this. Crystalline, with crystals <1mm – 3mm and usually uniform in texture.
Heat treated chert	Chert that has been heated deliberately in a fire, creating heating fractures throughout the artefact, and/or pot lid fractures.
Fossil wood	Fossilised plant material, usually trees. It is distinctive in hand specimen as, though the fossilised material is usually identical to chert in the appearance of the crystal structure, it usually preserves the original growth rings of the tree as coloured bands and/or as structural cracks or weaknesses following the tree rings. Not common in the main survey area, but prevalent in other areas of the park.

Table A7.1: Raw material types identified in the artefact analysis.

Dating the artefacts

Assigning artefacts to chronological periods was based on knowledge of the typical technologies used in each period in this area of Africa (outlined in chapter 2). Assigning artefacts in this manner inherently engenders a large error margin, for example, the Early Stone Age technology of hard hammer knapping (Mode 1) results in a limited typological suite of artefacts, usually cores and flakes. These Mode 1 artefacts continue to be manufactured throughout archaeological time as it is an expedient technology, with the result that Middle Stone Age core and flake material could be misclassified as Early Stone Age, and vice versa. A similar scenario could occur with Middle (Mode 3) and Later Stone Age (Mode 5) technology, where prepared cores characteristic of Mode 3 technology are produced throughout the two periods. The primary difference is in size as Later Stone Age artefacts are smaller on average, but there is an overlap and a number of artefacts could be classified as coming from either period. To screen out artefacts that could be misclassified in this way, independent categories were created for them, such as Early/Middle Stone Age, as well as a general 'unknown' category for stone artefacts that are clearly worked, but cannot be assigned to any period. Using these categories provided a safety net and meant that only artefacts that definitively originated from one period would be classified as coming from that period. The risk existed that substantial material would be removed from the analysis, but by including only technologically and typologically certain artefacts in each period, the sample, no matter how residual, ensured a higher degree of integrity. A second risk to consider was the skewing of results in favour of large Early Stone Age, 'typical' Middle Stone Age, or small Later Stone Age artefacts. However, as no comparisons are to be made with other artefact collections, and the creation of a typological sequence for the area was not the purpose of this research, the potential clustering should provide a shield for the results, as only residues are considered that certainly derive from one archaeological period.

Appendix 7.2 general description of the archaeological data

Basic parameters of the sample

A total of 1,845 artefacts were recovered from the 61 transects in the 168km^2 main survey area, weighing a total of 80.25 kg (table A7.2). As noted above, there was a possibility that sample sizes could be low due to uncertainties regarding identifying an artefact's typological association. It transpired that large numbers were classified as potentially deriving from any point within two periods e.g. Early/Middle Stone Age (table A7.2). These boundary classifications do not have much use in analysing patterns in the landscape, and they should not be assumed to be intermediate periods of technological development between the groups.

	Typology									
	Early Stone Age		Early/Middle Stone Age		Middle Stone Age		Middle/Later Stone Age		Later Stone Age	
Number of artefacts	86	4.6%	234	12.7%	29	1.6%	228	12.4%	346	18.8%
Weight (g)	26813.0		23190.0		1487.0		5775.5		2788.0	

	Typology										
	Early Iron Age		Iron Age		Later Iron Age		Unknown		20th Century		Total
Number of artefacts	1	.1%	668	36.2%	61	3.3%	191	10.4%	1	.1%	1845
Weight (g)	13.0		12622.0		471.0		6957.5		133.0		80250.0

Table A7.2: A table of the total number of artefacts recovered, the percentage of the collection they represent, and their total weight from the main random survey of 61 transects, organised by typological association (percentages are rounded to one decimal point).

The primary raw materials used in stone tool manufacture throughout the survey area are quartzite or quartz of some description (table A7.3), totalling 950 artefacts that represent 85.8% of the total. The remainder is mainly chert (13.4%), with small quantities of other materials such as glass and ore (0.8% others). Of the artefacts recorded in the sample the greatest proportion are pottery fragments (37.3%) (table A7.4). Table A7.2 shows that the majority of these fragments are undiagnostic of either of the phases of the Iron Ages, with 36.2% of the total artefact collection attributed to undifferentiated Iron Age, opposed to a combined total of 3.4% for the Early and Late Iron Age categories, where only one artefact could be classified as Early Iron Age. Simple cores and flakes, when totalled, make the greatest proportion of artefacts present (44.1%). The remainder are more technologically complex and can be classified more formally: prepared cores and blade cores comprise 0.8% of the sample, and stone artefacts such as segments, crested flakes, flaked flakes, blades, or prepared core flakes total 1.7% of the sample (table A7.4).

Raw material	Number of artefacts	
Fine Quartzite	265	23.9%
Medium Quartzite	166	15.0%
Coarse Quartzite	70	6.3%
Vein Quartz	12	1.1%
Chert	146	13.2%
Granite	1	.1%
Quartz	428	38.7%
Pink Quartzite	1	.1%
Tabular Quartzite	8	.7%
Black Rock	2	.2%
Heat Treated Chert	2	.2%
Fossil Wood	0	.0%
Silcrete	0	.0%
Ore with streak	4	.4%
Glass	0	.0%
Unknown	2	.2%
Total	1107	100.0%

Table A7.3 (above): The totals for artefact raw material.

Table A7.4 (right): Total numbers of tool types collected and percentage of total.

Tool type	Number of artefacts	
	Total	% of Total
Core	351	19.4%
Flake	308	17.0%
Broken Flake	118	6.5%
Split Cobble	113	6.2%
Chunk	102	5.6%
Pot	675	37.3%
Core Fragment	22	1.2%
Prepared Core Flake	14	.8%
Prepared Core	11	.6%
Shatter/debris	41	2.3%
Unutilised pigment	13	.7%
Hammerstone	5	.3%
Flaked flake	5	.3%
Retouched flake	5	.3%
Top stone/pestle	3	.2%
Blade Core	2	.1%
Blade	2	.1%
Blade fragment	2	.1%
Utilised pigment	2	.1%
Unknown	6	.3%
Misc 20th C	3	.2%
Crested Flake	1	.1%
Wall fragment (daga)	3	.2%
Segment	1	.1%
Bottom stone/mortar	1	.1%
Bipolar core	1	.1%
Total	1810	100.0%

The majority of the stone artefacts are in a fresh condition (table A7.5), and overall 68.4% are unrolled, 22% slightly rolled, and only 9.5% of the total are moderately or heavily rolled. There are some notable differences to the general trend, interpretations for which are presented in the following sections and chapter 8. Zones 3, 4, and 13, have fresh artefacts representing 90%, 96.3%, and 88.6% of the total for each zone respectively, which is much higher than the overall pattern. This at first glance appears high as it may be expected on these recent surfaces that the artefacts present might have been damaged by the activity of the Luangwa. In contrast zones 7, 12, and 14, which all contain the cobble mound topography in some form, have relatively few fresh artefacts (60.6%, 43.8%, and 55.7% respectively), and a greater number of slightly and moderately rolled artefacts. This was expected because as these zones all contain cobbles that have been deposited by an alluvial fan system that has subsequently been eroded, making it likely that any archaeological material would have been damaged (see section 6.6). In fact, the number of fresh artefacts present in these zones may seem high, and the reasons for this are explored in chapter 7, but essentially relates to the relatively large number of more recent Later Stone Age artefacts found in these zones.

It is possible that raw material could have an influence on artefact condition (Foley 1981:122); weaker raw material found in the Luangwa Valley, such as the coarse quartzite, could abrade to a higher degree more rapidly than a strong homogenous material such as quartz or fine quartzite. This relationship was examined by cross-tabulating the degree of abrasion exhibited by the artefacts with their typological affinity and raw material (table A7.6). Typological affinity was introduced to the analysis as the age of an artefact will clearly also have an effect on the amount of abrasion an artefact might have experienced, and therefore comparisons should only be made with artefacts of a similar age. Later Stone Age artefacts were excluded from the analysis as the majority are in fresh condition (see section 8.4). The relationship is complex, but the results indicate that raw material has not been a factor in the abrasion of the artefacts. For the Early Stone Age, the percentage (11% - 14%) of heavily rolled artefacts is similar for quartz, and medium and coarse quartzite. There are no heavily rolled fine quartzite artefacts, but there is a similar percentage (21% - 30%) of moderately rolled fine quartzite artefacts as medium and coarse quartzites. The absence of fine quartzite in the heavily rolled category, and quartz in the moderately rolled category, might be an artefact of the sample size rather than a genuine reflection on the amount of rolling experienced by artefacts made of different raw material. The indeterminate classifications were included as a comparison in table A7.6. Though this does not increase the sample size for individual periods, it still provides a useful comparison in this circumstance. In the Early/Middle Stone Age sample, despite the fact that there are few heavily rolled artefacts, the percentage of artefacts in each class of rolling is similar for each raw material type and is in a range of 12% – 20% of each other, which suggests that raw material did not have an influence in abrasion throughout the two periods. In the case of the Middle Stone Age and the Middle/Later Stone Age the pattern is more difficult to discern; the Middle Stone Age sample size is small and there is only one coarse quartzite artefact to provide a comparison to the harder raw materials. It is, however, fresh, indicating that despite its relative softness it has not been subject to any abrasion greater than the other artefacts of the period. In fact, the lack of coarse quartzite in both categories is a result of raw material selection on the part of Middle Stone Age hominins (see section 8.3), so it is not actually possible to compare the raw materials more susceptible to erosion, as they were simply not used. In general, it is apparent that raw material type does not influence the degree of abrasion that an artefact might have experienced, and therefore, the abrasion states of artefacts from any one period will be a direct function of geomorphological processes.

Condition (degree of rolling)	Numbers of artefacts and percentage sorted by geomorphological zone											
	3. Recent cut-offs		4. Sand and silt nw		5. Sand and silt sw		6. Nchindeni Hills		7. Cobble zone		8. Nchindeni cobbles	
Fresh	9	90.0%	52	96.3%	126	75.9%	145	78.8%	260	60.6%	11	73.3%
Slightly Rolled	1	10.0%	2	3.7%	24	14.5%	36	19.6%	113	26.3%	4	26.7%
Moderately Rolled	0	.0%	0	.0%	16	9.6%	3	1.6%	52	12.1%	0	.0%
Very Rolled	0	.0%	0	.0%	0	.0%	0	.0%	4	.9%	0	.0%

Condition (degree of rolling)	Numbers of artefacts and percentage sorted by geomorphological zone													
	9. Black rock		10. Chichele Hill		11. Sand Bank		12. Chipembele Ridge		13. Chipembele Ridge Sands		14. Cathedral Mopane zone		Table Totals	
Fresh	9	45.0%	10	76.9%	2	100.0%	7	43.8%	39	88.6%	78	55.7%	748	68.4%
Slightly Rolled	7	35.0%	3	23.1%	0	.0%	7	43.8%	5	11.4%	39	27.9%	241	22.0%
Moderately Rolled	4	20.0%	0	.0%	0	.0%	2	12.5%	0	.0%	15	10.7%	92	8.4%
Very Rolled	0	.0%	0	.0%	0	.0%	0	.0%	0	.0%	8	5.7%	12	1.1%

Table A7.5: Tables illustrating the numbers and percentages of artefacts in each abrasion class by geomorphological zone. Zones 1 and 2 have been excluded as these are the Luangwa River channel and its sand banks, which were not surveyed.

Typology	Raw material	Fresh		Slightly Rolled		Moderately Rolled		Very Rolled		Total	
		Total N	% of row	Total N	% of row	Total N	% of row	Total N	% of row	Total N	% of row
Early Stone Age	Fine Quartzite	15	65.2%	3	13.0%	5	21.7%	.	.0%	23	100.0%
	Medium Quartzite	5	17.2%	13	44.8%	7	24.1%	4	13.8%	29	100.0%
	Coarse Quartzite	5	19.2%	10	38.5%	8	30.8%	3	11.5%	26	100.0%
	Quartz	6	85.7%	.	.0%	.	.0%	1	14.3%	7	100.0%
Early/Middle Stone Age	Fine Quartzite	39	47.6%	28	34.1%	15	18.3%	.	.0%	82	100.0%
	Medium Quartzite	24	30.8%	27	34.6%	23	29.5%	4	5.1%	78	100.0%
	Coarse Quartzite	11	40.7%	11	40.7%	5	18.5%	.	.0%	27	100.0%
	Vein Quartz	1	33.3%	1	33.3%	1	33.3%	.	.0%	3	100.0%
	Chert	4	50.0%	4	50.0%	.	.0%	.	.0%	8	100.0%
	Quartz	13	40.6%	13	40.6%	6	18.8%	.	.0%	32	100.0%
	Tabular Quartzite	2	50.0%	1	25.0%	1	25.0%	.	.0%	4	100.0%
Middle Stone Age	Fine Quartzite	4	50.0%	4	50.0%	.	.0%	.	.0%	8	100.0%
	Medium Quartzite	.	.0%	2	66.7%	1	33.3%	.	.0%	3	100.0%
	Coarse Quartzite	1	100.0%	.	.0%	.	.0%	.	.0%	1	100.0%
	Chert	10	83.3%	2	16.7%	.	.0%	.	.0%	12	100.0%
	Quartz	4	80.0%	.	.0%	1	20.0%	.	.0%	5	100.0%
Middle/Later Stone Age	Fine Quartzite	38	59.4%	20	31.3%	6	9.4%	.	.0%	64	100.0%
	Mcdium Quartzite	9	37.5%	12	50.0%	3	12.5%	.	.0%	24	100.0%
	Coarse Quartzite	3	75.0%	1	25.0%	.	.0%	.	.0%	4	100.0%
	Vein Quartz	3	100.0%	.	.0%	.	.0%	.	.0%	3	100.0%
	Chert	52	89.7%	6	10.3%	.	.0%	.	.0%	58	100.0%
	Quartz	42	57.5%	28	38.4%	3	4.1%	.	.0%	73	100.0%
	Pink Quartzite	.	.0%	1	100.0%	.	.0%	.	.0%	1	100.0%

Condition (degree of rolling)

Table A7.6: Cross-tabulated data illustrating the relationship between artefact raw material used in the Early and Middle Stone Age against the degree of abrasion recorded. The indeterminate periods are also displayed to provide extra data, as the sample size in the Early and Middle Stone Age is relatively small.

General description of archaeological material by period

Table A7.7 summarises data for the average length, thickness, and weight of the key flake and core artefact types. Data for the indeterminate periods has been included in some of the cases analysed in this section, as they provide a general overview of the data and a comparison for the known period categories of the Early and Middle Stone Age, which have relatively low numbers of artefacts. It is apparent that in later periods, the artefacts produced are smaller than in those of the earlier periods. The indeterminate artefact classifications, which will include Mode 1 and 3, as well as Mode 3 and 5 technologies, also follow this general pattern (table A7.7). This suggests that the classification technique was consistent, with the indeterminate classes genuinely representing artefacts that might have been produced during either period.

The data from the cores in the sample is presented in table A7.8 and A7.9; the Early Stone Age (Mode 1) artefacts predominately have three to five removals, performed in one to three episodes. Middle Stone Age (Mode 3) cores typically have four to six removals, most frequently undertaken in one episode. The difference is presumably a result of the prepared core technique used in the Middle Stone Age, for which a core is knapped with the aim of producing one or two predetermined flakes (see section 2.2). This predetermination and planning has resulted in fewer visible episodes on the final core form, though presumably in many cases a larger number of episodes were necessary throughout the knapping procedure, evidence of which was removed in the final episode. Later Stone Age (Mode 5) cores frequently have two to five removals though the highest proportion has just three or four removals, as was the case with Mode 1 Early Stone Age material. This similarity between the scar counts on Mode 1 and 5 cores is superficial, and is due to the practice in the Later Stone Age of removing microlithic flakes from bipolar split cobble cores (a specialist operation absent in mode 1 and 2 repertoires); these are often relatively small and constrain the knapping procedure (Barham 1987:48).

Table A7.7: Average length, thickness (t) (in cm), and weight (grams), of key artefact types by typological association. Artefacts such as broken flakes, broken cores, split cobble cores, and pot are excluded here.

Tool type	Typology											
	Early Stone Age				Early/Middle Stone Age				Middle Stone Age			
	Mean length cm	Mean t cm	Total N	Mean weight	Mean length cm	Mean t cm	Total N	Mean weight	Mean length cm	Mean t cm	Total N	Mean weight
Flake	7.2	2.9	18	200.6	5.0	1.8	78	49.6	3.9	1.2	5	17.0
Core	8.5	5.1	62	360.8	6.5	3.4	118	148.8	6.4	1.9	6	75.0
Prepared Core Flake	.	.	0	.	.	.	0	.	4.2	1.4	7	27.1
Prepared Core	.	.	0	.	.	.	0	.	5.2	2.4	5	64.0
Blade	.	.	0	.	.	.	0	.	.	.	0	.
Blade Core	.	.	0	.	.	.	0	.	6.5	5.0	1	168.0
Crested Flake	.	.	0	.	.	.	0	.	.	.	0	.
Bipolar core	.	.	0	.	.	.	0	.	.	.	0	.

Tool type	Typology							
	Middle/Later Stone Age				Later Stone Age			
	Mean length cm	Mean t cm	Total N	Mean weight	Mean length	Mean t cm	Total N	Mean weight
Flake	3.3	1.2	88	13.9	2.2	.7	104	4.2
Core	4.8	2.3	68	47.7	3.1	1.6	68	19.5
Prepared Core Flake	3.4	1.0	5	10.8	2.1	.6	2	2.0
Prepared Core	5.4	1.5	1	58.0	3.5	1.6	5	19.0
Blade	4.9	1.3	1	20.0	2.7	.6	1	1.0
Blade Core	4.1	3.3	1	54.0	.	.	0	.
Crested Flake	.	.	0	.	2.2	.7	1	1.0
Bipolar core	.	.	0	.	2.1	1.1	1	7.0

Cores: Number of removals	Early Stone Age		Middle Stone Age		Later Stone Age	
	No, of artefacts	Column %	No, of artefacts	Column %	No, of artefacts	Column %
1	.	.0%	.	.0%	.	.0%
2	5	8.2%	.	.0%	14	16.7%
3	16	26.2%	.	.0%	27	32.1%
4	18	29.5%	2	15.4%	22	26.2%
5	11	18.0%	2	15.4%	15	17.9%
6	5	8.2%	4	30.8%	2	2.4%
7	4	6.6%	1	7.7%	3	3.6%
8	2	3.3%	1	7.7%	1	1.2%
9	.	.0%	1	7.7%	.	.0%
11	.	.0%	1	7.7%	.	.0%
12	.	.0%	1	7.7%	.	.0%

Table A7.8: Data for the number of removals from cores in the collection, and the percentage of each by period.

Cores: number of episodes	Early Stone Age		Middle Stone Age		Later Stone Age	
	No. of artefacts	Column %	No. of artefacts	Column %	No. of artefacts	Column %
1	30	54.5%	7	63.6%	61	79.2%
2	15	27.3%	4	36.4%	13	16.9%
3	9	16.4%	.	.0%	3	3.9%
5	1	1.8%	.	.0%	.	.0%

Table A7.9: Data for the number of episodes undertaken in the knapping procedure, and percentage period.

The flakes in the collection were analysed in relation to their typological affiliation and platform/butt types (table A7.10), scar counts (table A7.11), and cortex index (table A7.12). Due to the uncertainty associated with classifying artefacts to any one period a relatively small proportion of the sample appears in the Early and Middle Stone Age columns. Platform types for 336 flakes could be recorded, of which 17, 12, and 115 flakes were assigned to the Early, Middle, and Later Stone Age respectively, with the remainder classified in the indeterminate classes. The Early Stone Age has the highest proportion of natural or cortical platform types (58.8%) (table A7.10); the Middle Stone Age has the lowest (8.3%), but has the highest proportion of dihedral platform types (50%). The high number of dihedral platform types reflects the development of prepared core technology in the Middle Stone Age. It is probable though that the statistics here are distorted by the small sample sizes, as well as the fact that some Mode 1 flakes produced in the Middle Stone Age were likely to be classified in the indeterminate period categories, and are therefore excluded from the analysis. The larger sample size for the Later Stone Age gives a more representative view of the lithic technology used in this period. Half of the flakes have plain platforms (53.9%), with fewer dihedral platforms than for those of the Middle Stone Age (13%) and a higher proportion of cortical platforms (19.1%), as well as an overall higher proportion of marginal platforms than the earlier periods (13.9%). The latter probably represents a genuine technological constraint due to the smaller size of the split cobbles used in the Later Stone Age providing small core platforms to strike, or alternatively, it could be the result of crushing the platform that sometimes occurs during bipolar flaking (Barham 1987:48).

The maximum number of scar counts on Early Stone Age flakes is four (table A7.11), whereas in the Middle and Later Stone Ages there are up to six and seven scars on the dorsal surface. This indicates a general increase in the complexity of the knapping procedures used over time. There is an overall reduction in scar counts in the Later Stone Age where one or two scars are most common (47.5% in total), compared to the Middle Stone Age where three scars are most common (41.7%). This reduction of scar counts in the Later Stone Age is probably related to the smaller size of the cores and flakes placing limitations on extensive core reduction, as cores would rapidly become too small to manipulate easily and flake.

The Cortex index data for the flakes (table A7.12) indicates an overall pattern in the sample of a decrease in the amount of cortex on the artefacts. Wholly cortical flakes comprise 12.3% of the total, whilst 15.4% of the flakes retain greater than 50% of their cortex, 24.9% retain less than 50% of their cortex, and flakes with no cortex comprise 47.4% of the flakes in the total sample. The Early Stone Age has the highest proportion of cortical flakes (13.3%), but the majority (46.7%) retain less than 50% cortex. All the Middle Stone Age flakes have no cortex, due to the technology used for core preparation, which usually results in the cortex of the core being removed before the preferential flake removal. It is also a result of there being no way of distinguishing simpler Middle Stone Age flakes from Mode 1 core reduction still practiced in the Middle Stone Age, or flakes produced during the initial stages of complex core reduction. The majority of Later Stone Age flakes also have no cortex (53.5%), which is also likely to be a result of removal of the cortex during core preparation. Flakes with less than 50% cortex represent 24.8% of flakes, a figure virtually identical to the overall total for all the flakes. Flakes with greater than 50% cortex, or wholly cortical flakes are slightly less frequent than the overall totals, at 12.9%, and 8.9% respectively.

Platform / butt Type.	Typology: Early Stone Age		Early/Middle Stone Age		Middle Stone Age		Middle/Later Stone Age		Later Stone Age		Total	
	Count	Column %	Count	Column %	Count	Column %	Count	Column %	Count	Column %	Count	Column %
Plain	6	35.3%	44	52.4%	5	41.7%	61	56.5%	62	53.9%	178	53.0%
Dihedral	1	5.9%	4	4.8%	6	50.0%	13	12.0%	15	13.0%	39	11.6%
Cortical / Natural	10	58.8%	34	40.5%	1	8.3%	29	26.9%	22	19.1%	96	28.6%
Marginal	0	.0%	2	2.4%	0	.0%	5	4.6%	16	13.9%	23	6.8%
Soft hammer	0	.0%	0	.0%	0	.0%	0	.0%	0	.0%	0	.0%
Total	17	100.0%	84	100.0%	12	100.0%	108	100.0%	115	100.0%	336	100.0%

Table A7.10: Flakes tabulated by typological affiliation and platform/butt type.

Number of scars on dorsal	Typology: Early Stone Age		Early/Middle Stone Age		Middle Stone Age		Middle/Later Stone Age		Later Stone Age		Total	
	Count	Column %	Count	Column %	Count	Column %	Count	Column %	Count	Column %	Count	Column %
0	2	13.3%	15	19.7%	0	.0%	11	11.8%	8	7.9%	36	12.1%
1	5	33.3%	16	21.1%	1	8.3%	21	22.6%	22	21.8%	65	21.9%
2	3	20.0%	20	26.3%	1	8.3%	21	22.6%	26	25.7%	71	23.9%
3	2	13.3%	18	23.7%	5	41.7%	19	20.4%	31	30.7%	75	25.3%
4	3	20.0%	6	7.9%	1	8.3%	16	17.2%	9	8.9%	35	11.8%
5	0	.0%	0	.0%	2	16.7%	3	3.2%	3	3.0%	8	2.7%
6	0	.0%	0	.0%	1	8.3%	2	2.2%	2	2.0%	5	1.7%
7	0	.0%	1	1.3%	1	8.3%	0	.0%	0	.0%	2	.7%
Total	15	100.0%	76	100.0%	12	100.0%	93	100.0%	101	100.0%	297	100.0%

Table A7.11: Flakes tabulated by typological affiliation and number of flake scars on the dorsal surface of the flakes.

Flakes: Cortex index	Early Stone Age		Early/Middle Stone Age		Middle Stone Age		Middle/Later Stone Age		Later Stone Age		Total	
	Typology											
	Count	Column %	Count	Column %	Count	Column %	Count	Column %	Count	Column %	Count	Column %
Wholly cortical	2	13.3%	14	19.2%	0	.0%	11	11.8%	9	8.9%	36	12.3%
> 50% cortex	4	26.7%	19	26.0%	0	.0%	9	9.7%	13	12.9%	45	15.4%
< 50% cortex	7	46.7%	19	26.0%	0	.0%	22	23.7%	25	24.8%	73	24.9%
No cortex	2	13.3%	21	28.8%	11	100.0%	51	54.8%	54	53.5%	139	47.4%
Total	15	100.0%	73	100.0%	11	100.0%	93	100.0%	101	100.0%	293	100.0%

Table A7.12: Flakes tabulated by typological affiliation, and percentage of dorsal surface covered by cortex (cortex index).

Appendix 7.3: Total artefact counts sorted by transect and typological grouping

Transect number	Typology				
	Early Stone Age	Middle Stone Age	Later Stone Age	Iron Age	Total
CH1	.	.	3	.	3
CH2	5	.	1	.	6
CI1	.	1	8	.	9
GA1.1	.	.	3	.	3
GA1.2	3	.	2	.	5
GA1.3	.	1	.	.	1
GA1.4	3	1	25	109	138
GA1.5	2	.	3	.	5
GA1.6	6	1	6	1	14
GA1.7	1	.	6	15	22
GA1.8	.	.	1	.	1
GA1.9	.	2	6	10	18
GC2	.	.	44	1	45
GR1	2	.	1	.	3
GR2	16	2	22	.	40
GR3	5	.	6	.	11
GR4	11	4	9	.	24
GR5	3	.	13	.	16
GR6	9	1	16	.	26
GR7	5	4	34	.	43
MI1	.	.	19	28	47
MI2	1	.	5	1	7
MI3	.	.	.	1	1
MM1	.	.	1	.	1
MM2	5	4	6	.	15
MM3	3	2	2	.	7
MM4	.	2	4	.	6
MM5	1	1	4	.	6
MM6	.	1	1	47	49
NG1	1	.	10	.	11
NH1	.	.	6	.	6
NH3	.	.	23	.	23
NH4	.	.	2	.	2
NH5	.	.	2	.	2
NH6	.	.	3	.	3
NH7	.	.	34	.	34
NH9	2	.	.	.	2
NH10	.	1	1	31	33
NH11	1	1	1	.	3
NH13	1	.	6	.	7
NH14	.	.	2	.	2
NH16	.	.	5	.	5
RN3	.	.	.	181	181
RN4	.	.	.	217	217
RS1	.	.	.	86	86
TW2	.	.	.	1	1

Appendix 7.4: Lithic raw materials cross-referenced by transect number

Percentages used in figure 7.2 not displayed as the table would be too large.

Transect number	Total raw materials used on each transect.													
	Fine Quartzite	Medium Quartzite	Coarse Quartzite	Vein Quartz	Chert	Granite	Quartz	Pink Quartzite	Tabular Quartzite	Black Rock	Heat Treated Chert	Ore with streak	Unknown	TOTAL
CH1	1	.	.	.	.	.	2	.	.	.	.	.	.	3
CH2	7	4	.	.	.	.	5	.	.	.	.	.	.	16
CI1	1	.	.	.	1	.	8	.	.	.	.	.	.	10
GA1.1	3	2	.	.	5	.	1	.	.	.	.	.	.	11
GA1.2	4	3	.	.	.	.	2	.	.	.	.	.	.	9
GA1.3	3	1	.	.	.	.	2	.	.	.	.	.	.	6
GA1.4	13	2	2	.	1	.	46	.	2	1	.	.	.	67
GA1.5	8	1	2	.	.	.	3	.	.	.	.	.	.	14
GA1.6	11	5	2	.	2	.	21	.	2	.	1	.	.	44
GA1.7	1	1	.	.	.	.	7	.	.	.	.	.	1	10
GA1.8	1	.	1	.	.	.	1	.	.	.	.	.	.	3
GA1.9	4	2	1	.	4	.	13	.	.	.	.	.	.	24
GC2	1	.	.	.	8	.	43	.	1	.	1	.	.	54
GR1	3	3	3	.	2	.	2	.	.	.	.	.	.	13
GR2	40	21	6	.	12	.	37	.	.	.	.	.	.	116
GR3	5	2	4	.	.	.	6	.	.	.	.	.	.	17
GR4	15	19	9	1	8	.	20	.	1	.	.	.	.	72
GR5	11	5	2	.	10	.	15	.	.	.	.	.	.	43
GR6	7	10	.	.	3	.	17	.	.	.	.	.	.	37
GR7	17	28	4	.	50	.	31	.	.	.	.	.	.	130
MI1	2	.	.	.	.	.	18	.	.	.	.	.	.	20
MI2	7	1	1	.	1	.	14	.	.	.	.	.	.	24
MM1	1	.	.	.	3	.	1	.	.	.	.	.	.	5
MM2	11	12	4	.	14	.	8	.	1	.	.	.	.	50
MM3	8	1	2	.	8	.	8	.	.	.	.	.	.	27
MM4	6	4	1	.	6	.	13	.	.	.	.	.	.	30
MM5	3	8	1	.	.	.	9	.	.	.	.	.	.	21
MM6	4	2	.	.	.	.	3	.	.	.	.	.	.	9
NG1	4	1	.	.	1	.	9	.	.	.	.	.	1	16
NH1	5	.	1	2	1	.	.	.	.	.	.	.	.	9
NH2	2	1	1	1	.	1	.	.	.	.	.	.	.	6
NH3	4	1	.	1	4	.	19	.	.	.	.	.	.	29
NH4	8	1	3	.	.	.	.	1	.	.	.	.	.	13
NH5	8	2	2	.	.	.	.	.	.	.	.	.	.	12
NH6	6	.	3	.	.	.	.	.	.	.	.	4	.	13
NH7	3	3	.	.	2	.	34	.	1	.	.	.	.	43
NH8	1	.	.	1	.	.	.	.	.	.	.	.	.	2
NH9	3	1	1	1	.	.	2	.	.	.	.	.	.	8
NH10	1	3	1	.	.	.	5	.	.	.	.	.	.	10
NH11	2	7	7	.	.	.	1	.	.	.	.	.	.	17
NH12	.	1	.	.	.	.	.	.	.	.	.	.	.	1
NH13	4	3	1	2	.	.	.	.	.	.	.	.	.	10
NH14	2	.	1	3	.	.	1	.	.	.	.	.	.	7
NH16	7	5	1	.	.	.	1	.	.	.	.	.	.	15
RN3	2	.	.	.	.	.	.	.	.	.	.	.	.	2
RN4	2	.	3	.	.	.	.	.	.	1	.	.	.	6
RS1	1	.	.	.	.	.	.	.	.	.	.	.	.	1
TW2	2	.	.	.	.	.	.	.	.	.	.	.	.	2

Appendix 7.5: Total artefact counts and percentages for artefact condition cross-referenced by transect number

Transect number	Condition (degree of rolling)									
	Fresh		Slightly Rolled		Moderately Rolled		Very Rolled		Total	
CH1	2	66.7%	1	33.3%	.	.0%	.	.0%	3	100.0%
CH2	7	43.8%	7	43.8%	2	12.5%	.	.0%	16	100.0%
CI1	8	80.0%	2	20.0%	.	.0%	.	.0%	10	100.0%
GA1.1	9	81.8%	2	18.2%	.	.0%	.	.0%	11	100.0%
GA1.2	.	.0%	5	55.6%	4	44.4%	.	.0%	9	100.0%
GA1.3	6	100.0%	.	.0%	.	.0%	.	.0%	6	100.0%
GA1.4	55	84.6%	8	12.3%	2	3.1%	.	.0%	65	100.0%
GA1.5	6	42.9%	1	7.1%	7	50.0%	.	.0%	14	100.0%
GA1.6	30	68.2%	9	20.5%	5	11.4%	.	.0%	44	100.0%
GA1.7	8	80.0%	1	10.0%	1	10.0%	.	.0%	10	100.0%
GA1.8	1	33.3%	1	33.3%	1	33.3%	.	.0%	3	100.0%
GA1.9	20	83.3%	4	16.7%	.	.0%	.	.0%	24	100.0%
GC2	52	96.3%	2	3.7%	.	.0%	.	.0%	54	100.0%
GR1	9	69.2%	3	23.1%	1	7.7%	.	.0%	13	100.0%
GR2	67	57.8%	31	26.7%	17	14.7%	1	.9%	116	100.0%
GR3	10	58.8%	5	29.4%	2	11.8%	.	.0%	17	100.0%
GR4	32	43.8%	25	34.2%	16	21.9%	.	.0%	73	100.0%
GR5	31	72.1%	11	25.6%	1	2.3%	.	.0%	43	100.0%
GR6	20	54.1%	9	24.3%	6	16.2%	2	5.4%	37	100.0%
GR7	91	70.0%	29	22.3%	9	6.9%	1	.8%	130	100.0%
MI1	19	95.0%	1	5.0%	.	.0%	.	.0%	20	100.0%
MI2	20	83.3%	4	16.7%	.	.0%	.	.0%	24	100.0%
MM1	.	.0%	5	100.0%	.	.0%	.	.0%	5	100.0%
MM2	29	58.0%	9	18.0%	8	16.0%	4	8.0%	50	100.0%
MM3	15	57.7%	6	23.1%	3	11.5%	2	7.7%	26	100.0%
MM4	19	65.5%	7	24.1%	3	10.3%	.	.0%	29	100.0%
MM5	9	42.9%	9	42.9%	1	4.8%	2	9.5%	21	100.0%
MM6	6	66.7%	3	33.3%	.	.0%	.	.0%	9	100.0%
NG1	11	73.3%	4	26.7%	.	.0%	.	.0%	15	100.0%
NH1	7	77.8%	2	22.2%	.	.0%	.	.0%	9	100.0%
NH10	6	60.0%	3	30.0%	1	10.0%	.	.0%	10	100.0%
NH11	6	35.3%	11	64.7%	.	.0%	.	.0%	17	100.0%
NH12	1	100.0%	.	.0%	.	.0%	.	.0%	1	100.0%
NH13	9	90.0%	1	10.0%	.	.0%	.	.0%	10	100.0%
NH14	5	71.4%	2	28.6%	.	.0%	.	.0%	7	100.0%
NH16	11	78.6%	3	21.4%	.	.0%	.	.0%	14	100.0%
NH2	4	66.7%	2	33.3%	.	.0%	.	.0%	6	100.0%
NH3	24	96.0%	1	4.0%	.	.0%	.	.0%	25	100.0%
NH4	8	61.5%	5	38.5%	.	.0%	.	.0%	13	100.0%
NH5	9	90.0%	1	10.0%	.	.0%	.	.0%	10	100.0%
NH6	5	55.6%	3	33.3%	1	11.1%	.	.0%	9	100.0%
NH7	43	100.0%	.	.0%	.	.0%	.	.0%	43	100.0%
NH8	2	100.0%	.	.0%	.	.0%	.	.0%	2	100.0%
NH9	5	62.5%	2	25.0%	1	12.5%	.	.0%	8	100.0%
RN3	2	100.0%	.	.0%	.	.0%	.	.0%	2	100.0%
RN4	6	100.0%	.	.0%	.	.0%	.	.0%	6	100.0%
RS1	1	50.0%	1	50.0%	.	.0%	.	.0%	2	100.0%
TW2	2	100.0%	.	.0%	.	.0%	.	.0%	2	100.0%

Appendix 7.6: Artefact abrasion counts and percentages per transect tabulated by typological association

Transect number	Early Stone Age Condition (degree of rolling)							
	Fresh		Slightly Rolled		Moderately Rolled		Very Rolled	
CH2	3	60.0%	1	20.0%	1	20.0%	.	.0%
GA1.2	.	.0%	.	.0%	3	100.0%	.	.0%
GA1.4	.	.0%	1	33.3%	2	66.7%	.	.0%
GA1.5	.	.0%	.	.0%	2	100.0%	.	.0%
GA1.6	2	33.3%	2	33.3%	2	33.3%	.	.0%
GA1.7	.	.0%	.	.0%	1	100.0%	.	.0%
GR1	1	50.0%	1	50.0%	.	.0%	.	.0%
GR2	11	68.8%	2	12.5%	2	12.5%	1	6.3%
GR3	1	20.0%	3	60.0%	1	20.0%	.	.0%
GR4	5	45.5%	3	27.3%	3	27.3%	.	.0%
GR5	1	33.3%	2	66.7%	.	.0%	.	.0%
GR6	3	33.3%	2	22.2%	2	22.2%	2	22.2%
GR7	2	40.0%	3	60.0%	.	.0%	.	.0%
MI2	.	.0%	1	100.0%	.	.0%	.	.0%
MM2	.	.0%	1	20.0%	2	40.0%	2	40.0%
MM3	1	33.3%	.	.0%	.	.0%	2	66.7%
MM5	.	.0%	.	.0%	.	.0%	1	100.0%
NG1	.	.0%	1	100.0%	.	.0%	.	.0%
NH11	.	.0%	1	100.0%	.	.0%	.	.0%
NH13	1	100.0%	.	.0%	.	.0%	.	.0%
NH9	.	.0%	2	100.0%	.	.0%	.	.0%

Transect number	Middle Stone Age Condition (degree of rolling)							
	Fresh		Slightly Rolled		Moderately Rolled		Very Rolled	
CI1	1	100.0%	.	.0%	.	.0%	.	.0%
GA1.3	1	100.0%	.	.0%	.	.0%	.	.0%
GA1.4	1	100.0%	.	.0%	.	.0%	.	.0%
GA1.6	1	100.0%	.	.0%	.	.0%	.	.0%
GA1.9	1	50.0%	1	50.0%	.	.0%	.	.0%
GR2	2	100.0%	.	.0%	.	.0%	.	.0%
GR4	2	50.0%	1	25.0%	1	25.0%	.	.0%
GR6	.	.0%	1	100.0%	.	.0%	.	.0%
GR7	3	75.0%	1	25.0%	.	.0%	.	.0%
MM2	3	75.0%	1	25.0%	.	.0%	.	.0%
MM3	2	100.0%	.	.0%	.	.0%	.	.0%
MM4	1	50.0%	1	50.0%	.	.0%	.	.0%
MM5	.	.0%	.	.0%	1	100.0%	.	.0%
MM6	.	.0%	1	100.0%	.	.0%	.	.0%
NH10	1	100.0%	.	.0%	.	.0%	.	.0%
NH11	.	.0%	1	100.0%	.	.0%	.	.0%

Appendix 7.6 (continued)

Transcect number	Later Stone Age Condition (degree of rolling)							
	Fresh		Slightly Rolled		Moderately Rolled		Very Rolled	
CH1	1	50.0%	1	50.0%	.	.0%	.	.0%
CH2	1	100.0%	.	.0%	.	.0%	.	.0%
CI1	7	87.5%	1	12.5%	.	.0%	.	.0%
GA1.1	3	100.0%	.	.0%	.	.0%	.	.0%
GA1.2	.	.0%	2	100.0%	.	.0%	.	.0%
GA1.4	24	100.0%	.	.0%	.	.0%	.	.0%
GA1.5	3	100.0%	.	.0%	.	.0%	.	.0%
GA1.6	5	83.3%	1	16.7%	.	.0%	.	.0%
GA1.7	6	100.0%	.	.0%	.	.0%	.	.0%
GA1.8	1	100.0%	.	.0%	.	.0%	.	.0%
GA1.9	6	100.0%	.	.0%	.	.0%	.	.0%
GC2	44	100.0%	.	.0%	.	.0%	.	.0%
GR1	1	100.0%	.	.0%	.	.0%	.	.0%
GR2	20	90.9%	2	9.1%	.	.0%	.	.0%
GR3	6	100.0%	.	.0%	.	.0%	.	.0%
GR4	6	66.7%	3	33.3%	.	.0%	.	.0%
GR5	11	84.6%	2	15.4%	.	.0%	.	.0%
GR6	13	81.3%	2	12.5%	1	6.3%	.	.0%
GR7	27	79.4%	7	20.6%	.	.0%	.	.0%
MI1	19	100.0%	.	.0%	.	.0%	.	.0%
MI2	5	100.0%	.	.0%	.	.0%	.	.0%
MM1	.	.0%	1	100.0%	.	.0%	.	.0%
MM2	6	100.0%	.	.0%	.	.0%	.	.0%
MM3	1	100.0%	.	.0%	.	.0%	.	.0%
MM4	4	100.0%	.	.0%	.	.0%	.	.0%
MM5	4	100.0%	.	.0%	.	.0%	.	.0%
MM6	1	100.0%	.	.0%	.	.0%	.	.0%
NG1	8	88.9%	1	11.1%	.	.0%	.	.0%
NH1	6	100.0%	.	.0%	.	.0%	.	.0%
NH10	1	100.0%	.	.0%	.	.0%	.	.0%
NH11	1	100.0%	.	.0%	.	.0%	.	.0%
NH13	6	100.0%	.	.0%	.	.0%	.	.0%
NH14	2	100.0%	.	.0%	.	.0%	.	.0%
NH16	4	80.0%	1	20.0%	.	.0%	.	.0%
NH3	23	100.0%	.	.0%	.	.0%	.	.0%
NH4	1	50.0%	1	50.0%	.	.0%	.	.0%
NH5	1	50.0%	1	50.0%	.	.0%	.	.0%
NH6	3	100.0%	.	.0%	.	.0%	.	.0%
NH7	34	100.0%	.	.0%	.	.0%	.	.0%

Appendix 7.7: Raw material artefact counts and percentages per transect, tabulated by typological association

Transect number	Early Stone Age Raw material									
	Fine Quartzite		Medium Quartzite		Coarse Quartzite		Quartz		Unknown	
CH2	2	40.0%	3	60.0%	.	.0%	.	.0%	.	.0%
GA1.2	1	33.3%	2	66.7%	.	.0%	.	.0%	.	.0%
GA1.4	1	33.3%	.	.0%	2	66.7%	.	.0%	.	.0%
GA1.5		.0%		.0%	2	100.0%		.0%		.0%
GA1.6	2	33.3%	2	33.3%	1	16.7%	1	16.7%	.	.0%
GA1.7	.	.0%	.	.0%	.	.0%	.	.0%	1	100.0%
GR1	.	.0%	.	.0%	2	100.0%	.	.0%	.	.0%
GR2	8	50.0%	1	6.3%	4	25.0%	3	18.8%	.	.0%
GR3	1	20.0%	.	.0%	4	80.0%	.	.0%	.	.0%
GR4	4	36.4%	5	45.5%	2	18.2%	.	.0%	.	.0%
GR5	1	33.3%	2	66.7%	.	.0%	.	.0%	.	.0%
GR6	2	22.2%	5	55.6%	.	.0%	2	22.2%	.	.0%
GR7	1	20.0%	3	60.0%	.	.0%	1	20.0%	.	.0%
MI2	.	.0%	.	.0%	1	100.0%	.	.0%	.	.0%
MM2	.	.0%	2	40.0%	3	60.0%	.	.0%	.	.0%
MM3	.	.0%	1	33.3%	2	66.7%	.	.0%	.	.0%
MM5	.	.0%	1	100.0%	.	.0%	.	.0%	.	.0%
NG1	.	.0%	1	100.0%	.	.0%	.	.0%	.	.0%
NH11	.	.0%	.	.0%	1	100.0%	.	.0%	.	.0%
NH13	.	.0%	.	.0%	1	100.0%	.	.0%	.	.0%
NH9	.	.0%	1	50.0%	1	50.0%	.	.0%	.	.0%

Transect number	Middle Stone Age Raw material									
	Fine Quartzite		Medium Quartzite		Coarse Quartzite		Chert		Quartz	
CI1	.	.0%	.	.0%	.	.0%	1	100.0%	.	.0%
GA1.3	1	100.0%	.	.0%	.	.0%	.	.0%	.	.0%
GA1.4	1	100.0%	.	.0%	.	.0%	.	.0%	.	.0%
GA1.6	.	.0%	.	.0%	.	.0%	1	100.0%	.	.0%
GA1.9	.	.0%	.	.0%	.	.0%	1	50.0%	1	50.0%
GR2	1	50.0%	.	.0%	.	.0%	1	50.0%	.	.0%
GR4	2	50.0%	1	25.0%	.	.0%	.	.0%	1	25.0%
GR6	1	100.0%	.	.0%	.	.0%	.	.0%	.	.0%
GR7	.	.0%	.	.0%	.	.0%	3	75.0%	1	25.0%
MM2	.	.0%	1	25.0%	.	.0%	3	75.0%	.	.0%
MM3	.	.0%	.	.0%	.	.0%	2	100.0%	.	.0%
MM4	1	50.0%	.	.0%	.	.0%	.	.0%	1	50.0%
MM5	.	.0%	.	.0%	.	.0%	.	.0%	1	100.0%
MM6	1	100.0%	.	.0%	.	.0%	.	.0%	.	.0%
NH10	.	.0%	.	.0%	1	100.0%	.	.0%	.	.0%
NH11	.	.0%	1	100.0%	.	.0%	.	.0%	.	.0%

Appendix 7.7 (continued)

Transect number	Later Stone Age Raw material																	
	Fine Quartzite		Medium Quartzite		Coarse Quartzite		Vein Quartz		Chert		Quartz		Tabular Quartzite		Heat Treated Chert		Unknown	
CH1	.	.0%	.	.0%	.	.0%	.	.0%	.	.0%	2	100.0%	.	.0%	.	.0%	.	.0%
CH2	.	.0%	.	.0%	.	.0%	.	.0%	.	.0%	1	100.0%	.	.0%	.	.0%	.	.0%
CI1	.	.0%	.	.0%	.	.0%	.	.0%	.	.0%	8	100.0%	.	.0%	.	.0%	.	.0%
GA1.1	1	33.3%	.	.0%	.	.0%	.	.0%	2	66.7%	.	.0%	.	.0%	.	.0%	.	.0%
GA1.2	1	50.0%	.	.0%	.	.0%	.	.0%	.	.0%	1	50.0%	.	.0%	.	.0%	.	.0%
GA1.4	.	.0%	.	.0%	.	.0%	.	.0%	1	4.0%	24	96.0%	.	.0%	.	.0%	.	.0%
GA1.5	2	66.7%	.	.0%	.	.0%	.	.0%	.	.0%	1	33.3%	.	.0%	.	.0%	.	.0%
GA1.6	.	.0%	.	.0%	1	16.7%	.	.0%	1	16.7%	3	50.0%	.	.0%	1	16.7%	.	.0%
GA1.7	1	16.7%	.	.0%	.	.0%	.	.0%	.	.0%	5	83.3%	.	.0%	.	.0%	.	.0%
GA1.8	.	.0%	.	.0%	.	.0%	.	.0%	.	.0%	1	100.0%	.	.0%	.	.0%	.	.0%
GA1.9	.	.0%	.	.0%	1	16.7%	.	.0%	1	16.7%	4	66.7%	.	.0%	.	.0%	.	.0%
GC2	.	.0%	.	.0%	.	.0%	.	.0%	6	13.6%	37	84.1%	.	.0%	1	2.3%	.	.0%
GR1	.	.0%	.	.0%	.	.0%	.	.0%	.	.0%	1	100.0%	.	.0%	.	.0%	.	.0%
GR2	2	9.1%	.	.0%	.	.0%	.	.0%	3	13.6%	17	77.3%	.	.0%	.	.0%	.	.0%
GR3	1	16.7%	.	.0%	.	.0%	.	.0%	.	.0%	5	83.3%	.	.0%	.	.0%	.	.0%
GR4	1	11.1%	.	.0%	.	.0%	.	.0%	1	11.1%	7	77.8%	.	.0%	.	.0%	.	.0%
GR5	2	15.4%	.	.0%	.	.0%	.	.0%	1	7.7%	10	76.9%	.	.0%	.	.0%	.	.0%
GR6	1	6.3%	1	6.3%	.	.0%	.	.0%	1	6.3%	13	81.3%	.	.0%	.	.0%	.	.0%
GR7	5	14.7%	1	2.9%	.	.0%	.	.0%	13	38.2%	15	44.1%	.	.0%	.	.0%	.	.0%
MI1	1	5.3%	.	.0%	.	.0%	.	.0%	.	.0%	18	94.7%	.	.0%	.	.0%	.	.0%
MI2	.	.0%	.	.0%	.	.0%	.	.0%	1	20.0%	4	80.0%	.	.0%	.	.0%	.	.0%
MM1	.	.0%	.	.0%	.	.0%	.	.0%	.	.0%	1	100.0%	.	.0%	.	.0%	.	.0%
MM2	.	.0%	1	16.7%	.	.0%	.	.0%	1	16.7%	4	66.7%	.	.0%	.	.0%	.	.0%
MM3	.	.0%	.	.0%	.	.0%	.	.0%	1	50.0%	1	50.0%	.	.0%	.	.0%	.	.0%
MM4	1	25.0%	.	.0%	.	.0%	.	.0%	.	.0%	3	75.0%	.	.0%	.	.0%	.	.0%
MM5	1	25.0%	.	.0%	.	.0%	.	.0%	.	.0%	3	75.0%	.	.0%	.	.0%	.	.0%
MM6	.	.0%	.	.0%	.	.0%	.	.0%	.	.0%	1	100.0%	.	.0%	.	.0%	.	.0%
NG1	1	10.0%	.	.0%	.	.0%	.	.0%	.	.0%	8	80.0%	.	.0%	.	.0%	1	10.0%
NH1	4	66.7%	.	.0%	.	.0%	1	16.7%	1	16.7%	.	.0%	.	.0%	.	.0%	.	.0%
NH10	1	100.0%	.	.0%	.	.0%	.	.0%	.	.0%	.	.0%	.	.0%	.	.0%	.	.0%
NH11	.	.0%	.	.0%	.	.0%	.	.0%	.	.0%	1	100.0%	.	.0%	.	.0%	.	.0%
NH13	2	33.3%	2	33.3%	.	.0%	2	33.3%	.	.0%	.	.0%	.	.0%	.	.0%	.	.0%
NH14	.	.0%	.	.0%	.	.0%	1	50.0%	.	.0%	1	50.0%	.	.0%	.	.0%	.	.0%
NH16	4	80.0%	.	.0%	.	.0%	.	.0%	.	.0%	1	20.0%	.	.0%	.	.0%	.	.0%
NH3	3	13.6%	.	.0%	.	.0%	1	4.5%	4	18.2%	14	63.6%	.	.0%	.	.0%	.	.0%
NH4	2	100.0%	.	.0%	.	.0%	.	.0%	.	.0%	.	.0%	.	.0%	.	.0%	.	.0%
NH5	2	100.0%	.	.0%	.	.0%	.	.0%	.	.0%	.	.0%	.	.0%	.	.0%	.	.0%
NH6	3	100.0%	.	.0%	.	.0%	.	.0%	.	.0%	.	.0%	.	.0%	.	.0%	.	.0%
NH7	2	5.9%	1	2.9%	.	.0%	.	.0%	1	2.9%	29	85.3%	1	2.9%	.	.0%	.	.0%

Appendix 7.8: Artefact types sorted by typological association and transect number

Transect number	Early Stone Age Tool type							
	Flake		Broken Flake		Core		Core Fragment	
CH2	1	20.0%	1	20.0%	3	60.0%	.	.0%
GA1.2	.	.0%	.	.0%	3	100.0%	.	.0%
GA1.4	.	.0%	.	.0%	3	100.0%	.	.0%
GA1.5	1	50.0%	.	.0%	1	50.0%	.	.0%
GA1.6	3	50.0%	.	.0%	3	50.0%	.	.0%
GA1.7	1	100.0%	.	.0%	.	.0%	.	.0%
GR1	.	.0%	.	.0%	2	100.0%	.	.0%
GR2	4	25.0%	2	12.5%	10	62.5%	.	.0%
GR3	.	.0%	.	.0%	5	100.0%	.	.0%
GR4	4	36.4%	.	.0%	7	63.6%	.	.0%
GR5	1	33.3%	.	.0%	2	66.7%	.	.0%
GR6	.	.0%	.	.0%	9	100.0%	.	.0%
GR7	.	.0%	.	.0%	4	80.0%	1	20.0%
MI2	.	.0%	.	.0%	1	100.0%	.	.0%
MM2	1	20.0%	2	40.0%	2	40.0%	.	.0%
MM3	1	33.3%	.	.0%	2	66.7%	.	.0%
MM5	1	100.0%	.	.0%	.	.0%	.	.0%
NG1	.	.0%	.	.0%	1	100.0%	.	.0%
NH11	.	.0%	.	.0%	1	100.0%	.	.0%
NH13	.	.0%	.	.0%	1	100.0%	.	.0%
NH9	.	.0%	.	.0%	2	100.0%	.	.0%

Transect number	Middle Stone Age Tool type															
	Flake		Broken Flake		Core		Blade Core		Prepared Core		Prepared Core Flake		Split Cobble		Flaked flake	
CI1	1	100.0%	.	.0%	.	.0%	.	.0%	.	.0%	.	.0%	.	.0%	.	.0%
GA1.3	.	.0%	1	100.0%	.	.0%	.	.0%	.	.0%	.	.0%	.	.0%	.	.0%
GA1.4	1	100.0%	.	.0%	.	.0%	.	.0%	.	.0%	.	.0%	.	.0%	.	.0%
GA1.6	.	.0%	1	100.0%	.	.0%	.	.0%	.	.0%	.	.0%	.	.0%	.	.0%
GA1.9	1	50.0%	.	.0%	.	.0%	.	.0%	1	50.0%	.	.0%	.	.0%	.	.0%
GR2	1	50.0%	.	.0%	1	50.0%	.	.0%	.	.0%	.	.0%	.	.0%	.	.0%
GR4	.	.0%	.	.0%	3	75.0%	1	25.0%	.	.0%	.	.0%	.	.0%	.	.0%
GR6	.	.0%	.	.0%	.	.0%	.	.0%	.	.0%	.	.0%	1	100.0%	.	.0%
GR7	.	.0%	.	.0%	1	25.0%	.	.0%	.	.0%	3	75.0%	.	.0%	.	.0%
MM2	.	.0%	.	.0%	.	.0%	.	.0%	1	25.0%	3	75.0%	.	.0%	.	.0%
MM3	.	.0%	.	.0%	.	.0%	.	.0%	.	.0%	1	50.0%	.	.0%	1	50.0%
MM4	.	.0%	.	.0%	.	.0%	.	.0%	2	100.0%	.	.0%	.	.0%	.	.0%
MM5	.	.0%	.	.0%	.	.0%	.	.0%	.	.0%	.	.0%	1	100.0%	.	.0%
MM6	.	.0%	.	.0%	1	100.0%	.	.0%	.	.0%	.	.0%	.	.0%	.	.0%
NH10	.	.0%	.	.0%	.	.0%	.	.0%	1	100.0%	.	.0%	.	.0%	.	.0%
NH11	1	100.0%	.	.0%	.	.0%	.	.0%	.	.0%	.	.0%	.	.0%	.	.0%

Appendix 7.8 (continued)

Transect number	Later Stone Age														
	Flake	Broken Flake	Core	Blade	Prepared Core	Prepared Core Flake	Hammerst-one	Chunk	Split Cobble	Core Fragment	Blade fragment	Crested Flake	Segment	Bipolar core	Total
CH1	1	.	1	.	.	.	.	1	.	.	.	.	.	.	3
CH2	.	.	.	.	.	.	.	.	1	.	.	.	.	.	1
CI1	.	1	2	.	.	.	.	5	.	.	.	.	.	.	8
GA1.1	2	.	.	.	.	.	.	.	.	.	1	.	.	.	3
GA1.2	.	.	.	.	2	.	.	.	.	.	.	.	.	.	2
GA1.4	11	1	4	.	.	.	.	5	4	.	.	.	.	.	25
GA1.5	1	.	1	.	.	.	.	.	1	.	.	.	.	.	3
GA1.6	1	1	3	.	1	.	.	.	.	.	.	.	.	.	6
GA1.7	2	.	3	.	.	.	.	.	1	.	.	.	.	.	6
GA1.8	.	.	1	.	.	.	.	.	.	.	.	.	.	.	1
GA1.9	1	2	2	.	1	.	.	.	.	.	.	.	.	.	6
GC2	10	4	1	.	.	1	.	5	8	.	.	.	.	.	29
GR1	1	.	.	.	.	.	.	.	.	.	.	.	.	.	1
GR2	7	3	3	.	.	.	.	.	8	1	.	.	.	.	22
GR3	.	.	2	.	.	.	.	.	4	.	.	.	.	.	6
GR4	1	.	2	.	.	.	.	.	6	.	.	.	.	.	9
GR5	2	1	7	.	.	.	.	.	2	.	.	.	.	1	13
GR6	4	2	5	.	.	.	.	.	5	.	.	.	.	.	16
GR7	11	5	11	.	.	.	.	2	4	1	.	.	.	.	34
MI1	5	1	.	.	.	.	.	1	10	.	.	.	1	.	18
MI2	2	1	.	.	.	.	.	.	2	.	.	.	.	.	5
MM1	1	.	.	.	.	.	.	.	.	.	.	.	.	.	1
MM2	.	1	2	.	.	1	.	1	1	.	.	.	.	.	6
MM3	1	.	.	.	.	.	.	.	1	.	.	.	.	.	2
MM4	2	.	.	.	.	.	.	.	2	.	.	.	.	.	4
MM5	2	.	.	.	.	.	.	.	2	.	.	.	.	.	4
MM6	.	.	.	.	.	.	.	1	.	.	.	.	.	.	1
NG1	3	.	6	.	.	.	.	.	1	.	.	.	.	.	10
NH1	2	1	2	.	.	.	.	.	.	1	.	.	.	.	6
NH3	10	3	2	.	.	.	.	.	.	1	1	.	.	.	17
NH4	1	.	1	.	.	.	.	.	.	.	.	.	.	.	2
NH5	.	.	2	.	.	.	.	.	.	.	.	.	.	.	2
NH6	1	.	1	1	.	.	.	.	.	.	.	.	.	.	3
NH7	10	.	.	.	1	.	1	.	21	.	.	1	.	.	34
NH10	1	.	.	.	.	.	.	.	.	.	.	.	.	.	1
NH11	1	.	.	.	.	.	.	.	.	.	.	.	.	.	1
NH13	3	2	1	.	.	.	.	.	.	.	.	.	.	.	6
NH14	1	.	1	.	.	.	.	.	.	.	.	.	.	.	2
NH16	3	.	2	.	.	.	.	.	.	.	.	.	.	.	5

Appendix 7.9: Total number of Middle Stone Age artefact types cross-referenced by the transect they were recovered from, raw material, and degree of abrasion.

Transect number	Raw Material	Condition (degree of rolling)	Middle Stone Age Tool type							
			Flake	Broken Flake	Core	Blade Core	Prepared Core	Prepared Core Flake	Split Cobble	Flaked flake
CI1	Chert	Fresh	1	.	.	.	.	.	.	.
GA1.3	Fine Quartzite	Fresh	.	1	.	.	.	.	.	.
GA1.4	Fine Quartzite	Fresh	1	.	.	.	.	.	.	.
GA1.6	Chert	Fresh	.	1	.	.	.	.	.	.
GA1.9	Chert	Slightly Rolled	1	.	.	.	.	.	.	.
	Quartz	Fresh	.	.	.	.	1	.	.	.
GR2	Fine Quartzite	Fresh	.	.	1	.	.	.	.	.
	Chert	Fresh	1	.	.	.	.	.	.	.
GR4	Fine Quartzite	Fresh	.	.	1	.	.	.	.	.
		Slightly Rolled	.	.	1	.	.	.	.	.
	Medium Quartzite	Moderately Rolled	.	.	1	.	.	.	.	.
	Quartz	Fresh	.	.	.	1	.	.	.	.
GR6	Fine Quartzite	Slightly Rolled	.	.	.	.	.	.	1	.
GR7	Chert	Fresh	.	.	.	.	.	2	.	.
		Slightly Rolled	.	.	.	.	.	1	.	.
	Quartz	Fresh	.	.	1	.	.	.	.	.
MM2	Medium Quartzite	Slightly Rolled	.	.	.	.	1	.	.	.
	Chert	Fresh	.	.	.	.	.	3	.	.
MM3	Chert	Fresh	.	.	.	.	.	1	.	1
MM4	Fine Quartzite	Slightly Rolled	.	.	.	.	1	.	.	.
	Quartz	Fresh	.	.	.	.	1	.	.	.
MM5	Quartz	Moderately Rolled	.	.	.	.	.	.	1	.
MM6	Fine Quartzite	Slightly Rolled	.	.	1	.	.	.	.	.
NH10	Coarse Quartzite	Fresh	.	.	.	.	1	.	.	.
NH11	Medium Quartzite	Slightly Rolled	1	.	.	.	.	.	.	.

Appendix 7.10: Artefact counts for the Later Stone Age and Iron Age cross-referenced by individual transect block codes

Block A is the southern most block of each transect and block E is the northern most (Empty blocks are excluded).

Typology								
Block code	Later Stone Age	Iron Age	Block code	Later Stone Age	Iron Age	Block code	Later Stone Age	Iron Age
CH1D	2	.	GR6C	2	.	NH3B	2	.
CH1E	1	.	GR7A	14	.	NH3C	5	.
CH2B	1	.	GR7B	11	.	NH4C	1	.
CI1E	8	.	GR7D	2	.	NH4D	1	.
GA11C	1	.	GR7E	7	.	NH5B	1	.
GA11D	2	.	MI1A	19	28	NH5D	1	.
GA12C	1	.	MI2E	5	1	NH6B	1	.
GA12D	1	.	MI3A	.	1	NH6C	2	.
GA14A	2	.	MM1C	1	.	NH7B	1	.
GA14C	18	3	MM2A	3	.	NH7C	1	.
GA14E	5	107	MM2E	3	.	NH7E	32	.
GA15A	2	.	MM3A	2	.	RN3C	.	159
GA15B	1	.	MM4A	1	.	RN3D	.	22
GA16B	3	.	MM4B	2	.	RN4A	.	217
GA16C	3	1	MM4C	1	.	RS1B	.	86
GA17A	.	4	MM5B	2	.	TW2A	.	1
GA17C	6	.	MM5C	1	.			
GA17E	.	11	MM5E	1				
GA18A	1	.	MM6C	.	1			
GA19A	1	.	MM6D	.	38			
GA19B	5	8	MM6E	1	8			
GA19D	.	2	NG1A	1	.			
GC2A	1	.	NG1B	3	.			
GC2D	41	1	NG1C	3	.			
GC2E	2	.	NG1D	2	.			
GR1C	1	.	NG1E	1	.			
GR2A	4	.	NH10C	.	30			
GR2B	5	.	NH10E	1	1			
GR2D	3	.	NH11D	1	.			
GR2E	10	.	NH13A	5	.			
GR3A	3	.	NH13B	1	.			
GR3B	1	.	NH14B	2	.			
GR3E	2	.	NH16A	3	.			
GR4A	1	.	NH16B	1	.			
GR4E	8	.	NH16C	1	.			
GR5D	7	.	NH1A	1	.			
GR5E	6	.	NH1D	3	.			
GR6A	12	.	NH1E	2	.			
GR6B	2	.	NH3A	16	.			

Appendix 7.11: t-test to compare mean artefact counts

Independent samples t-test to compare mean artefact counts on transects in the Cobble Zone (7) and the Nchindeni Cobbles (zone 8) to mean artefact counts elsewhere in the survey area. A detailed description of the test is given in appendix 6.2. In this case the Levene test is greater than 0.05 so equal variances are assumed. There is a low significance for the t-test (<0.05) and the confidence interval does not equal zero so the difference between the number of artefacts found in the cobble zones and the number of artefacts found elsewhere in the survey area is significant.

Group Statistics

Location of transects with LSA artefacts	N (transects)	Mean	Std. Deviation	Std. Error Mean
On cobble zones 7 & 8	8	13.88	10.302	3.642
Off cobble zones	53	4.57	8.759	1.203

Independent Samples Test

Number of LSA	Levene's Test for Equality of Variances		t-test for Equality of Means						
	F	Sig.	t	df	Sig. (2-tailed)	Mean Difference	Std. Error Difference	95% Confidence Interval of the Difference Lower	Upper
Equal variances assumed	0.814	0.371	2.740	59	0.008	9.309	3.397	2.512	16.106
Equal variances not assumed			2.427	8.597	0.039	9.309	3.836	0.569	18.048

Appendix 7.12: t-test to compare mean artefact counts

Independent samples t-test to compare mean artefact counts on transects in the Cobble Zone (7), the Nchindeni Cobbles (zone 8), and the Cathedral Mopane zone (14), to the mean artefact counts elsewhere in the survey area. The Levene test is high so equal variances are assumed, and the significance is above 0.05. The result of this test demonstrates that when zone 14 is included with the other zones with cobbles there is no significant difference in the number of artefacts found in areas with cobbles.

Group Statistics

Location of transects with LSA artefacts	N (transects)	Mean	Std. Deviation	Std. Error Mean
On cobble zones 7, 8, & 14	14	9.21	9.480	2.534
Off cobble zones	47	4.77	9.269	1.352

Independent Samples Test

Number of LSA	Levene's Test for Equality of Variances		t-test for Equality of Means						
	F	Sig.	t	df	Sig. (2-tailed)	Mean Difference	Std. Error Difference	95% Confidence Interval of the Difference Lower	Upper
Equal variances assumed	0.357	0.552	1.568	59	0.122	4.448	2.837	-1.228	10.124
Equal variances not assumed			1.549	20.978	0.136	4.448	2.872	-1.524	10.421

Appendix 7.13: t-test to compare mean artefact counts

Independent samples t-test to compare mean artefact counts on transects in the Nchindeni Hills (zone 6) against the rest of the survey sample. The results indicate that there is no significant increase in the number of artefacts in the Nchindeni Hills.

Group Statistics

Location of transects with LSA artefacts	N	Mean	Std. Deviation	Std. Error Mean
On Nchindeni Hills (6)	16	5.31	9.499	2.375
Off Nchindeni Hills	45	5.96	9.506	1.417

Independent Samples Test

Number of LSA	Levene's Test for Equality of Variances		t-test for Equality of Means						
								95% Confidence Interval of the Difference	
	F	Sig.	t	df	Sig. (2-tailed)	Mean Difference	Std. Error Difference	Lower	Upper
Equal variances assumed	0.033	0.856	-.232	59	0.817	-.643	2.766	-6.179	4.892
Equal variances not assumed			-.233	26.441	0.818	-.643	2.765	-6.323	5.037

Appendix 7.14: t-test to compare mean artefact counts

Independent samples t-test to compare mean artefact counts on transects in the two sand and silt zones and the Recent Cut-off zone (zones 4, 5, &3 respectively) to the rest of the survey sample. The results indicate that there is no significant difference in the number of artefacts discovered in zones 3, 4, and 5 when compared to the rest of the sample.

Group Statistics

Location of transects with LSA artefacts	N	Mean	Std. Deviation	Std. Error Mean
On sand and silt	18	5.06	11.425	2.693
Off sand and silt	43	6.09	8.594	1.310

Independent Samples Test

Number of LSA	Levene's Test for Equality of Variances		t-test for Equality of Means						
								95% Confidence Interval of the Difference	
	F	Sig.	t	df	Sig. (2-tailed)	Mean Difference	Std. Error Difference	Lower	Upper
Equal variances assumed	0.189	0.666	-.389	59	0.699	-1.037	2.666	-6.372	4.297
Equal variances not assumed			-.346	25.429	0.732	-1.037	2.995	-7.200	5.125

Appendix 7.15: t-test to compare mean artefact counts

A repeat of the independent samples t-test in appendix 8.9 With the Recent Cut-off zone (3) excluded from the comparative sample and placed in the rest of the survey sample; so zones 4 and 5 were compared to the rest of the survey sample. The results indicate that there is no significant difference in the number of artefacts discovered in zones 4, and 5 when compared to the rest of the sample.

Group Statistics

Location of transects with LSA artefacts	N	Mean	Std. Deviation	Std. Error Mean
On sand and silt	10	9.10	14.341	4.535
Off sand and silt	51	5.14	8.188	1.147

Independent Samples Test

Number of LSA	Levene's Test for Equality of Variances		t-test for Equality of Means						
	F	Sig.	t	df	Sig. (2-tailed)	Mean Difference	Std. Error Difference	95% Confidence Interval of the Difference	
								Lower	Upper
Equal variances assumed	3.984	0.051	1.220	59	0.227	3.963	3.248	-2.536	10.461
Equal variances not assumed			.847	10.180	0.416	3.963	4.678	-6.435	14.360

Appendix 7.16: Fragments of shatter retrieved on the transects for all periods. Shatter was only found in association with Later Stone Age artifact scatters.

		Typology		
		Early Stone Age	Middle Stone Age	Later Stone Age
		Shatter/debris	Shatter/debris	Shatter/debris
		Number of artefacts	Number of artefacts	Number of artefacts
Block code	GC2D	.	.	15
	NH3A	.	.	6

Appendix 7.17: Binomial test to determine whether or not distance to streams is a factor affecting Late Stone Age artefact distribution in the landscape.

A binomial test is suitable where variables can only assume two values, which in this case is the presence or absence of artefacts. The hypothesized distribution used is the percentage of the total number of survey blocks in the landscape with and without artefacts, to which the survey blocks located within each buffer are compared. As 27% of blocks in the survey area overall contain artefacts, this forms the hypothesised proportion of blocks containing artefacts that should occur within each buffer if distance to streams is not a factor affecting artefact distribution. The binomial test is then used to compare the observed values to the hypothesised values. Small significance values (< 0.05) indicate that the observed distribution differs significantly from the hypothesised distribution.

		Category	N	Observed Prop.	Test Prop.	Asymp. Sig. (1-tailed)
Total survey	Group 1	Presence	27	0.27	0.27	0.552(b,a)
	Group 2	Absence	73	0.73		
	Total		100	1.00		
Buffer500	Group 1	Presence	29	0.29	0.27	0.362(a)
	Group 2	Absence	71	0.71		
	Total		100	1.00		
Buffer250	Group 1	Presence	40	0.40	0.27	0.003(a)
	Group 2	Absence	60	0.60		
	Total		100	1.00		

a Based on Z Approximation.
b Alternative hypothesis states that the proportion of cases in the first group < .27.

Appendix 7.18: Breakdown of artefact counts by block code and typological association for the Kasweta Road transects.

Block code	Typology					
	Early Stone Age	Early/Middle Stone Age	Middle Stone Age	Middle/Later Stone Age	Later Stone Age	Iron Age
	Number of artefacts	Number of artefacts	Number of artefacts	Number of artefacts	Number of artefacts	Number of artefacts
K2A	2	.	.	.	.	.
K2B	2	1	.	.	.	.
K2C	1	.	.	.	.	.
K2D	.	2	.	.	.	.
K2E	9	12	2	5	1	.
K4C	2	5	.	5	1	.
K4E	.	2	1	.	.	.
K5B	.	1	1	.	.	.
K5C	1	.	3	.	.	.
K5D	13	38	1	15	12	6
K5E	2	1	.	.	.	.

Appendix 7.19: Artefact counts for the O5 road transects, tabulated by the individual block code.

		Typology			
		Early Stone Age	Middle Stone Age	Later Stone Age	Total
		Number of artefacts	Number of artefacts	Number of artefacts	Number of artefacts
Block code	FS1C	1	3	7	11
	FS1J	4	.	.	4
	FS1K	3	3	2	8
	FS1L	2	1	.	3
	FS1O	2	.	1	3
	FW2C	2	3	2	7
	FW2D	.	.	5	5
	FW2L	1	1	.	2
	FW2N	2	1	1	4
	FW2O	6	.	6	12

Appendix 7.20: Tool types identified on the O5 road transects cross-referenced by typological association.

Transect number	Tool type	Typology						
		Early Stone Age	Early / Middle Stone Age	Middle Stone Age	Middle / Later Stone Age	Later Stone Age	Unknown	Total
FS1	Flake	3	.	1	1	1	.	6
	Broken Flake	.	.	.	.	1	.	1
	Core	2	3	.	.	2	.	7
	Blade Core	.	.	.	.	1	.	1
	Blade	.	.	1	1	.	.	2
	Prepared Core	.	.	1	.	.	.	1
	Split Cobble	.	.	.	.	2	.	2
FW2	Flake	1	5	.	5	2	.	13
	Broken Flake	.	.	1	1	.	.	2
	Core	7	6	.	6	1	1	21
	Prepared Core	.	.	1	.	.	.	1
	Prepared Core Flake	.	.	1	.	.	.	1
	Chunk	.	.	.	.	.	1	1
	Split Cobble	.	.	.	.	4	.	4
	Shatter/debris	.	.	.	.	.	6	6
	Flaked flake	1	.	.	.	.	.	1

Appendix 7.21: Artefact abrasion on the O5 road transects

Artefact abrasion is cross-referenced by typological association and transect block code, numerical data is presented, along with the percentages of artefacts present on each survey block, which is listed by each abrasion category for each typological grouping. The survey blocks form two groups relating to geomorphological features, one comprising C and D on each transect, and the other comprising J, K, L, and O (see text). The essential point to note is that transects in the second group of survey blocks contain the majority of abraded artefacts, whilst blocks C and D contain the majority of un-abraded artefacts.

Block code	Typology																							
	Early Stone Age								Middle Stone Age								Later Stone Age							
	Condition (degree of rolling)								Condition (degree of rolling)								Condition (degree of rolling)							
	Fresh		Slightly Rolled		Moder-ately Rolled		Very Rolled		Fresh		Slightly Rolled		Moder-ately Rolled		Very Rolled		Fresh		Slightly Rolled		Moder-ately Rolled		Very Rolled	
	Sum	Column Sum %	Sum	Column Sum %	Sum	Column Sum %	Sum	Column Sum %	Sum	Column Sum %	Sum	Column Sum %	Sum	Column Sum %	Sum	Column Sum %	Sum	Column Sum %	Sum	Column Sum %	Sum	Column Sum %	Sum	Column Sum %
FS1C	.	.0	1	16.7	.	.0	.	.0	3	42.9	.	.0	.	.0	.	.0	7	33.3	.	.0	.	.0	.	.0
FS1J	4	80.0	.	.0	.	.0	.	.0	.	.0	.	.0	.	.0	.	.0	.	.0	.	.0	.	.0	.	.0
FS1K	.	.0	2	33.3	1	20.0	.	.0	1	14.3	2	66.7	.	.0	.	.0	2	9.5	.	.0	.	.0	.	.0
FS1	.	.0	2	33.3	.	.0	.	.0	.	.0	1	33.3	.	.0	.	.0	.	.0	.	.0	.	.0	.	.0
FS1O	.	.0	.	.0	.	.0	2	40.0	.	.0	.	.0	.	.0	.	.0	.	.0	1	100	.	.0	.	.0
FW2	.	.0	.	.0	.	.0	.	.0	1	14.3	.	.0	.	.0	.	.0	.	.0	.	.0	.	.0	.	.0
FW2D	.	.0	.	.0	.	.0	.	.0	.	.0	.	.	.	.0	.	.0	5	23.8	.	.0	.	.0	.	.0
FW2L	.	.0	1	16.7	.	.0	.	.0	1	14.3	.	.0	.	.0	.	.0	.	.0%	.	.0	.	.0	.	.0
FW2N	.	.0	.	.0	.	.0	2	40.0	1	14.3	.	.0	.	.0	.	.0	1	4.8	.	.0	.	.0	.	.0
FW2O	1	20.0	.	.0	4	80.0	1	20.0	.	.0	.	.0	.	.0	.	.0	6	28.6	.	.0	.	.0	.	.0

Appendix 7.22: Raw data for the artefact counts for the first season's transects LU1 and CHO.

The data is presented by the 100x100m block code, and cross-referenced by typological association.

Block code	Typology					
	Early Stone Age	Middle Stone Age	Later Stone Age	Iron Age	Late Iron Age	Unknown
	Number of artefacts	Number of artefacts	Number of artefacts	Number of artefacts	Number of artefacts	Number of artefacts
CH11E	.	.	.	1	.	.
CH12B	.	.	5	.	.	.
CH13C	1	.	4	.	.	.
CH13E	.	.	1	5	.	.
CH15C	.	1	6	.	.	.
LU11B	.	.	26	1	2	49
LU14A	.	.	3	1	.	2
LU14B	.	.	4	.	.	1
LU14C	.	.	4	1	.	13
LU15A	.	.	120	3	.	247
LU15D	.	.	3	.	.	.
LU15E	5	1	4	.	.	.
LU15J	3	3	3	.	.	.
LU15M	.	.	5	.	.	.
LU15N	1	.	.	.	.	.
LU15O	3	.	4	.	.	.
LU16A	.	.	11	.	.	.
LU16B	.	.	10	.	.	.
LU16C	.	.	2	.	.	.
LU17A	.	.	.	1	.	.
LU17D	.	.	14	35	.	.
LU17N	.	1	4	.	.	.
LU17S	8	6	13	.	.	4
LU17X	3	3	5	.	.	2
LU18C	.	.	11	.	.	.
LU18D	.	.	8	1	.	.
LU33C	.	.	.	.	.	1
LU33E	.	.	6	6	.	8

10. References Cited

10.1. Texts Cited

Abe, Y. Marean, C.W. Nilssen, P.J. Assefa, Z. and Stone, E.C. 2002. The analysis of cutmarks on archaeofauna: a review and critique of quantification procedures, and a new image-analysis GIS approach. *American Antiquity.* **Vol:67 (4).** pp 642-663.

Adhikari, D.P. and Koshimizu, S. 2005. Debris flow disaster at Larcha, upper Bhotekoshi Valley, central Nepal. *The Island Arc.* **Vol:14.** pp 410-428.

Aldenderfer, M. 1996. Introduction. In Aldenderfer, M. and Maschner, H.D.G. (eds.) *Anthropology, Space and Geographic Information Systems.* Oxford University Press: Oxford.

Aldenderfer, M. and Maschner, H.D.G. (eds.) 1996. *Anthropology, Space and Geographic Information Systems.* Oxford University Press: Oxford.

Allen, K.M. Green, S.W. and Zubrow, E.B.W. (eds.) 1990. *Interpreting Space: GIS and Archaeology.* Taylor and Francis: London.

Ammerman, A.J. 1981. Surveys and archaeological research. *Annual Review of Anthropology.* **Vol:10.** pp 63-88.

Andrefsky, W. 1998. *Lithics: Macroscopic Approaches to Analysis.* Cambridge University Press: Cambridge.

Archer, D.R. 1971a. Rainfall. In Davies, D.H. (ed.) *Zambia in Maps.* University of London Press: London.

Archer, D.R. 1971b. Temperature, humidity, sunshine and winds. In Davies, D.H. (ed.) *Zambia in Maps.* University of London Press: London.

Asch, D.L. 1975. On sample size problems and the uses of non-probabilistic sampling. In Mueller, J.W. (ed.) *Sampling in Archaeology.* The University of Arizona Press: Tucson.

Asfaw, B. Beyene, Y. Suwa, G. Walter, R.C. White, T.D. WoldeGabriel, G. Yemane, T. 1992. The earliest Acheulean from Konso-Gardula. *Nature.* **Vol:360.** pp 732-735.

Ashton, N. and McNabb, J. 1996. The flint industries from the Waechter excavations. In Conway, B. McNabb, J. and Ashton, N. (eds.) *Excavations at Barnfield Pit, Swanscombe 1968 – 1972: British Museum Occasional Paper 94.* British Museum Press; London.

Asombang, R.N. 2004. Interpreting standing stones in Africa: a case study in north-west Cameroon. *Antiquity.* **Vol:78.** pp 294-305.

Astle, W.L. 1995. *South Luangwa National Park – Vegetation Map.* Geography Department, University of Maryland, College Park USA.

Astle, W.L. Phiri, P.S.M. and Prince, S.D. 1997a. Annotated checklist of the flowering plants and ferns of the South Luangwa National Park, Zambia. *Kirkia.* **Vol:16 (2).** pp 109-160.

Astle, W.L. Phiri, P.S.M. and Prince, S.D. 1997b. A dictionary of the vernacular-scientific names of the plants of the Luangwa Valley, Zambia. *Kirkia.* **Vol:16 (2).** pp 161-203.

Astle, W.L. Webster, R. and Lawrance, C.J. 1969. Land classification for management planning in the Luangwa Valley of Zambia. *Journal of Applied Ecology.* **Vol:6 (2).** pp 143-169.

Avery, D.M. 2003. Early and Middle Pleistocene environments and hominid biogeography; micromammalian evidence from Kabwe, Twin Rivers and Mumbwa Caves in central Zambia. *Palaeogeography, Palaeoclimatology, Palaeoecology.* **Vol: 189.** 55-69

Banning, E.B. 2002. *Archaeological Survey.* Kluwer Academic: New York.

Barber, M. 2006. Palaeolithic contrivance over Stonehenge. *British Archaeology.* **Vol:89**- July/August. pp 19-23.

Barham, L.S. 1987. The bipolar technique in southern Africa: a replication experiment. *South African Archaeological Bulletin.* **Vol:42 (145).** pp 45-50.

Barham, L.S. 2000. *The Middle Stone Age of Zambia.* Western Academic and Specialist Press: Bristol.

Barham, L.S. 2001. Central Africa and the emergence of regional identity in the Middle Pleistocene. In Barham, L.S. and Robson-Brown, K. (eds.) *Human Roots: Africa and Asia in the Middle Pleistocene.* Western Academic and Specialist Press: Bristol.

Barham, L.S. 2002. Backed tools in middle Pleistocene central Africa and their evolutionary significance. *Journal of Human Evolution.* **Vol: 43.** pp 585-603.

Barham, L.S. 2007. Modern is as modern does? Technological trends and thresholds in the south-central African record. In Mellars, P. Boyle, K. Bar-Yosef, O. and Stringer, C. (eds.) *Rethinking the Human Revolution: New Behavioural and Biological Perspectives on the Origin and Dispersal of Modern Humans.* (McDonald

Institute Monographs.) McDonald Institute for Archaeological Research: Cambridge.

Barham, L.S. and Jarman, C.L., 2005. New radiocarbon dates for the Early Iron Age in the Luangwa Valley, eastern Zambia. *Azania.* **Vol:XL.** pp114-121.

Barham, L.S. Phillips, W. Maher, B. Karloukovski, V. Duller, G. Jain, M. and Wintle, A. In Prep. *The Dating and Interpretation of a Mode 1 Site in the Luangwa Valley, Zambia.*

Barham, L.S. and Robson-Brown, K. (eds.) 2001. *Human Roots: Africa and Asia in the Middle Pleistocene.* Western Academic and Specialist Press: Bristol.

Barker, G. 1995. *A Mediterranean Valley: Landscape Archaeology and Annales History in the Biferno Valley.* Leicester University Press: London.

Barker, P. 1998. *Techniques of Archaeological Excavation. 3rd edition.* Routledge: London.

Barnett, T.F. 2005. Patterns on the rocks: report on recent work to survey rock art sites in the Wadi al-Hayat, Fezzan. *Libyan Studies.* **Vol:36.** pp 121-134.

Barton, C.M. Bernabeu, J.J. Aura, J.E. and Garcia, O. 1999. Land-use dynamics and socio-economic change: an example from the Polop Alto Valley. *American Antiquity.* **Vol:64 (4).** pp 609-634.

Barton, C.M., Bernabeu. J. Aura, J.E. Garcia, O. and La Roca, N. 2002. Dynamic landscapes, artefact taphonomy, and landuse modelling in the Western Mediterranean. *Geoarchaeology: An International Journal.* **Vol:17 (2).** pp 155-190

Bar-Yosef, O. and Kuhn, S.L. 1999. The big deal about blades: Laminar technologies and human evolution. *American Anthropologist.* **Vol:101 (2).** pp 322-338.

Bell, T. Wilson, A. and Wickham, A. 2002. Tracking the Samnites: landscape and communications routes in the Sangro Valley, Italy. *American Journal of Archaeology.* **Vol:106.** pp 169-86.

Bevan, A. and Conolly, J. 2004. GIS, archaeological survey, and landscape archaeology on the island of Kythera, Greece. *Journal of Field Archaeology.* **Vol:29.** pp 123-138.

Bevan, A. and Harlan, D. 2003. Kythera Island Project: archaeological survey: methods and preliminary results. *http://www.ucl.ac.uk/kip/survey.php* accessed 11/2007.

Binford, L.R. 1964. A consideration of archaeological research design. *American Antiquity.* **Vol:29 (4).** pp 425-441.

Bintliff, J. 1985. The Boeotia survey. In Macready, S. and Thompson, F.H. (eds.) *Archaeological Field Survey in Britain and Abroad.* Occasional Paper (new series) VI. The Society of Antiquaries of London: London.

Bintliff, J. 2000. The concepts of 'site' and 'offsite' archaeology in surface artefact survey. In Pasquinucci, M., Trement, F. (eds.) *The Archaeology of Mediterranean Landscapes 4: Non-destructive Techniques Applied to Landscape Archaeology.* Oxbow: Oxford.

Bisson, M.S. 1990. Lithic reduction sequences as an aid to the analysis of Later Stone Age quartz assemblages from the Luano Spring, Chingola, Zambia. *African Archaeological Review.* **Vol:8.** pp 103-138.

Bisson, M.S. 1992. A survey of Late Stone Age and Iron Age sites at Luano, Zambia. *World Archaeology.* **Vol:24 (2).** pp 234-248.

Blair, T.C. 1999. Cause of dominance by sheet flood vs. debris-flow processes on two adjoining alluvial fans, Death Valley, California *Sedimentology.* **Vol:46.** pp 1015-1028.

Bower, J. 1986. A survey of surveys: aspects of surface archaeology in sub-Saharan Africa. *The African Archaeological Review.* **Vol:4.** pp 21-40

Boismier, W.A. 1991. The role of research design in surface collection: an example from Broom Hill, Braishfield, Hampshire. In Schofield, A.J. (ed.) *Oxbow Monograph 4: Interpreting Artefact Scatters: Contributions to Ploughzone Archaeology.* Oxbow Books: Oxford.

Brandt, R. Groenewoudt, B.J. and Kvamme, K.L. 1992. An experiment in archaeological site location: modelling in the Netherlands using GIS techniques. *World Archaeology.* **Vol:24 (2).** pp 268-282.

Cairncross, B. 2001. An overview of the Permian (Karoo) coal deposits of southern Africa. *African Earth Sciences.* **Vol:33** pp 529-562.

Cannon, S.H. 1997. *Evaluation of the Potential for Debris and Hyperconcentrated Flows in Capulin Canyon as a Result of the 1996 Dome Fire, Bandelier National Monument, New Mexico.* USGS Open File Report 97-136 at *http://pubs.usgs.gov/of/1997/ofr-97-0136/* accessed on 31/08/07.

Castaing, C. 1991. Post-Pan-African tectonic evolution of South Malawi in relation to the Karroo and Recent East African Rift Systems. *Tectonophysics.* **Vol:191.** pp 55-73.

Challis, K. Priestnall, G. Gardner, A. Henderson, J. O'Hara, S. 2002. Corona remotely sensed imagery in dryland archaeology: the Islamic city of al-Raqqa, Syria. *Journal of Field Archaeology.* **Vol:29.** pp 139-153.

Chapman, H. 2006. *Landscape Archaeology and GIS.* Tempus: Stroud.

Chazan, M. 1997. Redefining Levallois. *Journal of Human Evolution.* **Vol:33.** pp 719-735.

Chenhall, R.G. 1975. A Rational for archaeological sampling. in Mueller, J.W. (ed.) *Sampling in Archaeology.* Tuscon: University of Arizona Press.

Cherry, J.F. 1982. A preliminary definition of site distribution on Melos. In Renfrew, C. and Wagstaff, M. (eds.) *An Island Polity: The Archaeology of Exploitation in Melos.* Cambridge University Press: Cambridge.

Cherry, J.F. Gamble, C. Shennan, S. (eds.) 1978. *Sampling in Contemporary British Archaeology.* BAR British series 50. BAR: Oxford.

Church, M. and Mark, D.M. 1980. On size and scale in geomorphology. *Progress in Physical Geography.* **Vol:4.** pp 342-390.

Clark, D. 1982. *The Cambridge History of Africa.Vol.1: From the Earliest Times to c.500 BC.* Cambridge University Press: Cambridge.

Clark, G. 1971. *World Prehistory: A New Outline.* Cambridge University Press: Cambridge.

Clark, J.D. 1950. The newly discovered Nachikufu culture of Northern Rhodesia and the possible origin of certain elements of the South African Smithfield Culture. *South African Archaeological Bulletin.* **Vol:5.** pp 86-98.

Clark, J.D. 1954. Upper Sangoan industries from northern Nyasaland and the Luangwa Valley: A case of environmental differentiation? *South African Journal of Science.* **Vol:50 (8).** pp 201-208.

Clark, J.D. 1967. *Atlas of African Prehistory.* University of Chicago: Chicago and London.

Clark, J.D. 1970. *The Prehistory of Africa.* Thames and Hudson: Southampton.

Clark, J.D. 2001. *Kalambo Falls Prehistoric Site: Volume III.* Cambridge University Press: Cambridge.

Clark, J.D. and Brown K.S. 2001. The Twin Rivers kopje, Zambia: Stratigraphy, fauna, and artefact assemblages from the 1954 and 1956 excavations. *Journal of Archaeological Science.* **Vol: 28.** pp 305-330

Clarke, D.L. 1972. (ed.) *Models in Archaeology.* Methuen and co. Ltd: London.

Clarke, W.G. 1922. *Our Homeland Prehistoric Antiquities and How to Study Them.* The Homeland Association: London.

Cohen, A.S. Stone, J.R. Beuning, K.R.M. Park, L.E. Reinthal, P.N. Dettman, D. Scholz, C.A. Johnson, T.C. King, J.W. Talbot, M.R. Brown, E.T. and Ivory, S.J. 2007. Ecological consequences of early Late Pleistocene megadroughts in tropical Africa. *Proceedings of the National Academy of Sciences.* **Vol:104 (42).** pp 16422-16427.

Collatz, J.G. Berry, J.A. and Clark, J.S. 1998. Effects of climate and atmospheric CO2 partial pressure on the global distribution of C4 grasses: present, past, and future. *Ocealogica.* **Vol: 144.** pp 441-454.

Cole, M.M. 1963. Vegetation and geomorphology in northern Rhodesia: an aspect of the distribution of the savannah of central Africa. *Geographical Journal.* **Vol:129 (3).** pp 290-305

Conolly, J. and Lake, M. 2006. *Geographical Information Systems in Archaeology.* Cambridge University Press: Cambridge.

Cowgill, G.L. 1975. A selection of samplers: comments on archaeo-statistics. In Mueller, J.W. (ed.) *Sampling in Archaeology.* The University of Arizona Press: Tucson.

Crawford, O.G.S. 1928. *Air Survey and Archaeology. Ordnance Survey Professional Papers, new series No. 7 (2nd ed.).* H.M stationary office: London

Crawford, O.G.S. 1929. *Air-photography for Archaeologists.* Ordnance Survey, H.M. Stationary Office: London.

Crumley, C. 1993. Analyzing historic ecotonal shifts. *Ecological Applications.* **Vol:3 (3).** pp 377-384

Crumley, C. and Marquardt, W.H. 1990. Landscape: a unifying concept in regional analysis. In Allen, K.M., Green, S.W. and Zubrow, E.B.W. (eds.). *Interpreting Space: GIS and Archaeology.* Taylor and Francis: London.

Dalla Bona, L. 2000. Protecting cultural resources through forest management planning in Ontario using archaeological predictive modelling. In Wescott, K.L. and Brandon, R.J. (eds). *Practical applications of GIS for archaeologists: a predictive modelling kit.* Taylor and Francis: London.

D'Andrea, A., Gallotti, R, and Piperno, M. 2002. Taphonomic interpretation of the developed Oldowan site of Garba IV (Melka Kunture, Ethiopia) through GIS application. *Antiquity.* **Vol:76.** pp 991-1001.

deMenocal, P.B. 1995. Plio-Pleistocene African Climate. *Science.* **Vol:270.** pp 53-59.

deMenocal, P.B. Ortiz, J. Guilderson, T. and Sarnthein, M. 2000. Coherent high- and low- latitude climate variability during the Holocene warm period. *Science.* **Vol:288.** pp 2198-2202.

De Waele, B, and Mapani, B. 2002. Geology and correlation of the central Irumide belt. *Journal of African Earth Sciences.* **Vol:35.** pp 385-397.

Delanges, A. and Roche, H. 2005. Late Pliocene hominid knapping skills: The case of Lokalalei 2C, West Turkana, Kenya. *Journal of Human Evolution.* **Vol:48.** pp 435-472.

Dennis, J.C. Ungar, P.S. Teaford, M.F. and Glander, K.E. 2004. Dental topography and molar wear in *Alouatta palliata* from Costa Rica. *American Journal of Physical Anthropology.* **Vol:125.** pp 152-161.

Derricourt, R. 1980. *People of the Lakes: Archaeological Studies in Northern Zambia.* Manchester University Press: Manchester.

Derricourt, R. 1985. *Man on the Kafue: Archaeology and History of the Itezhitezhi Area of Zambia.* Ethnographica: London.

Dickson, B.D. Pearl, F.B. Gang, G-U. Kahinju, S. and Wandibba, S. 2004. Site reconnaissance in the Kipsing and Tol river watersheds of central Kenya: implications for Middle and later Stone Age land-use patterns. *African Archaeological Review.* **Vol:21 (3).** pp 153-191.

Dixey, F. 1937a. The Geology of part of the upper Luangwa Valley, northeastern Rhodesia. *Quarterly Journal of the Geological Society London.* **Vol:93.** pp 52-76.

Dixey, F. 1937b. The pre-Karoo landscape of the lake Nyasa region, and a comparison of the Karoo structural directions with those of the rift valley. *Quarterly Journal of the Geological Society London.* **Vol:93.** pp 77-93.

Drysdall, A.R. and Weller, R.K. 1966. Karoo sedimentation in Northern Rhodesia. *Transactions and Proceedings of the Geological Society of South Africa.* **Vol:69.** pp 39-69.

Duncan, R.B. and Beckman, K.A. 2000. The application of GIS predictive site location models within Pennsylvania and West Virginia. In Wescott, K.L. and Brandon, R.J. (eds). *Practical Applications of GIS for Archaeologists: a Predictive Modelling Kit.* Taylor and Francis: London.

Dunnell, R.C. and Dancey, W.S. 1983. The siteless survey: a regional scale data collection strategy. In Schiffer, M.B. (ed.) *Advances in Archaeological Method and Theory Volume 6.* Academic Press: New York.

Dupont, L.D. Jahns, S. Marret, F. and Ning, S. 2000. Vegetation change in equatorial west Africa: time slices for the last 150 Ka. *Palaeogeography, Palaeoclimatology, Palaeoecology.* **Vol: 155.** pp 95-122.

Dupont, L.M. Donner, B. Schneider, R. and Wefer, G. 2001. Mid-Pleistocene environmental change in tropical Africa began as early as 1.05 Ma. *Geological Society of America.* **Vol:29 (3).** pp 195 – 198.

Ebert, J.I. 2000. The state of the art in "inductive" predictive modelling: seven big mistakes (and lots of smaller ones). In Wescott, K.L. and Brandon, R.J. (eds.) *Practical Applications of GIS for Archaeologists: a Predictive Modelling Kit.* Taylor and Francis: London.

Ebinger, C.J. Deino, A.L. Tesha, A.L. Becker, T. and Ring, U. 1993. Tectonic controls on rift basin morphology – evolution of the northern Malawi (Nyasa) Rift. *Journal of Geophysical Research – Solid Earth.* **Vol:98 (B10).** pp 17821-17836.

Elton, S. Barham, L. Andrews, P. and Sambrook Smith, G.H. 2003. Pliocene femur of *Theropithecus* from the Luangwa Valley, Zambia. *Journal of Human Evolution.* **Vol:44.** pp 133-139.

Fagan, B.M. 1967. *Iron Age Cultures in Zambia: volume one.* Chatto and Windus: London.

Fanning, P.C. and Holdaway, S.J. 2004. Artifact visibility at open sites in Western New South Wales, Australia. *Journal of Field Archaeology.* **Vol:29.** pp 255-271.

Favory, F. and Raynaud, C. 2000. Definition ou hierarchisation des sites? Approche integree en Gaule mediterraneenne. In Pasquinucci, M. Trement, F. (eds.) *The Archaeology of Mediterranean Landscapes 4: Non-destructive Techniques Applied to Landscape Archaeology.* Oxbow: Oxford.

Field, J.S. and Lahr, M.M. 2005. Assessment of the southern dispersal: GIS-based analyses of potential routes at Oxygen Isotope Stage 4. *Journal of World Prehistory.* **Vol:19 (1).** pp 1-43.

Fitzjohn, M. 2007. Viewing places: GIS applications for examining the perception of space in the mountains of Sicily. *World Archaeology.* **Vol:39 (1).** pp 36-50.

Foley, R. 1978. Incorporating sampling into initial research design: some aspects of spatial archaeology. In Cherry, J.F., Gamble, C., Shennan, S. (eds.) *Sampling in Contemporary British Archaeology.* BAR British series 50. BAR: Oxford.

Foley, R. 1981. *Off-site Archaeology and Human Adaptation in Eastern Africa.* BAR International series 97. BAR: Oxford.

Forte, M. 1995. Scientific visualisation and archaeological landscape: the case study of a terramara, Italy. In Lock, G and Stancic, Z. (eds.) *Archaeology and Geographic Information Systems.* Taylor and Francis: London.

Fox, C.F. 1932. *The personality of Britain: its Influence on Inhabitant and Invader in Prehistoric and Early Historic Times.* National Museum of Wales: Cardiff.

Fry, G.L.A. Skar, B. Jerspasen, G. Bakkestuen, V. and Erikstad, L. 2004. Locating archaeological sites in the landscape: a hierarchical approach based on landscape indicators. *Landscape and Urban Planning.* **Vol:67.** pp 97-107.

Gaffney, V. 1995. GIS and ED: a trap for the unwary? In Lock, G and Stancic, Z. (eds.) *Archaeology and Geographic Information Systems.* Taylor and Francis: London.

Gaffney, V. Stancic, Z. and Watson, H. 1995. The impact of GIS on archaeology: a personal perspective. In Lock, G and Stancic, Z. (eds.) *Archaeology and Geographic Information Systems.* Taylor and Francis: London.

Gamble, C. 1997. *Timewalkers: The Prehistory of Global Colonisation.* Penguin: St. Ives.

Gasse, F. Barker, P. and Johnson, T.C. 2002. A 24,000yr diatom record from the northern basin of Lake Malawi. In Odada, E.O. and Olago, D.O. (eds.) *The East African Great Lakes: Limnology, Palaeolimnology and Biodiversity.* Kluwer Academic Publishers: Netherlands.

Genty, D. Blamart, D. Ouahdi, R. Gilmour, M. Baker, A. Jouzal, J. and Van Exter, S. 2003. Precise dating of Dansgaard-Oeschger climate oscillations in western Europe from stalagmite data. *Nature.* **Vol: 421.** pp 833-837.

Gillings, M. Mattingley, D. and van Dalen, J. (eds.) 1999. *Geographic Information Systems and Landscape Archaeology.* Oxbow Books: Oxford.

Gillings, M. and Sbonias, K. 1999. Regional survey and GIS: The Boeotia Project. In Gillings, M., Mattingley, D. and van Dalen, J. (eds.) *Geographic Information Systems and Landscape Archaeology.* Oxbow Books: Oxford.

Gilvear, D. Winterbottom, S. and Sichingabula, H. 2000. Character of channel planform change and meander development: Luangwa River, Zambia. *Earth Surface Processes and Landforms.* **Vol:25.** pp 421-436.

Gingele, F.X. Muller, P.M. and Schneider, R.R. 1998. Orbital forcing of freshwater input in the Zaire Fan area, clay mineral evidence form the last 200Kyr. *Palaeogeography, Palaeoclimatology, Palaeoecology.* **Vol: 138.** pp 17-26

Golberg, P. and Macphail, R.I. 2006. *Practical and Theoretical Geoarchaeology.* Blackwell: Oxford.

Gourad, K. Accessed March 2005. GIS Archaeology Survey. *http://www.gisdevelopment.net/application/archaeology/general/archg0001.htm*

Graham, D.J. Reid, I. Rice, S.P. 2005. Automated sizing of coarse-grained sediments: image-processing procedures. *Mathematical Geology.* **Vol:37 (1).** pp 1-28.

Graham, J. 1988. Collection and analysis of field data. In Tucker, M.E. (ed.). *Techniques in Sedimentology.* Blackwells: Oxford.

Gruen, A. and Sauerbier, M. 2005. Nasca – Palpa; Photogrammetric Reconstruction of the Geoglyphs of Nasca and Palpa. *http://www.photogrammetry.ethz.ch/research/eru/index.html.* Accessed March 2005.

Gupta, S. 1997. Himalayan drainage patterns and the origin of fluvial megafans in the Ganges foreland basin. *Geology.* **Vol:25 (1).** pp 11-14.

Guzzetti, F. Marchetti, M. Reichenbach, P. 1997. Large alluvial fans in the north-central Po Plain (Northern Italy). *Geomorphology.* **Vol:18.** pp 119-136.

Harvey, A.M. 1997. The role of alluvial fans in arid zone fluvial systems. In Thomas, D.S.G. (ed.) *Arid Zone Geomorphology: Process, Form, and Change in Drylands 2nd Ed.* John Wiley & Sons Ltd: New York.

Herz, N. and Garrison, E.G. 1998. *Geological Methods for Archaeology.* Oxford University Press: New York.

Heward, A.P. 1978. Alluvial fan and lacustrine sediments from the Stephanian A and B (La Magdalena, Cinera – matallana and Sabero) coalfields, northern Spain. *Sedimentology.* **Vol:25** pp 451-488.

Hodder, I. and Orton, C. 1976. *Spatial Analysis in Archaeology.* Cambridge University Press: Cambridge.

Hopley, P.J. Weeden, G.P. Marshall, J.D. Herries, A.I.R. Latham, A.G. Kuykendall, K.L. 2007. High- and low-latitude orbital forcing of early hominin habitats in South Africa. *Earth and Planetary Science Letters.* **Vol:256.** pp 419-432.

Howard, J.L. 1993. The statistics of counting clasts in rudites: a review, with examples from the upper Palaeogene of southern California, USA. *Sedimentology.* **Vol:40.** pp 157-174.

Huckabay, J.D. 1989. Man and the disappearance of the Zambezian dry evergreen forest. In Kadomura, H. (ed.) *A Preliminary Report of the Tropical African Geomorphology and Late-Quaternary Research Project (TAGELAQP) 1987/88.* Department of Geography Tokyo Metropolitan University: Tokyo.

Huggett, J. 2000. Looking at intra-site GIS. In Lockyear, K., Sly, T.J.T., Mihailescu-Bîrliba, V. (eds.) *CAA 96: Computer Applications and Quantitative Methods in Archaeology.* BAR International Series: 845. BAR Publishing: Oxford.

Huggett, R.J. 2007. *Fundamentals of Geomorphology 2nd ed.* Routledge: online book

Huntley, B.J. 1982. Southern African Savannas. In Huntley, B.J. and Walker, B.H. (eds.) *Ecology of Tropical Savannas.* Springer Verlag Berlin: Heidelberg.

Ingold, T. 1993. The temporality of the landscape. *World Archaeology.* **Vol:25 No.2.** pp 152-174.

Iwasaki, K. and Shinoda, M. 1989. Climatic environments of Zambia. In Kadomura, H. (ed.) *A preliminary report of the Tropical African Geomorphology and Late-Quaternary Research Project (TAGELAQP) 1987/88.* Department of Geography Tokyo Metropolitan University: Tokyo.

Kadomura, H. (ed.) 1989. *A Preliminary Report of the Tropical African Geomorphology and Late-Quaternary Research Project (TAGELAQP) 1987/88.* Department of Geography Tokyo Metropolitan University: Tokyo.

Keay, S.J. and Millet, M. 1991. Surface survey and site recognition in Spain: the Ager Tarraconensis survey and its background. In Schofield, A.J. (ed.) *Oxbow Monograph 4: Interpreting Artefact Scatters: Contributions to Ploughzone Archaeology.* Oxbow Books: Oxford.

Keesstra, S.D. Huissteden, J. Vandenberghe, J. Van Dam, O. de Gier, J. and Plezier, I.D. 2005. Evolution of the morphology of the river Dragonja (SW Slovenia) due to land-use changes. *Geomorphology.* **Vol:69.** pp 191-207.

Klein, R.J. 1999. *The Human Career. 2nd ed.* University of Chicago Press. London.

Knighton, D. 1998. *Fluvial Forms and Processes: A New Perspective.* Hodder Arnold: London.

Kuman, K. and Clarke, R.J. 2000. Stratigraphy, artefact industries and hominid associations for Sterkfontein, Member 5. *Journal of Human Evolution.* **Vol:38.** pp 827-847.

Kvamme, K.L. 1990. One-sample tests in regional archaeological analysis: new possibilities through computer technology. *American Antiquity.* **Vol:55 (2).** pp 367-381.

Leakey, M.D. 1966. A review of the Oldowan culture from Olduvai Gorge Tanzania. *Naure.* **Vol:210.** pp 462-466.

Leeder, M. 1999. *Sedimentology and Sedimentary Basins: From Turbulence to Tectonics.* Blackwell Science: Bodmin.

Leuschner, D.C. and Sirocko, F. 2000. The low-latitude monsoon climate during Dansgaard-Oeschger cycles and Heinrich events. *Quaternary Science Reviews.* **Vol:19.** pp 243-254.

Llobera, M. 1996, Exploring the topography of mind: GIS, social space and archaeology. *Antiquity.* **Vol:70.** pp 612-622

Llobera, M. 2003. Extending GIS-based visual analysis: the concept of *visualscapes. International Journal of Geographical Information Science.* **Vol:17 (1).** pp 25-48.

Llobera, M. 2007. Reconstructing visual landscapes. *World Archaeology.* **Vol:39 (1).** pp 51-69.

Lock, G and Stancic, Z. (eds.). 1995. *Archaeology and Geographic Information Systems.* Taylor and Francis: London.

Lock, G. Bell, T. and Lloyd, J. 1999. Towards a methodology for modelling surface survey data: the Sangro Valley Project. In Gillings, M. Mattingly, D. and van Dalen, J. (eds.) *Geographic Information Systems and Landscape Archaeology.* Oxbow Books: Oxford.

Lock, G. and Harris, T. 2006. Enhancing predictive archaeological modelling: integrating location, landscape, and culture. In Mehrer, M.W. and Wescott, K.L. (eds.) *GIS and Archaeological Site Location Modelling.* Taylor and Francis: London.

MacCrae, F.B. and Lancaster, D.G. 1937. Stone Age sites in Nothern Rhodesia. *Man.* **Vol:37.** pp 62-64.

Mäckel, R. 1971. Vegetation and the forest estate. In Davies, D.H. (ed.) *Zambia in Maps.* University of London Press: London.

Madry, S.L.H. 1990. The realities of hardware. In Allen, K.M., Green, S.W. and Zubrow, E.B.W. (eds.) *Interpreting Space: GIS and Archaeology.* Taylor and Francis: London.

Mazel, A.D. Frink, D.S. Muscarella, O.W. Joukowsky, M.S. 1984. Archaeological survey of the Natal Drakensberg, Natal, South Africa. *Journal of Field Archaeology.* **Vol:11 (3).** pp 345-364.

McBrearty, S. and Brooks, A. 2000. The revolution that wasn't: a new interpretation of the origin of modern human behaviour. *Journal of Human Evolution.* **Vol: 39.** pp 453-563.

McNabb, J. and Ashton, N. (eds.) 1996. *Excavations at Barnfield Pit, Swanscombe 1968 – 1972: British Museum Occasional Paper 94.* British Museum Press; London.

McConnel, R.B. 1972. Geological development of the rift system of eastern Africa. *Geological Society of America Bulletin.* **Vol:83.** pp 2249-2572.

McIntosh, S.K. and McIntosh, R.J. 1993. Field survey in the Tumulus zone of Senegal. *The African Archaeological Review.* **Vol:11.** pp 73-107.

McIntyre, C. 2004. *Zambia: The Bradt Travel Guide 3rd ed.* Bradt: Chalfont St Peter.

Mellars, P. Boyle, K. Bar-Yosef, O. and Stringer, C. (eds.) 2007. *Rethinking the Human Revolution: New Behavioural and Biological Perspectives on the Origin and Dispersal of Modern Humans.* (McDonald Institute Monographs.) McDonald Institute for Archaeological Research: Cambridge.

Miller, S.F. 1969a. *The Nachikufan Industries of the Later Stone Age in Zambia.* Ph.D. thesis. Berkeley: California.

Miller, S.F. 1969b. Contacts between the Later Stone Age and the Early Iron Age in southern central Africa. *Azania.* **Vol:4.** pp 81-90.

Milliken, S. and Cook, J. (eds.) 2001. *A Very Remote Period Indeed. Papers on the Palaeolithic Presented to Derek Roe.* Oxbow Books: Oxford.

Mitchell, P. 1994. The archaeology of the Phuthiatsana-ea-Thaba Bosiu Basin, Lesotho, southern Africa: changes in the Later Stone Age regional demography. *Antiquity.* **Vol:68.** pp 83-96.

Mitchell, P. 1996. The late Quaternary landscape at Sehonghong in the Lesotho highlands, southern Africa. *Antiquity.* **Vol 70.** pp 623-38.

Mitchell, P. 2002. *The Archaeology of Southern Africa.* Cambridge University Press: Cambridge

Moore, J.A. and Keene, A.S. 1983. *Archaeological Hammers and Theories.* Academic Press: London.

Mueller, J.W. (ed.) 1975. *Sampling in Archaeology.* The University of Arizona Press: Tucson.

Mueller, J.W. (ed.) 1975. *Sampling in Archaeology.* The University of Arizona Press: Tucson.

Musonda, F.B. 1985. *Aspects of the Prehistory of the Lusemfwa Drainage Basin, Zambia, during the last 20,000 years.* Ph.D. Thesis. University of California, Berkeley.

Musonda, F.B. 1987. The significance of pottery in Zambian Later Stone Age contexts. *The African Archaeological Review.* **Vol:5.** pp 147-158.

Odada, E.O. and Olago, D.O. (eds.) 2002. *The East African Great Lakes: Limnology, Palaeolimnology and Biodiversity.* Kluwer Academic Publishers: Netherlands.

Oesterlen, P.M. and Blenkinsop, T.G. 1994. Extension directions and strain near the failed triple junction of the Zambezi and Luangwa Rift zones, southern Africa. *Journal of African Earth Sciences.* **Vol:18 (2).** pp175-180.

Oguchi, T. and Ohmori, H. 1994. Analysis of relationships among alluvial fan area, source basin area, basin slope, and sediment yield. Zeitschrift fur Geomorphologie. Vol:38 (4). pp 405-420.

Orton, C. 2000. *Sampling in Archaeology.* Cambridge University Press: Cambridge.

Partridge, T.C. deMenocal, P.B. Lorentz, S.A. Paiker, M.J. and Vogel, J.C. 1997. Orbital forcing of climate over south Africa: a 200,000 year rainfall record from the Pretoria Saltpan. *Quaternary Science Reviews.* **Vol: 16.** pp 1125-1133.

Pasquinucci, M., Trement, F. (eds.) 2000. *The Archaeology of Mediterranean Landscapes 4: Non-destructive Techniques Applied to Landscape Archaeology.* Oxbow: Oxford.

Pavitt, N. 2001. *Africa's Great Rift Valley.* Harry Abrams: New York.

Pearl, F, 2004. *Archaeology and Geoarchaeology of the Mukogodo Hills and Ewaso Ng'iro Plains, Central Kenya.* BAR International Series 1247, BAR Publishing: Oxford.

Pederson, J. Smith, G. Pazzaglia, F. 2001. Comparing the modern, Quaternary, and Neogene records of climate-controlled hillslope sedimentation in southeast Nevada. *Geological Society of America Bulletin.* **Vol:103 (3).** pp 305-319.

Philip, G. Donoghue, D. Beck, A. and Galiatsatos, N. 2002. Corona satellite photography: an archaeological application from the Middle East. *Antiquity.* **Vol:76.** pp 109-118.

Phillips, C.W. 1980. *Archaeology in the Ordnance Survey 1791 – 1965.* The Council for British Archaeology: London.

Phillips, W.M. Barham, L.S. and Kubik, P.W. 2005. Dating Alluvial sediments with cosmogenic nuclides. *Geochimica et Comochimica Acta.* **Vol: 69 (10).** pp A166-A166.

Phillipson, D.W. 1973. The prehistoric succession in Eastern Zambia: a preliminary report. *Azania.* **Vol:8.** pp 3-24.

Phillipson, D.W. 1974. Iron Age history and archaeology in Zambia. *Journal of African History.* **Vol:15 (1).** pp 1-25.

Phillipson, D.W. 1976. *The Prehistory of Eastern Zambia.* Memoir number six of the British Institute in eastern Africa: Nairobi.

Phillipson, D.W. 2005. *African Archaeology 3rd Ed.* Cambridge University Press: Cambridge.

Phillipson, L. 1975. Survey of the Stone Age archaeology of the upper Zambezi Valley: 1, the Northern part of the Valley. *Azania.* **Vol:10.** pp 1-48.

Phillipson, L. 1976. Survey of the Stone Age archaeology of the upper Zambezi Valley: 2, excavations at Kandanda. *Azania.* **Vol:11.** pp 49-81.

Phillipson, L. 1977. Survey of the Stone Age archaeology of the Upper Zambezi Valley: 3. The Southern part of the Valley. *Azania.* **Vol:12.**

Plog, S. Plog, F. and Wait, W. 1978. Decision making in modern surveys. In Schiffer, M.B. (ed.) *Advances in Archaeological Method and Theory. Volume 1.* Academic Press: New York.

Potts, R. Jorstad, T. and Cole, D. 1996. The role of GIS in interdisciplinary investigations at Olorgesailie, Kenya, a Pleistocene archaeological locality. In Aldenderfer, M. and Maschner, H.D.G. (eds.) *Anthropology, Space and Geographic Information Systems.* Oxford university Press: Oxford.

Prothero, D.R. and Schwab, F. 1996. *Sedimentary Geology: An Introduction to Sedimentary Rocks and Stratigraphy.* W.H. Freeman and Company: New York.

Pwiti, G. 1996. *Continuity and Change: An Archaeological Study of Farming Communities in Northern Zimbabwe AD 500 – 1700.* Studies in African archaeology 13: Uppsala

Renfrew, C. and Wagstaff, M. (eds.) 1982. *An Island Polity: The Archaeology of Exploitation in Melos.* Cambridge University Press: Cambridge

Robertshaw, P. 1994. Archaeological survey, ceramic analysis, and state formation in western Uganda. *The African Archaeological Review.* **Vol:12.** pp 105-131.

Robertson, J.H. 2000. Early Iron Age archaeology in central Zambia. *Azania.* **Vol: 35.** pp 147-181.

Sampson, C.G. 1985. *Atlas of Stone Age Settlements in the Central and Upper Seacow Valley.* Memoirs van die nasionale museum Bloemfontein No. 20: Bloemfontein.

Sampson, C.G. 1988. *Stylistic Boundaries Among Mobile Hunter-Foragers.* Smithsonian Institution: Washington.

Sampson , C.G. 2001. An upper Acheulian settlement pattern in the Upper Karoo region of South Africa. In Milliken, S. and Cook, J. (eds.) *A Very Remote Period Indeed. Papers on the Palaeolithic Presented to Derek Roe.* Oxbow Books: Oxford.

Sampson, C.G. and Southard, M.D. 1973. Variability and change in the Nachikufan industry of Zambia. *South African Archaeological Bulletin.* **Vol:28.** pp78-89.

Schefuβ, E. Schouten, S. Fred Jansen, J.H. and Sinninghe Damste, J.S. 2003. African vegetation controlled by tropical sea surface temperatures in the mid-Pleistocene period. *Nature.* **Vol:422.** pp 416-421.

Schiffer, M.B. (ed.) 1983. *Advances in Archaeological Method and Theory Volume 6.* Academic Press: New York.

Schiffer, M.B. 1987. *Formation Processes of the Archaeological Record.* University of New Mexico Press: Albuquerque.

Schiffer, M.B. Sullivan, A.P. and Klinger, T.C. 1978. The design of archaeological surveys. *World Archaeology.* **Vol:10 (1).** pp 1-28.

Schofield, A.J. (ed.) 1991. *Oxbow Monograph 4: Interpreting Artefact Scatters: Contributions to Ploughzone Archaeology.* Oxbow Books: Oxford.

Scholz C.A. Johnsonc T.C. Cohend A.S. Kinge J.W. Peck J.A. Overpeck J.T. Talbot M.R. Brown E.T. Kalindekafeh L. Amoakoi P.Y.O. Lyonsa R.P. Shanahand T.M. Castaneda I.S. Heile C.W. Forman S.L. McHargue L.R. Beuning K.R. Gomez J. and Pierson J. 2007. East African megadroughts between 135 and 75 thousand years ago and bearing on early-modern human origins. *Proceedings of the National Academy of Sciences.* **Early Edition.** pp1-6.

Semaw, S., Renne, P., Harris, J.W.K., feibel, C.S., Bernor, R.L., Fesseha, N., and Mowbray, K. 1997. 2.5 million year old stone tools from Gona, Ethiopia. *Nature.* **Vol:385.** pp 333-336.

Sepulchre, P. Ramstein, G. Fluteau, F. Schuster, M. Tiercelin, J-J. and Brunet, M. 2006. Tectonic uplift and eastern Africa aridification. *Science.* **Vol:313.** pp 1419-1423.

Shaw, P.A. 1997. Africa and Europe. In Thomas, D. (ed.) *Arid Zone Geomorphology: Process, Form and Change in Drylands. 2nd Ed.* Wiley and Sons: Chichester.

Shennan, S. 1985. *Experiments in the collection and analysis of archaeological survey data; the East Hampshire survey.* Department of Archaeology and Prehistory University of Sheffield: Sheffield.

Shoko, D.S.M. 1996. Extension directions and strain near the failed triple junction of the Zambezi and Luangwa Rift zones, southern Africa. *Journal of African Earth Sciences.* **Vol:22 (4).** pp 617-620.

Sigurdsson, H. (ed.) 2000. *Encyclopedia of Volcanoes.* Academic Press: London.

Silvester, R.J. 1991. *The Fenland Project Number 4: the Wissey Embayment and the Fen Causeway, Norfolk.* East Anglian Archaeology Report No. 52: Hunstanton.

Smith, M.J. Rose, J. and Booth, S. 2006. Geomorphological mapping of glacial landforms from remotely sensed data: An evaluation of the principal data sources and an assessment of their quality. *Geomorphology.* **Vol:76.** pp 148-165.

Stankiewicz, J. and de Wit, M.J. 2006. A proposed drainage evolution model for Central Africa – Did the Congo flow east? *Journal of African Earth Sciences.* **Vol:44.** pp 75-84.

Steel, R.J. Maehle, S. Nilsen, H. Roe, S.L. and Spinnangr, A. 1977. Coarsening-upward cycles in the alluvium of Hornelen Basin (Devonian) Norway: sedimentary response to tectonic events. *Geological Society of America Bulletin.* **Vol:88.** pp 1124-1134.

Stokes, S. Haynes, G. Thomas, D.S.G. Horrocks, J.L. Higginson, M. and Malifa, M. 1998. Punctuated aridity in southern Africa during the last glacial cycle: The chronology of linear dune construction in the northeastern Kalahari. *Palaeogeography Palaeoclimatology Palaeoecology.* **Vol: 137.** pp 305-322

Stow, D.A.V. 2006. *Sedimentary Rocks in the Field.* Manson Publishing: London.

Sundstrom, L. 1993. A simple mathematical procedure for estimating the adequacy of site survey strategies. *Journal of Field Archaeology.* **Vol:20 (1).** pp 91-96.

Tappen, M. 1995. Savanna ecology and natural bone deposition. *Current Anthropology.* **Vol:36 (2).** pp 223-260.

Thieme, J.G. and Johnson, R.L. 1981. *Geological map of the Republic of Zambia.* Geological Survey Department: Lusaka.

Thomas, D.H. 1975. Nonsite sampling in archaeology: Up the creek without a site? In Mueller, J.W. (ed.) *Sampling in Archaeology.* The University of Arizona Press: Tucson.

Thomas, D.S.G. (ed.) 1997. *Arid Zone Geomorphology: Process, Form, and Change in Drylands 2nd Ed.* John Wiley & Sons Ltd: New York.

Thomas, D.S.G. and Shaw, P.A. 2002. Late Quaternary environmental change in central southern Africa: new data, synthesis, issues and prospects. *Quaternary Science Reviews.* **Vol:21.** pp 783-797.

Thomas, M.F. 1999. Evidence for high energy landforming events of the central African plateau: eastern province, Zambia. *Zeitschrift fur Geomorphologie.* **Vol:43 (3).** pp 273-297.

Thomas, M.F. and Murray, A.S. 1999. On the age and significance of Quaternary colluvium in eastern Zambia. *Palaeoecology of Africa and the Surrounding Islands.* **Vol:27.** pp 117-133.

Toth, N. 1985. The Oldowan reassessed: a close look at early stone artefacts. *Journal of Archaeological Science.* **Vol:12.** pp 101-120.

Trapnell, C.G. 1996. *The Soils, Vegetation, and Traditional Agriculture of Zambia, Volume II. North Eastern Zambia [Ecological Survey 1937 – 1942].* Redcliffe Press Ltd: Bristol.

Trauth, M.H. Deino, A.L. Bergner, A.G.N. and Strecker, M.R. 2003. African climate change and orbital forcing during the last 175kyr BP. *Earth and Planetary Science Letters.* **Vol:206.** pp 297-313.

Tryon, C.A. and McBrearty, S. 2006. Tephrostratigraphy of the Bedded Tuff Member (Kapthurin Formation, Kenya) and the nature of archaeological change in the later middle Pleistocene. *Quaternary Research.* **Vol:65.** pp 492-507.

Tucker, M.E. 1991. *Sedimentary Petrology: An Introduction to the Origin of Sedimentary Rocks (2nd Ed.).* Blackwell Science: Cambridge.

Ur, J. 2002. Settlement and landscape in northern Mesopotamia: the Tell Hamoukar Survey 2000-2001. *Akkadica.* **Vol:123.** pp 57-88.

Ur, J. 2003. Corona satellite photography and ancient road networks: a northern Mesopotamian case study. *Antiquity.* **Vol:77 (295).** pp 102-115.

Utting, J. 1988. *An Introduction to the Geological History of the South Luangwa National Park, Zambia. Geological Survey Department Special Publication no.1.* Geological Survey Department Zambia: Lusaka.

Vallance, J.W. 2000. Lahars. In Sigurdsson, H. (ed.). *Encyclopedia of Volcanoes.* Academic Press: London.

van de Velde, P. and De Waele, B. 1994. *Report Number 105: Geology of the Mupamadzi River Area, Explanation of Sheet 1231, SW quarter.* Geological Survey Department Zambia: Lusaka.

van Leusen, M. 1995. A defence of environmental determinism. In Lock, G and Stancic, Z. (eds.) *Archaeology and Geographic Information Systems.* Taylor and Francis: London.

Vogel, J.O. 1978. Recent archaeological survey in western Zambia. *Current Anthropology.* **Vol:19 (1).** pp 148-149.

Vogel, J.O. 1984. An early Iron Age settlement system in Southern Zambia. *Azania.* **Vol:19.** pp 61-78.

Vogel, J.O. 1987. Iron Age farmers in southwestern Zambia: some aspects of spatial organisation. *The African Archaeological Review.* **Vol:5.** pp 159-170.

Wallace, L.A. 1907. North-Eastern Rhodesia. *The Geographical Journal.* **Vol:29 (4).** pp 369-395.

Warren, R.E. and Asch, D.L. 2000. A predictive model of archaeological site location in the Eastern Prairie Peninsula. In Westcott, K.L. and Brandon, R.J. (eds.) *Practical applications of GIS for archaeologists: a predictive modelling kit.* Taylor and Francis: London.

Wells, S.G. and Harvey, A.M. 1987. Sedimentologic and geomorphic variations in strom-generated alluvial fans, Howgill Fells, northwest England. *Geological Society of America Bulletin.* **Vol:98** pp 182-198

Went, D.J. 2005. Pre-vegetation alluvial fan facies and processes: an example from the Cambro-Ordovician Rozel Conglomerate Formation. Jersey, Channel Islands. *Sedimentology.* **Vol:52.** pp 693-713.

Wescott, K.L. and Brandon, R.J. (eds.) 2000. *Practical Applications of GIS for Archaeologists: A Predictive Modelling Kit.* Taylor and Francis: London.

Wescott, K.L. and Kuiper, J.A. 2000. Using a GIS to model prehistoric site distributions in the Upper Chesapeake Bay. In Wescott, K.L. and Brandon, R.J. (eds.) *Practical Applications of GIS for Archaeologists: A Predictive Modelling Kit.* Taylor and Francis: London.

Wheatley, D. and Gillings, M. 2002. *Spatial Technology and Archaeology: The Archaeological Applications of GIS.* Taylor and Francis: London.

White, F. 1983. *The Vegetation of Africa.* UNESCO: Paris.

Whitley, T.G. 2004. Spatial variables as proxies for modelling cognition and descision-making in archaeological settings: a theoretical perspective. *Internet Archaeology.* **Vol:16** *http://intarch.ac.uk/journal/issue16/whitley_index.html*

Willey, G.R. 1953. Prehistoric settlement patterns in the Viru Valley, Peru. *Bureau of American Ethnology Bulletin 155.* Smithsonian Institution: Washington DC.

Willey, G.R. and Sabloff, J.A. 1974. *A History of American Archaeology.* Thames and Hudson: London.

Willoughby, P.R. 1993. The Middle Stone Age in East Africa and modern human origins. *African Archaeological Review.* **Vol:11.** pp 3-20.

Wobst, H.M. 1983. We can't see the forest for the trees: sampling and the shapes of archaeological distributions. In Moore, J.A. and Keene, A.S. (eds.) *Archaeological Hammers and Theories.* Academic Press: London.

Wohl, E.E. Anthony, D.J. Madsen, S.W. and Thompson, D.M. 1996. A comparison of surface sampling methods for coarse fluvial sediments. *Water Resources Management.* **Vol:32 (10).** pp 3219-3226.

10.2. List of Internet Resources Cited

Online articles

Bevan, A. and Harlan, D. 2003. Kythera Island Project: archaeological survey: methods and preliminary results. http://www.ucl.ac.uk/kip/survey.php accessed 11/2007.

Cannon, S.H. 1997. *Evaluation of the Potential for Debris and Hyperconcentrated Flows in Capulin Canyon as a Result of the 1996 Dome Fire, Bandelier National Monument, New Mexico.* USGS Open File Report 97-136 at http://pubs.usgs.gov/of/1997/ofr-97-0136/ accessed on 31/08/07.

Whitley, T.G. 2004. Spatial variables as proxies for modelling cognition and descision-making in archaeological settings: a theoretical perspective. *Internet Archaeology.* **Vol:16** http://intarch.ac.uk/journal/issue16/whitley_index.html

Online data sources

Lake Malawi Drilling Project. 2005. http://malawidrilling.syr.edu/ accessed 21/10/07. Information quoted accurate in May 2008.

ResMap. 2005. http://www.resmap.com./ accessed 16/11/05. Information still present, although prices are much higher in May 2008 for the basic products, *c.* $300 US.

ResMap. 2005. http://www.resmap.com/products/products_L7_m.html *accessed 16/11/05*). Information quoted accurate in May 2008.

Sunny Beach Technology LtD. 2007. www.sunny-beach.net/random_numbers/. Information accurate May 2008.

United States Geological Survey. 2005. http://landsat7.usgs.gov/ accessed 16/11/05. This product information has been deleted, but was found at http://landsat.usgs.gov/products_productinformation.php May 2008.

United States Geological Survey 2008 http://seamless.usgs.gov/website/seamless/products/srtm3arc.asp accessed 29/06/09.

United States Geological Survey 2007. Earth Explorer System (http://edcsns17.cr.usgs.gov/EarthExplorer/ accessed 10/2007. System running May 2008.

United States Geological Survey 2009 *http://eros.usgs.gov/products/satellite/declass1.php*. 29/06/09.

10.3. List of Map Resources Cited

Thieme, J.G., and Johnson, R.L. 1981. *Geological map of the Republic of Zambia.* Geological Survey Department: Lusaka.

Zambia Government Map Department. 1972. *Map number 1231B3*. The Surveyor General: Lusaka.

Zambia Government Map Department. 1972. *Map number 1331B1*. The Surveyor General: Lusaka.

Zambia Government Map Department. 1972. *Map number 1331B2*. The Surveyor General: Lusaka.

CAMBRIDGE MONOGRAPHS IN AFRICAN ARCHAEOLOGY

Series Editors: John Alexander, Laurence Smith and Timothy Insoll
Publication enquiries to Dr John Alexander, St John's College, Cambridge

No 1 BAR S75, 1980 **The Niger Delta** *Aspects of its Prehistoric Economy and Culture* by Nwanna Nzewunwa. ISBN 0 86054 083 9

No 2 BAR S89, 1980 **Prehistoric Investigations in the Region of Jenne, Mali** *A Study in the Development of Urbanism in the Sahel* by Susan Keech McIntosh and Roderick J. McIntosh ISBN 0 86054 103 7

No 3 BAR S97, 1981 **Off-Site Archaeology and Human Adaptation in Eastern Africa** *An Analysis of Regional Artefact Density in the Amboseli, Southern Kenya* by Robert Foley. ISBN 0 86054 114 2

No 4 BAR S114, 1981 **Later Pleistocene Cultural Adaptations in Sudanese Nubia** by Yousif Mukhtar el Amin. ISBN 0 86054 134 7

No 5 BAR S119, 1981 **Settlement Patterns in the Iron Age of Zululand** *An Ecological Interpretation* by Martin Hall. ISBN 0 86054 143 6

No 6 BAR S139, 1982 **The Neolithic Period in the Sudan, c. 6000-2500 B.C.** by Abbas S. Mohammed-Ali. ISBN 0 86054 170 3

No 7 BAR S195, 1984 **History and Ethnoarchaeology in Eastern Nigeria** *A Study of Igbo-Igala relations with special reference to the Anambra Valley* by Philip Adigwe Oguagha and Alex Ikechukwu Okpoko. ISBN 0 86054 249 1

No 8 BAR S197, 1984 **Meroitic Settlement in the Central Sudan** *An Analysis of Sites in the Nile Valley and the Western Butana* by Khidir Abdelkarim Ahmed. ISBN 0 86054 252 1

No 9 BAR S201, 1984 **Economy and Technology in the Late Stone Age of Southern Natal** by Charles Cable. ISBN 0 86054 258 0

No 10 BAR S207, 1984 **Frontiers** *Southern African Archaeology Today* edited by M. Hall, G. Avery, D.M. Avery, M.L. Wilson and A.J.B. Humphreys. ISBN 0 86054 268 8. £23.00.

No 11 BAR S215, 1984 **Archaeology and History in Southern Nigeria** *The ancient linear earthworks of Benin and Ishan* by P.J. Darling. ISBN 0 86054 275 0

No 12 BAR S213, 1984 **The Later Stone Age of Southernmost Africa** by Janette Deacon. ISBN 0 86054 276 9

No 13 BAR S254, 1985 **Fisher-Hunters and Neolithic Pastoralists in East Turkana, Kenya** by John Webster Barthelme. ISBN 0 86054 325 0

No 14 BAR S285, 1986 **The Archaeology of Central Darfur (Sudan) in the 1st Millennium A.D.** by Ibrahim Musa Mohammed. ISBN 0 86054 367 6.

No 15 BAR S293, 1986 **Stable Carbon Isotopes and Prehistoric Diets in the South-Western Cape Province, South Africa** by Judith Sealy. ISBN 0 86054 376 5.

No 16 BAR S318, 1986 **L'art rupestre préhistorique des massifs centraux sahariens** by Alfred Muzzolini.. ISBN 0 86054 406 0

No 17 BAR S321, 1987 **Spheriods and Battered Stones in the African Early and Middle Stone Age** by Pamela R. Willoughby. ISBN 0 86054 410 9

No 18 BAR S338, 1987 **The Royal Crowns of Kush** *A study in Middle Nile Valley regalia and iconography in the 1st millennia B.C. and A.D.* by Lázló Török.. ISBN 0 86054 432 X

No 19 BAR S339, 1987 **The Later Stone Age of the Drakensberg Range and its Foothills** by H. Opperman. ISBN 0 86054 437 0

No 20 BAR S350, 1987 **Socio-Economic Differentiation in the Neolithic Sudan** by Randi Haaland. ISBN 0 86054 453 2

No 21 BAR S351, 1987 **Later Stone Age Settlement Patterns in the Sandveld of the South-Western Cape Province, South Africa** by Anthony Manhire. ISBN 0 86054 454 0

No 22 BAR S365, 1987 **L'art rupestre du Fezzan septentrional (Libye) Widyan Zreda et Tarut (Wadi esh-Shati)** by Jean-Loïc Le Quellec. ISBN 0 86054 473 7

No 23 BAR S368, 1987 **Archaeology and Environment in the Libyan Sahara** *The excavations in the Tadrart Acacus*, 1978-1983 edited by Barbara E. Barich. ISBN 0 86054 474 5

No 24 BAR S378, 1987 **The Early Farmers of Transkei, Southern Africa Before A.D. 1870** by J.M. Feely. ISBN 0 86054 486 9

No 25 BAR S380, 1987 **Later Stone Age Hunters and Gatherers of the Southern Transvaal** *Social and ecological interpretation* by Lyn Wadley. ISBN 0 86054 492 3

No 26 BAR S405, 1988 **Prehistoric Cultures and Environments in the Late Quaternary of Africa** edited by John Bower and David Lubell. ISBN 0 86054 520 2

No 27 BAR S418, 1988 **Zooarchaeology in the Middle Nile Valley** *A Study of four Neolithic Sites near Khartoum* by Ali Tigani El Mahi. ISBN 0 86054 539 3

No 28 BAR S422, 1988 **L'Ancienne Métallurgie du Fer à Madagascar** by Chantal Radimilahy. ISBN 0 86054 544 X

No 29 BAR S424, 1988 **El Geili The History of a Middle Nile Environment, 7000 B.C.-A.D. 1500** edited by I. Caneva. ISBN 0 86054 548 2

No 30 BAR S445, 1988 **The Ethnoarchaeology of the Zaghawa of Darfur (Sudan) Settlement and Transcience** by Natalie Tobert. ISBN 0 86054 574 1

No 31 BAR S455, 1988 **Shellfish in Prehistoric Diet Elands Bay, S.W. Cape Coast, South Africa** by W.F. Buchanan. ISBN 0 86054 584 9

No 32 BAR S456, 1988 **Houlouf I** *Archéologie des sociétés protohistoriques du Nord-Cameroun* by Augustin Holl. ISBN 0 86054 586 5

No 33 BAR S469, 1989 **The Predynastic Lithic Industries of Upper Egypt** by Liane L. Holmes. ISBN 0 86054 601 2 (two volumes)

No 34 BAR S521, 1989 **Fishing Sites of North and East Africa in the Late Pleistocene and Holocene** *Environmental Change and Human Adaptation* by Kathlyn Moore Stewart. ISBN 0 86054 662 4

No 35 BAR S523, 1989 **Plant Domestication in the Middle Nile Basin** *An Archaeoethnobotanical Case Study* by Anwar Abdel-Magid. ISBN 0 86054 664 0

No 36 BAR S537, 1989 **Archaeology and Settlement in Upper Nubia in the 1st Millennium A.D.** by David N. Edwards. ISBN 0 86054 682 9

No 37 BAR S541, 1989 **Prehistoric Settlement and Subsistence in the Kaduna Valley, Nigeria** by Kolawole David Aiyedun and Thurstan Shaw. ISBN 0 86054 684 5

No 38 BAR S640, 1996 **The Archaeology of the Meroitic State** *New perspectives on its social and political organisation* by David N. Edwards. ISBN 0 86054 825 2

No 39 BAR S647, 1996 **Islam, Archaeology and History** *Gao Region (Mali) ca. AD 900 - 1250* by Timothy Insoll. ISBN 0 86054 832 5

No 40 BAR S651, 1996 **State Formation in Egypt**: *Chronology and society* by Toby A.H. Wilkinson. ISBN 0 86054 838 4

No 41 BAR S680, 1997 **Recherches archéologiques sur la capitale de l'empire de Ghana** *Etude d'un secteur d'habitat à Koumbi Saleh, Mauritanie. Campagnes II-III-IV-V (1975-1976)-(1980-1981)* by S. Berthier. ISBN 0 86054 868 6

No 42 BAR S689, 1998 **The Lower Palaeolithic of the Maghreb** *Excavations and analyses at Ain Hanech, Algeria* by Mohamed Sahnouni. ISBN0 86954 875 9

No 43 BAR S715, 1998 **The Waterberg Plateau in the Northern Province, Republic of South Africa, in the Later Stone Age** by Maria M. Van der Ryst. ISBN 0 86054 893 7

No 44 BAR S734, 1998 **Cultural Succession and Continuity in S.E. Nigeria** *Excavations in Afikpo* by V. Emenike Chikwendu. ISBN 0 86054 921 6

No 45 BAR S763, 1999 **The Emergence of Food Production in Ethiopia** by Tertia Barnett. ISBN 0 86054 971 2

No 46 BAR S768, 1999 **Sociétés préhistoriques et Mégalithes dans le Nord-Ouest de la République Centrafricaine** by Étienne Zangato. ISBN 0 86054 980 1

No 47 BAR S775, 1999 **Ethnohistoric Archaeology of the Mukogodo in North-Central Kenya** *Hunter-gatherer subsistence and the transition to pastoralism in secondary settings* by Kennedy K. Mutundu. ISBN 0 86054 990 9

No 48 BAR S782, 1999 **Échanges et contacts le long du Nil et de la Mer Rouge dans l'époque protohistorique (IIIe et IIe millénaires avant J.-C.)** *Une synthèse préliminaire* by Andrea Manzo. ISBN 1 84171 002 4

No 49 BAR S838, 2000 **Ethno-Archaeology in Jenné, Mali** *Craft and status among smiths, potters and masons* by Adria LaViolette. ISBN 1 84171 043 1

No 50 BAR S860, 2000 **Hunter-Gatherers and Farmers** *An enduring Frontier in the Caledon Valley, South Africa* by Carolyn R. Thorp. ISBN 1 84171 061 X

No 51 BAR S906, 2000 **The Kintampo Complex** *The Late Holocene on the Gambaga Escarpment, Northern Ghana* by Joanna Casey. ISBN 1 84171 202 7

No 52 BAR S964, 2000 **The Middle and Later Stone Ages in the Mukogodo Hills of Central Kenya** *A Comparative Analysis of Lithic Artefacts from Shurmai (GnJm1) and Kakwa Lelash (GnJm2) Rockshelters* by G-Young Gang. ISBN 1 84171 251 5

No 53 BAR S1006, 2001 **Darfur (Sudan) In the Age of Stone Architecture c. 1000 - 1750 AD** *Problems in historical reconstruction* by Andrew James McGregor. ISBN 1 84171 285 X

No 54 BAR S1037, 2002 **Holocene Foragers, Fishers and Herders of Western Kenya** by Karega-Mũnene. ISBN 1 84171 1037

No 55 BAR S1090, 2002 **Archaeology and History in Ìlàrè District (Central Yorubaland, Nigeria) 1200-1900 A.D.** by Akinwumi O. Ogundiran. ISBN 1 84171 468 2

No 56 BAR S1133, 2003 **Ethnoarchaeology in the Zinder Region, Republic of Niger: the site of Kufan Kanawa** by Anne Haour. ISBN 1 84171 506 9

No 57 BAR S1187, 2003 **Le Capsien typique et le Capsien supérieur** *Évolution ou contemporanéité. Les données technologiques* by Noura Rahmani. ISBN 1 84171 553 0

No 58 BAR S1216, 2004 **Fortifications et urbanisation en Afrique orientale** by Stéphane Pradines. ISBN 1 84171 576 X

No 59 BAR S1247, 2004 **Archaeology and Geoarchaeology of the Mukogodo Hills and Ewaso Ng'iro Plains, Central Kenya** by Frederic Pearl. ISBN 1 84171 607 3

No 60 BAR S1289, 2004 **Islamic Archaeology in the Sudan** by Intisar Soghayroun Elzein. ISBN 1 84171 639 1.

No 61 BAR S1308, 2004 **An Ethnoarchaeological Study of Iron-Smelting Practices among the Pangwa and Fipa in Tanzania** by Randi Barndon. ISBN 1 84171 657 X.

No 62 BAR S1398, 2005 **Archaeology and History in North-Western Benin** by Lucas Pieter Petit. ISBN 1 84171 837 8.

No 63 BAR S1407, 2005 **Traditions céramiques, Identités et Peuplement en Sénégambie** *Ethnographie comparée et essai de reconstitution historique* by Moustapha Sall. ISBN 1 84171 850 5

No 64 BAR S1446, 2005 **Changing Settlement Patterns in the Aksum-Yeha Region of Ethiopia: 700 BC – AD 850** by Joseph W. Michels. ISBN 1 84171 882 3.

No 65 BAR S1454, 2006 **Safeguarding Africa's Archaeological Past** *Selected papers from a workshop held at the School of Oriental and African Studies, University of London, 2001* edited by Niall Finneran. ISBN 1841718920

No 66 BAR S1537, 2006 **Excavations at Kasteelberg, and the Origins of the Khoekhoen in the Western Cape, South Africa** by Andrew B. Smith. ISBN 1 84171 969 2.

No 67 BAR –S1549, 2006 **Archéologie du Diamaré au Cameroun Septentrional** *Milieux et peuplements entre Mandara, Logone, Bénoué et Tchad pendant les deux derniers millénaires* by Alain Marliac ISBN 1 84171 978 1.

No 68 BAR S1602, 2007 **Chasse et élevage dans la Corne de l'Afrique entre le Néolithique et les temps historiques** by Joséphine Lesur. ISBN 978 1 4073 0019 1.

No 69 BAR S1617, 2007 **The Emergence of Social and Political Complexity in the Shashi-Limpopo Valley of Southern Africa, AD 900 to 1300** *Ethnicity, class, and polity* by John Anthony Calabrese ISBN 978 1 4073 0029 0.

No 70 BAR S1658, 2007 **Archaeofaunal remains from the past 4000 years in Sahelian West Africa** *Domestic livestock, subsistence strategies and environmental changes* by Veerle Linseele ISBN 978 1 4073 0094 8.

No 71 BAR S1667, 2007 **Il Sahara centro-orientale Dalla Preistoria ai tempi dei nomadi Tubu / The Central-Oriental Sahara. From Prehistory to the times of the nomadic Tubus** by Vanni Beltrami con le fotografie e i riassunti in inglese di Harry Proto / with English summaries and photographs by Harry Proto. ISBN 978 1 4073 0102 0.

No 72 BAR S1679, 2007 **Memory and the Mountain: Environmental Relations of the Wachagga of Kilimanjaro and Implications for Landscape Archaeology** by Timothy A. R. Clack. ISBN 978 1 4073 0117 4.

No 73 BAR S1736, 2008 **Archaeological Investigations of Iron Age Sites in the Mema Region, Mali (West Africa)** by Tereba Togola. ISBN 978 1 4073 0178 5.

No 74 BAR S1847, 2008 **Current Archaeological Research in Ghana** edited by Timothy Insoll. ISBN 978 1 4073 0334 5.

No 75 BAR S1860, 2008 **Holocene Prehistory of the Southern Cape, South Africa Excavations at Blombos Cave and the Blombosfontein Nature Reserve** by Christopher Stuart Henshilwood. ISBN 978 1 4073 0343 7.

No 76 BAR S1873, 2008 **The Archaeology of Tanzanian Coastal Landscapes in the 6th to 15th Centuries AD (the Middle Iron Age of the Region)** by Edward John David Pollard. ISBN 978 1 4073 0353 6.

No 77 BAR S1926, 2009 Using **Stone Tools: The Evidence from Aksum, Ethiopia** by Laurel Phillipson with an introduction by Professor Rodolfo Fattovich. ISBN 978 1 4073 0408 3

www.ingramcontent.com/pod-product-compliance
Lightning Source LLC
La Vergne TN
LVHW071207210926
848482LV00015B/742